AVIONIC SYSTEMS

DONALD H. MIDDLETON
GENERAL EDITOR

Longman
Scientific &
Technical

Longman Scientific & Technical,
Longman Group UK Limited,
Longman House, Burnt Mill, Harlow,
Essex CM20 2JE, England
and Associated Companies throughout the world.

© Longman Group UK Limited 1989

All rights reserved, no part of this publication may be reproduced, stored in a retrieval system, or transmitted in any form or by any means, electronic, mechanical, photocopying, recording, or otherwise without either the prior written permission of the Publishers or a licence permitting restricted copying in the United Kingdom issued by the Copyright Licensing Agency Ltd, 33–34 Alfred Place, London, WC1E 7DP.

First published 1989

British Library Cataloguing in Publication Data
Avionic systems.
 1. Aircraft. Electronic equipment
 I. Middleton, Don, *1921*–
629.135′5

ISBN 0-582-01881-1

Set in Linotron 202 9½/12 pt Ehrhardt Roman

Produced by Longman Singapore Publishers (Pte) Ltd.
Printed in Singapore

CONTENTS

Preface	vi
Glossary of acronyms used in text	vii
Glossary	x

1 **The Evolution of Avionics**   1
M. W. Wilson

Early development of avionics	1
Origins of radar	6
Transition from thermionic valves to solid state	14

2 **System Design Considerations**   19
S. J. K. Walker

The ethos of civil aircraft safety	19
ARINC specifications	22
Redundancy	25
Reliability	26
Built-in test equipment (BITE)	27
Automatic test equipment (ATE)	29

3 **Digital Technology**   31
I. Moir

Microprocessors	33
Memory devices	35
Data buses	37
Software development methodologies	44

4 **Flight Decks and Cockpits**   48
L. F. E. Coombs

The pilot's place	48
Instruments and displays	52
Avionics the only answer	55
Display and control input technologies	62
Flight deck systems	65
Flight deck examples	73
Military cockpits	76

5 Flight Control Systems — 84
I. Moir

Principles of flight control	84
Essential elements of control	91
Civil systems	93
UK military developments	97
Advanced developments	109

6 Aircraft Management Systems — 115
I. Moir

Engine and propulsion control	116
Flight performance management	120
Utilities management systems	126
EAP systems	131
Health and usage monitoring (HUM)	140
Stores management systems	142

7 Navigation Systems — 148
S. J. K. Walker

Radio wave propagation	148
Navigation equipment	149
Air traffic control	150
Navigation displays	150
Flight management systems	153
Point source aids	153
Hyperbolic and grid systems	161
Doppler navigation	164
Self-contained systems	165
Air data acquisition and air data computers	168
Attitude heading reference system (AHRS)	169
Laser technology	171

8 Communications Systems — 174
T. T. Brown

Design considerations	175
Frequency bands	176
HF transmission	177
UHF/VHF transmission	179
Real time channel evaluation	180
Aircraft antennas	181
Receivers	184
Frequency synthesis	186
Voice systems	189
Transmitter modulation	192
Information theory applications	193
Data links	195
Joint tactical information distribution system (JTIDS)	196
Satellite communications	200
Public correspondence	200

9 Airborne Radar   202
J. A. C. Kinnear

Propagation   203
Functional elements of a radar   205
The antenna   209
The transmitter   215
Types of radar   216
Pulse Doppler   218
Civil aviation applications   219
Military applications   222

10 Electronic Warfare – an Overview   228
B. R. Drake

The EW spectrum   230
Electronic support measures   232
Electronic countermeasures   235
Electro-optics and infra-red   242
The future   245

11 Future trends and developments   247
G. Warwick

Avionic integration   247
Cockpit integration   249
Sensor integration   253
Control integration   254
Database integration   255
Knowledge integration   256

Index   258
Index of names   262
Index of aircraft   263
Index of research aircraft   265
Index of gas turbines and propulsion units   266

PREFACE

It is beyond doubt that the origins of what we now know as avionics was a major factor in preserving the world from Nazi domination in the 1940s and it should not be forgotten that the first manifestations of its capability were demonstrated in February 1935 when Dr Watson Watt's team carried out the famous experiment at the BBC's Daventry transmitter and received a signal reflected back from a Heyford bomber approaching Daventry.

The five decades since this memorable event have seen the enterprise and brilliance of scientists and engineers at research centres and manufacturers' works combine with computer technology to produce systems of almost unbelievable capability and reliability.

This book is intended to fill a gap in the aviation technical bookshelf by discussing the present state of the art in a manner intelligible to both the student and the technician.

As General Editor it has been my privilege and pleasure to work with the authors of each chapter – each a leading practitioner in his particular subject. To them all I express my appreciation of their enthusiasm and cooperation.

To all the companies named in the text I owe a debt of gratitude for their cooperation, for the provision of photographs and information and, where appropriate, for their courtesy in permitting their engineers to contribute to this book.

Valuable advice has been given by Mr Mike Hirst, Mr Andy Hofton, Mr Harry Ratcliffe, Mr David Yeomans and Wing Commander Charles McClure of the College of Aeronautics, Cranfield, Mr A. H. Fox of Southall College, Mr Peter Drake of the CSE Oxford Air Training School, Wing Commander G. S. Bowden at Brunel College, Professor G. C. Bloodworth and Dr R. J. Patton of the University of York, Dr Bernstein and Mr David Sweeting of Queen Mary College, University of London. Mr John Burrows, Assistant Engineering Training Superintendent of British Airways and members of his staff were also extremely helpful.

My thanks for their assistance are also due to Mr John Saull, Director and Chief Surveyor of the Civil Aviation Authority and Mr D. Hawkes. From RAE Bedford, Mr R. G. White offered valuable advice as did Dr John Clarke at the Royal Signals and Radar Establishment at Malvern.

D. H. Middleton
General Editor

GLOSSARY OF ACRONYMS USED IN TEXT

AAD	automatic attitude director	BFO	beat frequency oscillator
ACARS	automatic communications and recording system	BIFU	bus interface unit
		BITE	built-in test equipment
ACCP	automatic configuration control processor	BLEU	Blind Landing Experimental Unit (RAE)
ACT	active control technology	BSI	British Standards Institution
ACU	antenna coupler unit	CAR	Civil Airworthiness Requirements
ADA	a computer language	CCS	communications control system
ADC	air data computer	CDI	course/deviation indicator
ADD	airstream direction detector	CDU	control and display unit
ADF	automatic direction finding	CMOS RAM	working memory (computer)
ADI	attitude direction indicator	CNI	communications, navigation and identification
ADIRS	air data inertial reference system		
ADIRU	air data computer and inertial reference unit	COHO STALO	coherent oscillator/stable local oscillator
ADM	air data module	CORE	controlled requirement expression
ADMC	actuator drive and monitor computer	CPU	central processing unit
ADS	automatic dependence surveillance	CRT	cathode ray tube
ADSEL	address selective (SSR system)	CSAS	control and stability augmentation
AEEC	Airline Electrical Engineering Committee	CS/MA-CA	carrier sense multiple access clash avoidance
AFCS	automatic flight control system	CVR	cockpit voice recorder
AFGS	automatic flight guidance system	DAFCS	digital automatic flight control
AHRS	attitude and heading reference system	DATAC	data-autonomous transmission and communication
AHRU	attitude and heading reference unit	DECS	digital engine control system
AI	airborne interception	DECU	digital engine control unit
AM	amplitude modulation	DFCS	digital flight control system
AMRICS	automatic management, receiver and intercom	DME	distance measuring equipment
		DOA	direction of arrival
AMSU	aircraft motion sensing unit	DSIC	Dowty & Smiths Industries Controls
APC	aeronautical public correspondence		
ARINC	Aeronautical Radio Inc	DTU	data transfer unit
ASPJ	airborne self-protection jammer	DVI	direct voice input
ASV	air-to-surface vessel (radar)	EAD	electronic attitude director
ASW	air-sea warfare (radar)	EAP	Experimental Aircraft Programme (BAe)
ATC	air traffic control		
ATE	automatic test equipment	EAW	airborne early warning radar
ATF	advanced tactical fighter	ECAM	electronic centralised aircraft maintenance
ATR	air transport racking		
ATU	antenna tuning unit	ECCM	electronic counter counter-measures
BC	bus controller	ECM	electronic counter-measures
BCAR	British Civil Airworthiness Requirements	EIS	electronic instrument system

vii

Glossary of acronyms used in text

EICAS	engine indicating and crew alert system	LRU	line replaceable unit
EPROM	electronically programmable read-only memory	LUF	lowest usable frequency
		MASS	master armament selector switch
EEPROM or E²PROM	electronically erasable programmable only memory	MAW	mission adaptive wing
		MDP	maintenance data panel
EFA	European Fighter Aircraft	MECU	main engine control unit
EAS	equivalent airspeed	MFCD	multi-function colour display
ELAC	elevator/aileron computer	MFD	multi-function display
ELINT	electronic intelligence	MIDS	management information and decision support
EMCS	energy monitoring and control system	MIT	Massachusetts Institute of Technology
EME	electromagnetic interference	MLS	microwave landing system
EOB	electronic order of battle	MOD	Ministry of Defence
ESM	electronic support measures	MOSFET	metal oxide silicon field effect transistor
EUROCAE	European Organisation for Civil-Aviation Electronics	MPCD	multi-purpose colour display
FAA	Federal Aviation Administration	MTBF	mean time between failures
FAC	flight augmentation computer	MTI	moving target indicator
FADEC	full authority digital engine control	NASA	National Aeronautics and Space Administration
FCS	flight control system		
FEC	forward error correction	NDB	non-directional beacon
FFT	fast Fourier transform (processor)	NPL	National Physical Laboratory
FLIR	forward looking infrared		
FMCS	flight management computer system	PCM	pulse code modulation
FMCU	flight management control unit	PFCU	powered flying control unit
FMS	flight management system	PPI	plan position indicator Alt. present position indicator
FSK	frequency shift keying		
GEOS	geostationary satellite	PRF	pulse repetition frequency
G-LOC	G-induced loss of consciousness	PSK	phase shift keying
GPS	global positioning system	RAE	Royal Aerospace Establishment
HOTAS	hands on throttle and stick	RAM	random access memory
HOL	high-level operating language	RBI	relative bearing indicator
HSI	horizontal situation indicator	RDDMI	radio direction/distance magnetic indicator
HUD	head-up display		
HUMS	health and usage monitoring system	RMI	radio magnetic indicator
		R NAV	area navigation
ICAO	International Civil Aviation Organisation	ROM	read-only memory
		RTT	round-trip timing
IEE	Institution of Electrical Engineers	RTTY	radio teletype
IEEE	Institution of Electrical and Electronic Engineers	RWR	rear warning radar
		SAE	Society of Automotive Engineers
IFF	identification, friend or foe	SAFRA	semi-automatic functional requirements analysis
ILS	instrument landing system		
IN	inertial navigation	SAW	surface acoustic wave
INEWS	integrated electronic warfare system	SID	standard instrument departure
I/OU	input/output unit	SIGINT	signals intelligence
ISA	instruction set architecture	SLAR	side-looking airborne radar
ITU	International Telecommunications Union	SMP	systems management processor
		SMS	stores management system
JAR	Joint Airworthiness Requirements	SMTD	STOL manoeuvre technology demonstrator
JTIDS	joint tactical information distribution system		
		SSEC	static source error correction
LATCC	London Air Traffic Control Centre	SSR	secondary surveillance radar
LCD	liquid crystal display	STAR	standard terminal arrival route
LCFC	low cycle fatigue counter	STOL	short-take-off-and-landing
LED	light-emiting diode	SUMS	structure usage monitoring
LOS	line of sight	TACAN	tactical air navigation

Glossary of acronyms used in text

TDMA	time division multiple access	**VHPIC**	very high performance integrated circuit
THS	trimmable horizontal surface (actuator)	**VLF**	very low frequency
TOA	time of arrival	**VLSI**	very large-scale integration
TWT	travelling wave tube (CRT)	**VNAV**	vertical navigation
UDF	unducted fan	**VOR**	VHF omni-directional radio range
UHF	ultra high frequency	**VOR/DME**	VOR directional guidance with DME distance measuring input
UMS	utilities management system	**VORTAC**	combination of VOR and TACAN
USAF	United States Air Force	**VOS**	voice-operated switch
USN	United States Navy	**VSWR**	voltage standing wave ratio
UV PROM	a computer programme memory	**WAMS**	weapon aiming mode selection
VERDAN	versatile digital analyser	**W/SMS**	weapon/stores management system
VHF	very high frequency		
VHLSI	very high large-scale integrated circuit		

GLOSSARY

Accelerometer	Device for the measurement of acceleration.
Aeroelastic effects	The result of interaction between aerodynamic forces and the structure of the aircraft. Usually de-stabilising and harmful.
Airframe	The complete assembly of the aircraft without its engine and removable items such as external stores and furnishings.
Airspeed	Velocity of aircraft in relation to the air through which it is flying.
Airspeed – Indicated (IAS)	Reading shown on airspeed indicator. This does not take into account instrument and error, position error and compressibility effect.
Airspeed – True (TAS)	The indicated airspeed corrected for the factors mentioned in definition of IAS.
Algorithm	An accepted method of computation either numerical or algebraic. A fundamental procedure in the design of software.
Alphanumeric	A precisely repeatable character, either a capital letter or numeral, produced by electronic means to be displayed or read rapidly by OCR (Optical Character Recognition) systems.
Analogue computer	A calculating device operating by the representation of numerical values. Input data can be processed from, for example, a control surface, so that all characteristics of the control can be analysed in real time.
Anti-skid	Device for preventing the wheels of an aircraft locking under braking conditions.
Artificial horizon	Primary flight instrument to inform pilot of attitude of aircraft in relation to the horizon ahead of the aircraft.
Authority – Total	The ability of the control system to control in its entirety the functions to which it is connected.
Authority – partial	The exercise of partial control of functions.
Autopilot	An electronic system which will automatically control the aircraft in pitch, roll and yaw modes. It may be part of the Automatic Flight Control System in which case it can be programmed to control the particular mode in which the aircraft is being operated.
Autothrottle	Propulsion engine control which may be linked to the AFCS and, if fitted, the automatic landing system, to ensure that the engine power is controlled in accordance with the operational requirements, such as maintenance of the glide path on the approach.
Azimuth	A bearing or direction in the horizontal plane.
Back beam	The reciprocal beam on the other side of the transmitter, for example, of an ILS localiser.
Bandwidth	Range of frequencies within which an antenna circuit, component, equipment or system will perform to its specification. Quoted in number of Hz between limits of frequency band.

Glossary

Bit	(From binary digit). Unit of data in digital systems comprising single character (0 or 1) in binary form.
Bit rate	Speed at which data can be processed in system. Quoted in bits, kbits or Mbits per second.
Brassboard	A functioning prototype of an avionic system.
Breadboard	A prototype, not necessarily capable of flight test, to prove the circuitry of an avionic system and its functional capability.
Burn-through range	The point in space where the target's signal amplitude exceeds the jamming signal on the radar display.
Byte	A group of bits processed as a single unit. An 8-bit electronic data processing word would be a sequence of eight consecutive bits.
Canard	An aircraft configuration with the horizontal stabiliser in front of the mainplane. The foreplane is also known as the canard.
Category 2 approach	An element of the bad-weather landing minima agreed by ICAO. A combination of a decision height 200 to 100 ft (60–30 m) and visibility runway visual range RVR of 2,600 to 1,300 ft (800–400 m).
Chaff/window	Metallic, radar-reflecting strips used to saturate or confuse enemy radars.
Chip	An electronic device sliced from a substrate of a single crystal semi-conductor.
Control algorithms	Control laws. (See ALGORITHM).
Control of bleed	Control of high-pressure bleed air from a gas turbine engine.
CORE	Controlled Requirements Expression – a computer software definition tool.
Cross-coupled	An aerodynamic destabilising interaction between two or more flight modes.
Data bank	Retrievable storage system for digital information.
Data bus	A system of inter-communication between different avionic units in the aircraft which, by using multiplexing techniques of transmitting consecutive pulses divided in frequency, phase and time along a very limited number of conductors, reduce substantially the number of wiring looms.
Delta, delta-sigma modulation	Both convert analogue signal waveforms to single-digit code bit-streams. Delta converting the rate-of-change of the wave-form and delta-sigma the instantaneous amplitude (by pre-integration before the delta comparator). This compares with PCM, pulse code modulation, converting instantaneous amplitude into n-digit code for amplitude levels. For example 4 digits give $2^4 = 16$ amplitude levels. Circuitry is simpler than in PCM.
Dedicated	Applicable for only one application.
Demodulation	The extraction of a modulated signal from its carrier.
Dielectric	A material capable of sustaining an electrical field and undergoing polarisation, a radome can be made of a composite material having dielectric properties.
Digital	Functioning with discrete numbers, bits or other parcels of data.
Dipole	An antenna (aerial) comprising two separate conductors in line and co-axially, fed at the mid-point.
Directional gyro	A free-gyro instrument used to indicate direction in azimuth.
Dutch roll	A phenomenon, associated primarily with the first generation of swept wing aircraft, in which a lateral oscillation with yawing and rolling elements occurred.
ECAM	Electronic Centralised Aircraft Monitoring.
EICAS	Engine Indication and Crew Alert System.
EIS	Electronic Instrumentation System.
Electronic countermeasures (ECM)	The actions taken to prevent, disrupt or reduce the enemy's effective use of the electromagnetic spectrum.

Glossary

Electronic support measures (ESM)	The interception, location and identification of radiated electromagnetic energy to provide emitter recognition and intelligence.
Electronic warfare (EW)	The means of determining or disrupting the enemy's use of the electromagnetic spectrum whilst safeguarding its use for friendly forces.
Empennage	The complete assembly of horizontal and vertical stabilisers of an aircraft.
Fan marker	A position fixing radio beacon, keyed for identification, radiating in a fan-shaped vertical pattern.
Fibre optics	A method of transmitting control signals along thin fibres of glass or suitable plastic material. Light entering at the end of the filament is transmitted by internal reflection. Fibres, generally of a few microns diameter, are assembled into bundles of tens of thousands to over 100,000.
Flag	Small, brightly coloured plate which emerges on the face of an electro-mechanical instrument to give visual warning of a malfunction. Also used on solid-state displays.
Flight envelope	A series of curves of aircraft performance and other factors defining the maximum limits within which the aircraft may be safely flown.
Flight profile	Specification of complete flight usually in terms of altitude plotted against distance.
Fluxvalve/fluxgate	Sensing element of a remotely indicating compass giving an electrical signal proportional to the intensity of external magnetic field acting along its axis.
Fly-by-light (Colloq)	Aircraft control system using fibre-optic filaments for the transmission of signals.
Fly-by-wire (Colloq)	Aircraft control system using electrical wiring for the transmission of signals.
Gain scheduling	The scheduling of the gain or amplification of a control circuit to modify the control characteristics.
Gated	An electromagnetic pulse functioning under the control of another pulse, usually synchronised with it.
Geostationary satellite	A satellite launched into a trajectory which ensures that it remains over a specific point on the surface of the earth.
Glass cockpit (Colloq)	A cockpit in which electronic displays have replaced the majority of conventional flight instruments.
Glideslope/glidepath	The descent path, at an angle from the horizontal, of an aircraft during the approach and landing phase.
Ground clutter	Returns on a radar screen caused by direct reflections from the ground. These degrade the value of the display by interference.
Hardover condition	A potentially hazardous situation when a control system unexpectedly develops a fault which moves the control to the limit of its operating range.
Heterodyne	The mixture of two AC electronic signals to generate a third one equal to the sum or difference of their frequencies.
Holographic combining glass	Special glass screens used in head-up displays which enhance the quality of the images.
Hyperbolic systems	Navaids based upon synchronised transmissions from ground stations – master and slave, the transmissions produce upon the aircraft screen lines of constant time difference to indicate range on the form of hyperbolic position lines as in the Decca system.
ICAO	International Civil Aviation Organisation.
IFF	Identification Friend or Foe.
ILS Localiser	Aerial and beam giving directional guidance in an instrument landing system.

Glossary

IN systems	Inertial navigation systems.
Intake ramp	A variable geometry wedge or wedges located in the air intakes of a supersonic aircraft engine to create oblique shock waves to achieve maximum efficiency. These are controlled automatically by the engine control system.
Integrated circuit	Microelectronic device consisting of a series of circuits upon a single crystal, usually silicon, substrate.
I/O card	A card serving an input/output function; interfacing aircraft signals with digital electronics.
Ionisation	Conversion of atoms to ions (electrically charged atoms or groups of atoms).
Ionogram	A graph plotting radio frequency against time of round trip of the pulse to the reflective layer.
Ionosphere	The whole of the ionised region of the atmosphere of the Earth, including Appleton E and F layers and the Kennelly-Heaviside layer.
Keypads	An array of push or press alphanumeric keys or buttons.
Laser	Light Amplification by Stimulated Emission of Radiation.
Load alleviation	A technique of electronic control, via sensors, which can alleviate, for example, an excessive up-load on the wing by upward deflection of the ailerons in unison.
Magnetic isotropy	Non-polarised magnetic field.
Manchester bi-phase	A means of encoding digital data by recording a change of state as the signal passes through zero.
Manoeuvre margin	A quantity which defines the stability and hence manoeuvrability of an aircraft by relating the centre of gravity distance from the centre of pressure in relation to aircraft length.
MOD	Ministry of Defence. (UK).
Monitoring devices	Devices capable of continuous surveillance of performance of sub-systems. Some are capable of initiating corrective action.
Multiplex	Combination of signals from different sources into a common channel in which they are separated by frequency, phase or time or a combination of these.
MPCD	Multi-Purpose Colour Display.
Nap of the earth flying	Flying as low as possible without hitting natural or man-made objects.
Non-volatile memory	Not capable of erasure.
PAVE PILLAR	A modular avionics architecture defined and developed by the US Air Force.
Phase	The time relationship between two or more electronic frequencies.
Photon	A parcel of magnetic energy emitted by a single electron.
Pi filter network	Filter for impedance transformation and/or frequency selection.
Potentiometer	A resistance with variable tap.
Pressure transducer	Device capable of generating an electrical signal proportional to fluid pressure or pressure difference.
Probe	External device upon the aircraft to obtain information about the air through which it is flying and the relationship between the aircraft attitude and the free stream flow.
Protocol	A library of messages held by operational organisations to suit that particular organisation's requirements.
PTT button	'Press to transmit' button.
Pulsed radar	The most common form of radar with signals emitted in the form of pulses.

Glossary

Racetrack pattern	Flight path of an airways holding pattern comprising two semi-circles joined by two parallel legs.
Radar warning receiver (RWR)	An ESM equipment providing threat warning, identification and direction of enemy radar transmissions.
Radio range	Early American radio navigation aid, land-based, broadcasting continuous coded signals, Morse N was heard to the right of the correct flight path, Morse A to the left of it.
Raster	Production of a CRT image by scanning in closely spaced horizontal lines either alternately or in sequence as in a domestic television receiver.
Reheat/afterburning	Injection and combustion of extra fuel in a specially designed jetpipe to give thrust augmentation.
Resonant cavity magnetron	A device comprising a closed hollow space with electrically conductive walls within which microwaves are generated by the excitation of an electron beam within a magnetic field.
Rho-theta	Radio navigation aid providing bearing and distance from a particular station.
SAFRA	Semi-Automated Functional Requirements Analysis – a software development tool.
Servo controls	System of powering aircraft controls to augment the strength of the pilot and to provide feedback to indicate degree of force required.
Skip distance	The distance from a transmitter at which the first reflected sky wave may be received.
Slat	Moveable part of wing leading edge which can be moved forwards or downwards mechanically to form a slot with the wing surface to delay airflow breakaway.
Slew Control (HOTAS)	Control for pilot to initiate rapid slew of weapon laying symbology.
Smart	(1) Particularly ingenious use of a device, ie smart skin – capable of being used as an antenna or for uses other than structural. (2) Devices with integral intelligence.
Softkey	A key whose function is programmed by software to change for differing modes of operation.
Spoiler	Moveable surface at upper rear surface of wing which can be deflected or opened to reduce lift and/or increase drag.
Stolport	Airport for the use of short take-off and landing aircraft. eg London City.
Stores	A load related to the mission of a military aircraft, carried either on external pylons or within the fuselage. eg bombs, sub-munitions, missiles, drop tanks.
Stovl	Short take-off vertical landing.
Strapdown	Device in aircraft not mounted upon gimbals.
Supercritical wing	Wing designed for efficient cruise performance above the critical Mach number, typically has a bluff leading edge, fairly flat top surface with a cambered undersurface.
Symbol generator	Computer for the production of symbols for CRT displays.
Synchro resolver (AC)	A three wire AC transmission system used for remote indication which can resolve and identify the difference in the current in the three wires.
Systems integrator	An avionic unit which integrates the outputs of two or more discrete systems.
Thermal imaging	Infra-red technique of producing pictorial displays or print-outs to show temperature variations in field of view.
Thermistor	A resistor, the value of which varies with temperature.
Thrust vectoring	Technique of varying direction of thrust of a jet or rocket motor by altering the direction of the jetpipe(s). eg Harrier.
Tilt switch	Switch activated by tilting its location.

Glossary

Touchscreen	Alphanumeric readout of commands and data on a screen which can be activated or interrogated by the pilot touching the relevant word or display.
Transistor	Electronic device consisting of a semiconductor which can amplify or control inputs to it.
Travelling wave tube	A type of microwave amplifier in which an interaction occurs between an electron beam and a radio frequency field travelling in the same or the opposite direction.
Triad	American deterrent philosophy embracing the simultaneous effect of submarine-launched ballistic missiles, orthodox manned bombers and silo based ballistic missiles on land.
Unducted fan	A propulsion unit comprising a gas turbine engine driving a multi-bladed propeller designed to give efficient cruise at high sub-sonic speeds.
USMS	Utility Systems Management System – the integrated system fitted to the EAP.
V_1	Speed at which pilot in command must decide to take-off or abort.
V_2	Take-off safety speed, the lowest at which the aircraft will meet the handling criteria of the climb with one engine inoperative.
V_r	Speed on the runway at which the PIC will commence take-off by rotating the aircraft into the lift-off attitude.
VOR beacon	A fixed beacon radiating a circular pattern horizontally. Upon this is superimposed a rotating directional pattern at 30 Hz the output phase modulation of which is unique to each heading established from the beacon which, itself, is part of the VOR – VHF omnidirectional radio range, the most common of radio navaids.
Voter monitoring	Binary logic system which can compare signals in several channels and detect a mismatch which may develop; the mismatched output is then 'voted out'.
Waypoint	A fixed, accurately defined location which forms the beginning or the end of a particular airline or military route segment.
Yaw	Movement of an aircraft about its Oz or normal axis as initiated by the rudder(s).
Yaw damper	Automatic sub-system of flying control system which senses the onset of yaw and initiates the correct control action to overcome it.

CHAPTER 1 THE EVOLUTION OF AVIONICS

MICHAEL W. WILSON BSc, CEng, MRAeS

Mr Wilson is a writer and consultant on technical aspects of aviation. He joined Vickers-Armstrong's Guided Weapons Department in 1956, later transferring to systems engineering on Vanguard, VC 10 and TSR 2 aircraft. In 1965 he was appointed Technical Editor of *Flight International*. From 1978 to 1981 Mr. Wilson was Public Relations executive, Carl Byoir. In 1981 he was invited to develop and launch *Jane's Avionics Yearbook*. Since 1985 Mr Wilson's activities have been equally shared between aviation writing and consultancy and editing a Christian magazine *Prophecy Today*.

EARLY DEVELOPMENT OF AVIONICS

Avionics is the now universally accepted shorthand for aeronautical or aviation electronics, an industry that began to blossom during the Second World War and has now come into astonishing flower. The term originated during the 1960s in the USA when the Western defence industry as a whole was undergoing rapid and radical change, in no small way spurred by the 1950–53 Korean War. Urgent and generous defence budgets pushed forward a host of new technologies to maintain national and international security in the face of growing and world-wide tensions.

The rapid advance in aviation electronics did not, of course, take place in isolation, but was in step with developments in this field for a wide range of industrial, commercial, scientific, defence and other applications. But while avionics represents a relatively small proportion of the total electronics market, it exercises a quite disproportionate influence as a result of aviation's special needs for reliability, low weight, small size and minimal power consumption. Strictly speaking, the term 'avionics' should be reserved only for equipment specially designed for airborne application and carried as part of that application. Thus a ground-based test or check-out system is not categorised as avionics, and neither is flight-test equipment installed for aircraft development and certification purposes. Again, to qualify, the electrical currents circulating in a system should substantially comprise information signals rather than heavy duty energy to activate electro-mechanical devices. Thus an electrical generator would not be classified as avionics. In practice, the subject has rather blurred edges, so that for example a turn and slip indicator (basically an electric motor with a spring attached) would now come under the term; the context is usually sufficient to show whether the description is appropriate or not.

The Wright brothers showed the practicability of powered flight in 1903 with their *Flyer*. For the first few years the tiny band of pioneers was too busy finding out about the fundamentals of flight to spend much time on the 'frills' that were to come later. But August 1910 saw the first recorded application of electronics, when Canadian designer and pilot J. D. A. McCurdy transmitted and received wireless signals via an H. M. Morton set in his Curtiss seaplane at Sheepshead Bay, New York State. The following month saw another demonstration, this time in England, when the pilot of a Bristol Boxkite over the British Army's firing ranges on Salisbury Plain signalled 'enemy in sight' to a ground-based receiver a quarter of a mile away.

Aeroplanes came into their own, as weapons, during the First World War. In that conflict, operated by armies of different nationalities, they were used for reconnaissance, noting enemy troop concentrations, and for calling fall-of-shot for the artillery. The latter task, a real-time application as we would now say, called for an air-to-ground radio link.

Of little use at the time, but curiously prophetic, was the demonstration in 1914 in Paris by Lawrence Sperry of a gyroscopically-based control system for keeping an aeroplane in level flight without the intervention of a human pilot (Fig. 1.1). Sperry had already developed the gyroscope to a high degree of perfection and saw the aeroplane as a new and potentially important business opportunity. Apart from being grossly under-powered, aircraft before 1914 generally exhibited very poor handling qualities and so could be difficult to fly. His equipment, fitted to a Curtiss flying boat, was an example of what would now be called an autostability system. Sperry was later to devise the turn-and-slip indicator, the simplest of all blind-flying devices. This gyro-based instrument was to launch Sperry into the flight-instrument (and, later, flight-control) field.

An advance on the turn-and-slip, the artificial horizon provides much more information and is easier to interpret; some of the very first examples provided the essential attitude guidance for crews of the big Gotha bombers during their long-range night raids on London in 1918. About this time, too, yet another gyro-based instrument appeared – the directional gyro; updated by a magnetic compass, it had a steady needle movement and so was much easier to use, particularly in the rapid establishment of a new heading. By the end of the war Britain also had a strategic night-attack force, and both sides had begun to get a feel for at least two of the problems that were to bug bomber pilots over Europe for the next thirty years – navigation and target recognition in poor weather and at night.

The First World War gave impetus to aircraft design and huge strides were made in structural efficiency, aerodynamics, engine power and reliability. But instrumentation was still simple: direct-reading engine RPM, oil temperature and pressure gauges, airspeed indicator, altimeter and compass continued for many years to be the principal guides to aircraft behaviour and progress. None of these instruments contained any electronic, or even electrical, elements; wireless continued as the sole forerunner of the avionics revolution to come.

Military aircraft development after 1918 slowed almost to a halt; the huge massacres of trench warfare had sickened people and turned governments away from re-armament. But new challenges appeared for aviation. America, Britain, France and Italy encouraged or supported long-range flights to open up new areas of wilderness or provide better links with empires or colonies. At the same time small airlines began to mushroom, sometimes serving just one or two routes. In particular, the government-regulated mail services were seen as tempting markets, with built-in subsidies; however they called for strict schedules, emphasising among other things accurate and reliable poor-weather navigation.

The possibilities of using radio for navigation as well as for communication had

Fig. 1.1 An attempt at aircraft stabilisation was made by Lawrence Sperry in 1914 when he fitted a cluster or gyroscopes to a Curtiss flying boat and flew over the Seine with his mechanic walking along the wing whilst Sperry flew the boat 'hands off' *British Aerospace Dynamics Bracknell*

Early development of avionics

The Evolution of Avionics

been appreciated quite early on; during the First World War wireless was used to 'fix' the position of Zeppelin airships on their way to bomb London. The method adopted was to broadcast signals to radio stations in Germany, which measured by triangulation the ship's position and then transmitted it to the crew. The same method was used in a remarkable, 3000-mile flight by a Zeppelin airship during 1917, from Bulgaria to the relief of a military unit in German East Africa.

Radio was soon recognised by the infant airline industry as an essential adjunct to the safe, efficient and profitable conduct of its business. It provided pilots with weather reports and surface-pressure measurements for altimeter setting, the former being particularly valuable in considering the availability of alternative airports in the event of fog or dangerously low cloud. Speech communication was employed for short to medium ranges, but Morse code was preferable at longer distances since the limited transmitter power could be channelled into a much smaller bandwidth, permitting the signals to punch their way through static or thunderstorm interference in a way not possible with speech under the same conditions.

Perhaps even more valuable to the new airlines was the radio range navigation aid. An American invention, radio range defined the position of an airport by four radio beams in the form of a cross, the beams being identified by the Morse letters for A and N. The pilot could find the centre-line of a beam by listening in on head-phones and steering until the letters were heard with equal strength. To help him fix his position (as opposed to direction), so-called 'fan-markers' were arranged to direct beams vertically upwards.

These radio rangers were the forerunners of today's sophisticated airways systems which are defined by radio beacons. But operating in the HF band, they were susceptible to interference by weather and could be 'bent' or deflected by natural or artificial obstacles on the ground. They could thus be tricky to use, and navigation was often a full-time occupation for the second pilot of the two-man

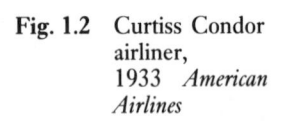

Fig. 1.2 Curtiss Condor airliner, 1933 *American Airlines*

Early development of avionics

crews which came to be seen as obligatory for the new multi-engined airliners of the late 1920s.

By the early 1930s the airlines had established themselves as a viable form of transport. Aircraft began to grow in size and complexity and be capable of longer flights, by day and by night. They were beginning to put considerable physical strain on their pilots. Airlines' staffs and aircraft designers put their heads together to work out ways by which pilots could be relieved of some of the many tasks they were now shouldering, and to transform them from 'one-armed paper hangers' into managers. The first fruits of these discussions appeared in 1933, when Eastern Airlines (then known as Eastern Air Transport) began accepting a fleet of Curtiss-Wright Condor twin-engined biplane airliners (Fig. 1.2) fitted with the world's first autopilots. The Sperry A1 electro-hydraulic systems did much to reduce the physical efforts of flying big transports on the popular New-York–Miami route, with its notoriously bad East Coast weather in winter.

In the same year Wiley Post had a Sperry A2 autopilot fitted to his Lockheed Vega monoplane *Winnie Mae* for his solo, round-the-world flight (Fig. 1.3). By the mid- to late 1930s autopilots were in general use to ease the workload in multi-engined aircraft that were now weighing up to 20,000 lb. By that time re-armament had begun in Europe and the USA, and the first examples of the new generation of four-engined bombers were being laid down. The performance of Britain's new

Fig. 1.3 Lockheed Vega. Wiley Post, in 'Winnie Mae', made the first solo flight around the world in 1933, flying 15,596 miles in 7 days 18 hours and $49\frac{1}{2}$ minutes. A Sperry A2 autopilot was fitted *Lockheed California Company*

Stirlings and Halifaxes and the B-17 Fortresses and B-24 Liberators from American stables, with twice the speed of the Douglas DC-3 and Boeing 247 commercial 'twins', called for substantially better autopilots. In the UK pneumatic systems continued to rule (in essence, offshoots of the air-driven attitude and turn indicators), but the newer Sperry and Minneapolis-Honeywell autopilots in the USA were electrically driven.

In addition, the new US systems could be 'slaved' to the secret and much-lauded Norden bomb-sight. Much more complex than their commercial aircraft predecessors, these bomber autopilots weighed up to 250 lb. Nevertheless, like every other item of electronic equipment on the aircraft, they were 'add-ons', designed in isolation from the aircraft; it would be another twenty years before the word 'integration' entered the designer's vocabulary.

ORIGINS OF RADAR

The prospect and advent of the heavy bomber gave impetus to considerations of detecting attacking fleets at a distance, giving time for defensive forces to be alerted and placed in position. Lord Trenchard, first leader of the Royal Air Force, had since the 1920s held the view that 'the bomber would always get through' and, notably in the UK, his convictions promoted the search for electronic means of long-range detection. Early experiments to detect aircraft from the infra-red emission of their engines failed because the detectors could not be made sufficiently sensitive – their time had not yet come. But British research in another waveband, around 30 MHz, showed that metal aircraft structures generated echoes when illuminated by a ground-based transmitter (Fig. 1.4). The discovery was put to good use during the Battle of Britain, when not only could German bomber fleets assembling over the French coast be detected while still out of visual range, but also an assessment of their strengths could be made so that the response of the waiting Hurricane and Spitfire squadrons could be carefully tailored to meet each situation.

The Chain Home early warning radar was invaluable in defeating the daylight raids, and the German Air Force turned its attention to night bombing. To counter the new threat, Britain's Telecommunications Research Establishment initiated the development of a family of airborne detection radars. They were fitted, initially, to Blenheim 1 night-fighters and later to Mosquitoes (Fig. 1.5) and Beaufighters. With a range of only a mile or so (ground clutter intervened at longer distance), the intercepting aircraft had to be directed, or 'vectored', to the vicinity of the target by ground radar before the latter could be picked up. The fighters had two-man crews – pilot and radar operator, and considerable skill was needed to interpret the crude, 'raw-echo' displays. Nearly simultaneous radar developments were afoot in Germany, and, when two years later the RAF began to launch large-scale night raids, many crews and aircraft fell victim to similar radar systems flying in Messerschmitt Bf 110 and Junkers Ju 88 twin-engined, two-seat night-fighters, also controlled from the ground, though later tactics involved independent operation with improved, longer-range equipment.

At the beginning of the Second World War new and revolutionary techniques appeared, known as electronic countermeasures (see Chapter Ten). The early activities involved the subtle deflection of German navigation beams so that the bombers 'riding' them by night would drop their bombs harmlessly in open countryside instead of on cities such as Coventry and Derby. The idea of 'spoofing' the enemy's interrogation transmissions was to catch on widely, one of the first

Origins of radar

Fig. 1.4 On 26 February 1935 Dr Robert Watson-Watt used the Royal Aircraft Establishment Handley Page Heyford bomber, K 6902, to demonstrate, with the cooperation of the BBC transmitter at Daventry and a primitive radar receiver using a cathode ray tube made by Cossor, that an aircraft flying in the vicinity of the transmitter could be detected. The experiment was repeated in December 1936 and later an airborne transmitter and receiver detected echoes from the coast near Harwich.

The equipment was designed by Dr. E. G. Bowen and his colleagues at the Telecommunications Research Establishment at Bawdsey in collaboration with Dr Watson-Watt who was the Superintendent at the National Physical Laboratory at Slough.

This was the beginning of development of AI and ASV radar – so valuable in the Second World War   *Handley Page Association.*

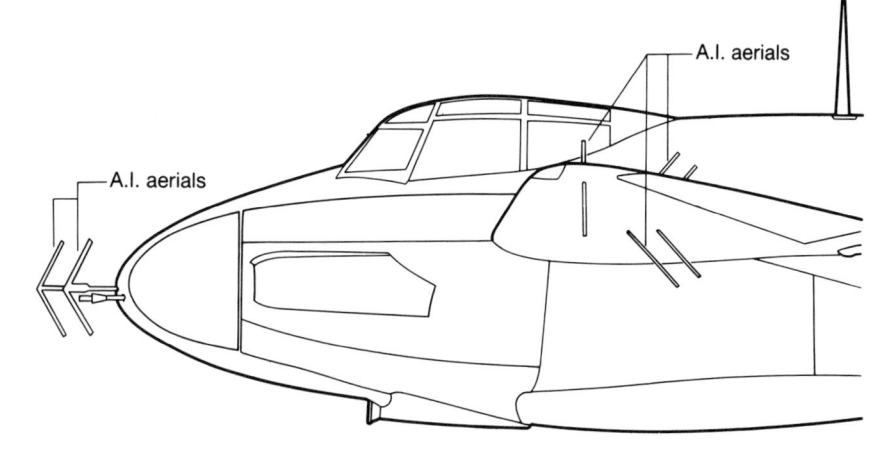

Fig. 1.5 De Havilland DH 98 Mosquito Mk II night fighter with early aircraft identification (AI) aerials on nose and wings   *British Aerospace, Hatfield*

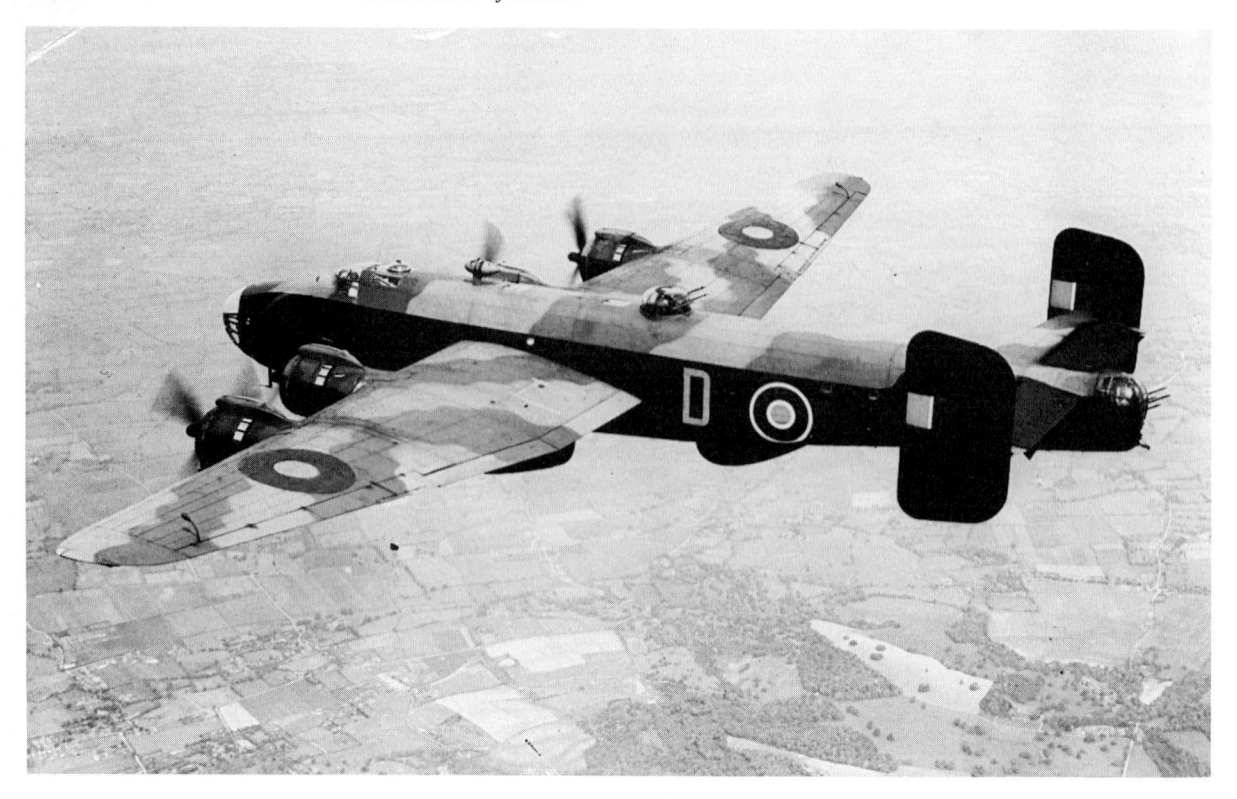

Fig. 1.6 Handley Page Halifax heavy bomber with H2S blister under rear fuselage *Handley Page Association*

airborne applications of the technique being code-named Window. Vast numbers of thin metal strips, acting as half-wave dipoles, were dropped by the bomber force on its way to the target to jam the German detection and tracking radars. Window, now called chaff, is still a greatly used and cheap method of jamming hostile radars.

The British night-bomber offensive over Germany highlighted the grave navigation and target-recognition deficiencies over a dark, rain-laden Europe that should have been appreciated years before. A combination of area navigation systems such as Gee, point navigation aids like Oboe and target indentifications devices like H2S (Fig. 1.6), however, greatly improved the effectiveness of the strategic night offensive. Gee was a system for broadcasting synchronised transmissions from three ground stations, resulting in an interference pattern over a very large area that could be measured and processed by an airborne receiver to provide a navigation fix up to distances of several hundred miles from the transmitters. Oboe was an intersecting beam system that, unlike Gee, could be used by only one aircraft at a time owing to the collision risk. Again based on ground transmitters, it was a precision device, and as such was used by individual aircraft to mark targets by means of flares to guide the main bomber force. H2S however was an airborne radar directed at the ground to 'paint' a map of the terrain, from which details of the target could be made out. An important derivative of H2S was the ASV (Air to Surface Vessel) radar carried by maritime patrol aircraft such as Whitleys, Sunderlands and Catalinas for detecting enemy submarines.

In all of the airborne radar development, the pacing factor was the need to boost the power level of the transmitted beam to increase range, and push up the frequency of the energy so that finer detail could be made out. The Chain Home ground-based early warning system worked at a wavelength of 50 metres and early airborne radars operated at around 1.5 metres. However, space limitations precluded large aerials, especially on night-fighters, and the need for better resolution with small antennas was a further spur to extend the frequency range. A major breakthrough came with the invention, in Britain, of the resonant-cavity magnetron, which could generate energy at around 10 centimetres, in what later came to be called the microwave region. Not until a crashed aircraft was examined did German engineers discover the new development. These centimetric radars had another advantage: they could use internally mounted dish-like reflectors rather than the external, drag-producing stick-type aerials.

Together with the development of radar (initially known by its British inventors as 'radiolocation', the more precise American term 'radar' as a shortened form of 'radio detection and ranging' soon came into universal use) came the first efforts in the twin technology of signal processing. Scientists were well aware that the raw radar echoes contained a large quantity of potentially valuable information locked up in them. In the UK the first efforts in this direction were made to improve the performance of radars for single-seat fighters, in which the pilot had little time to spare on radar management. Little headway was made in the UK, the German bomber offensive having petered out by the end of 1941. However US aircraft such as Navy Hellcats successfully employed pilot-only radars against the Japanese in the Pacific war. Signal-processing was eventually to become a huge activity, employing top brains and substantial budgets. Today it is recognised as the key in airborne radar performance. Much of the progress in today's systems is due to the work done by NASA, the US National Aeronautics and Space Administration, in processing the faint, blurred images of the planets returned by the first American deep-space probes such as the Mariners.

The new navigation and target-detection aids produced a flood of information that had to be digested as far as possible, and presented to the crew in a convenient way. Attention therefore focused on the improvisation of commercial, pre-war cathode-ray tubes for this purpose. Development over many years has produced a breed of tube with great resolution, and one moreover which is 'sunlight-visible', in other words, the picture or 'symbology' is bright enough to be seen even in the presence of direct sunlight, as may be the situation in the cockpit of a typical modern fighter with a transparent canopy permitting the pilot an all-round view.

Another development in the field of displays was the gyro gunsight, yet another application of the gyroscope principle. Pre-war fighters typically carried a traditional 'ring and bead' sight. While satisfactory in a stern chase, it provided no guidance to the pursuing pilot on how much lead angle to allow his guns in turning flight. Shooting proficiency was therefore based on pilot judgement. But with a gyroscopically compensated aiming point there was now much less room for error. Gyro gunsights continue to enjoy a large market among purchasers of light combat aircraft, but an even bigger one as aiming devices for ground-based anti-aircraft guns. With the advent of electronic data-processing, however, they were recognised as having great new possibilities. In the late 1950s and early 1960s they emerged as head-up displays, or HUDs. Continuing development, along with the introduction of fast analogue (and, later, digital) computing elements, led to devices of extraordinary versatility, with operational modes to suit every phase of flight-navigation, terrain-clearance or terrain following, combat or ground attack, and providing release and guidance information on every type of airborne weapon. Since the mid-1970s they have become the principal information interface between the pilot of the aircraft and his target. The first aircraft to be equipped with a HUD was probably the US Navy's A-5 Vigilante low-level carrier bomber.

In the decade 1945–55 aviation made the transition into the jet age. Almost overnight, fighters could fly half as fast again and reach much greater heights. By 1958 the first-generation jet airliners and military transports were flying. All had substantially greater performance than their piston-engined predecessors, in particular a much greater speed range. It was at this stage that some of the penalties of jet aviation became apparent. Aircraft became short of fuel very quickly, especially low down, and so more efficient navigation was needed to ensure accurate flying and prevent the pilot becoming lost. At the same time undesirable aerodynamic qualities began to manifest – Dutch roll in heavy transports and dangerous roll/pitch cross-coupling in the later Mach 1-plus fighters such as the F-100 Super Sabre.

These problems were not amenable to solution by changes or refinements in the design of the airframe itself. They were overcome, or at least minimised, by the introduction of control systems to provide a degree of 'artificial' stability to remove unpleasant motion in passenger aircraft and to protect the airframe from potentially catastrophic damage in the case of highly manoeuvrable fighters. The provision of artificial stability, or stability augmentation as it has come to be known, has been a requirement in most, if not all, swept-wing transport aircraft and fighters up to the present day. The simplest devices are known as yaw-dampers, comprising gyro-based motion sensors generating signals of suitable phase and magnitude that act on the rudder through an actuator to oppose any aircraft motion not commanded by the pilot. In some aircraft stability augmentation is needed in all phases of flight, in others flight may be continued with an inoperative system if speed and/or height are reduced. Nowadays yaw-dampers are fitted to many unswept-wing aircraft of quite low performance to improve their handling qualities.

The advent of yaw-dampers (some of which appeared as early as the latter years of the Second World War) marked the end of an age when electronic devices could be considered as optional, and not in any way associated with the performance or handling of an aircraft. From now on they would be given just as much consideration during the design stages as any other system.

The radio range beacon navigation system devised in the USA continued to be improved over the years, but was eventually replaced by another beacon system that in the 1960s was to attain world-wide acceptance: VOR, or VHF Omni-Range. Operating at much higher frequencies than radio range, it was more accurate and less susceptible to interference but – like its forebear – employed simple and cheap-to-build radio receivers virtually identical with those used for voice communication; indeed, the two functions for light-aircraft users are often packaged into a single box known as a nav-com.

The war in the Pacific, as the Americans chased the Japanese back to their homeland in a series of long-distance, island-hopping campaigns, showed up the need for a self-contained navigation system, one independent of ground beacons that could fade, be jammed or destroyed, and with global coverage rather than the local-area guidance of ground stations. American industry approached the problem by arranging a marriage between gyros (which by now had reached a high degree of perfection) and sensitive accelerometers to produce a new method known as inertial navigation. The first such system, built naturally enough by Sperry, emerged in 1950 and was flown experimentally in a Douglas DC-3 airliner. It was heavy, expensive, took several hours to warm up and reach a steady temperature, and exhibited severe drift. Ten years later IN systems (as they came to be called) were being fitted to a new generation of US high-performance military aircraft, and a decade later had been chosen as the principal navigation equipment on the

Fig. 1.7 Boeing 747 in British Caledonian Airlines livery *British Caledonian Airlines*

three wide-body transports – Boeing 747 (Fig. 1.7), Lockheed L-1011 TriStar and Douglas DC-10 – then in final design. They were certainly expensive: the three separate IN systems required on the 747 to meet US and UK airworthiness standards in respect of performance and ability to 'ride' failures cost around $300,000, but rapidly proved their worth.

Another self-contained navigation device developed more or less simultaneously with IN was Doppler, a radar that measures frequency differences between echoes from beams directed ahead of and behind an aircraft. Whereas an inertial navigation system measures acceleration, adds it up over a period of time to give speed, and a second time to provide present position, Doppler measures speed, from which one integration gives aircraft position. Doppler was considerably less expensive than IN because it employed a technology with virtually no moving parts; IN on the other hand used a host of delicate motors, synchros, slip-rings and costly gimbals and bearings. Doppler tended to be used in older, medium-range aircraft such as Boeing 707s and Douglas DC-8s. However IN did have the advantage that the attitude gimbals on which the accelerometers were fixed could supply pitch and roll information to flight instruments such as the artificial horizon (nowadays providing considerably more data than just aircraft attitude, and called the flight director) and autopilot or autostabilisation system.

In terms of performance, IN and Doppler tend to be complementary (see Chapter Seven). Inertial navigation equipment is very accurate over short periods of time, but drifts slowly owing to minute imperfections in the mechanical elements – gimbals, accelerometers and gyros. Doppler on the other hand can show quite large errors from one second to another as the aircraft flies over irregular ground and sea, but over a period they cancel out. Some long-range combat aircraft combine the two in a Doppler-monitored inertial navigation system, the two characteristics combined to provide a very accurate system over long ranges. For military aircraft with IN only, the drift problem is overcome by taking navigation 'fixes' from time to time *en route* to the target and resetting the system.

IN and Doppler are examples of what are now called 'area navigation' systems: provided the coordinates of departure are set into the equipment before take-off, they will compute the position of the host aircraft wherever it flies. The ground-based, so-called hyperbolic systems such as Gee also provide area-navigation performance within their coverage. By contrast, beacon systems show aircraft position with regard to fixed ground references, and external calculation is needed every time a fix is required. With the advent of microprocessors the computation can be run continuously and automatically, providing local area-navigation, or RNAV, performance.

Loran (Long-range navigation), basically a development of the Gee system used for navigation during the Second World War is now a popular and inexpensive navigation system. Based on ground-stations, it has the attraction, once again, of requiring only receiver technology. Another beacon-based system is Omega, which together with the US Navy's VLF Very Low Frequency network designed basically for submarines, is beginning to be widely used.

Meanwhile a satellite-based navigation system called GPS/Navstar is now undergoing operational trials in a variety of aircraft, ships and land vehicles. Initiated by the US Defense Department, Navstar is a network of 18 satellites in highly inclined, non-stationary orbits, each broadcasting time and position signals from which ground or airborne receivers may fix their position. Once again, as a passive (and militarily secure) receiving system, Navstar will be relatively inexpensive to the user. It will however, provide accuracy of a hitherto unknown order – errors of only a few feet are claimed. It is likely that the system will also be made available to commercial users such as the airlines and general aviation in the same way that Omega VLF was offered.

Communications technology has had something of a facelift (see Chapter Eight). Up to 1940 the standard communication frequencies were to be found in the 2–30 MHz band. But in that year the first VHF sets were produced. Operating at more than 100 MHz, they arrived just in time and had a new clarity of performance to benefit control of the RAF fighter squadrons during the Battle of Britain. Its line-of-sight coverage was a limitation at that time. Later still, UHF was developed, with even better performance. Meanwhile HF remained obligatory for long-distance flight, its exasperatingly unpredictable qualities calling for a full-time crew member in the form of a radio operator. But in the 1970s the advent of microprocessors opened up a way to process automatically the HF waveband and latch onto the purest signals at any one time. HF is therefore enjoying something of a revival, and what was once a moribund technology is now attracting a new generation of communication engineers to further improve the breed. A direct result of this development is that the radio operator has become an extinct, or nearly extinct, species, joining the navigator, who was deposed by automated navigation systems.

Airborne radar for combat aircraft has undergone intensive improvement and refinement. As related earlier, the world's first airborne interception radars became operational with the RAF in 1940. Its first recorded 'kill' occurred in November of that year as a result of the guidance provided to the pilot of a Beaufighter by his Mk IV set. The principal limitation was the need for a ground-control system (with a much more powerful radar) to guide the pilot to within a few miles of his target. The development of magnetrons to generate more power and the move to centimetric wavelengths produced a narrower beam with a greater power density, substantially increasing range.

These early pulsed radars had a major and fundamental limitation — they could not 'look down' to detect targets at a lower altitude because their relatively small echoes were lost amid the flood of returns from the ground. The way forward was through an application of the Doppler principle, in which echoes whose movement corresponded to the speed of the aircraft were suppressed. But for this operation to be possible the shape and phase of successive pulses had to be very closely similar, in other words, the generating device was required to produce pulse trains with great coherence. The magnetron could not do this, and radars had to await a new development, the travelling-wave tube TWT which has another advantage in being able to generate pulse trains with a much higher average power level, giving greater detection ranges. The current generation fighters, such as Panavia Tornado (Fig. 5.19) F-15 Eagle and F-16 Falcon can operate in air-to-air, air-to-ground, air-to-surface vessel modes for attack, ground-mapping, homing on to refuelling tankers and terrain-following, and scan the skies for perhaps 100 miles ahead while automatically tracking maybe a dozen targets and drawing the pilot's attention to those which pose the greatest risk or which represent the most attractive targets.

Whilst electronic countermeasures in the form of Window were being used more than forty years ago, general acceptance of the need for combat aircraft to have some form of defence against missiles and their tracking radars was long delayed. The aircraft flown by the late Gary Powers did not have such protection, and its destruction on a reconnaissance flight over the USSR in 1960 in what became known as the 'U-2 incident' alerted everyone to the deficiency. The first of many US combat aircraft to fall victim to Soviet missiles in Vietnam was downed five years later. US government and industry moved swiftly, and combat aircraft were equipped with receivers on wings, nose and tail that could detect radiation from missiles or missile control centres and give warning to the pilot of the direction of approach. The pilot could then either take evasive action or fire chaff (the modern equivalent of Window) to jam radar-guided missiles, or intense flares to fool weapons that home onto the infra-red radiation from the prospective target's engines (see Chapter 10).

The invention and subsequent development of lasers has made an impact on aviation in two areas (see Chapter Seven). The first application is as a replacement for conventional gyro-based attitude sensors in inertial navigation systems. Laser beams generated by a central source are projected in opposite directions around a small, triangular glass cavity. Any rotation of the cavity is revealed as a phase difference between the two beams since they now travel different distances. In a practical ring-laser gyro, or RLG as the devices are termed, three such cavities are mounted in a box at right-angles to one another to sense pitch, roll and yaw. Together with their corresponding accelerometers, they are rigidly mounted in the box and are therefore not space-stabilised, unlike a conventional gyro IN system. A major advantage is that there are no moving parts to go wrong. The second application is target-ranging. Accurate weapon delivery to a ground target is critically dependent on a knowledge of the distance of the target from the attacking aircraft so that the weapon can be established on an appropriate trajectory. While range is usually measured now by means of radar, the beam width is quite wide at the commonly used, x-band (10 cm) wavelengths. But laser energy, typically with a wavelength of 10^{-4} cm, can be directed into a much narrower beam so that the measurement error, when the beam illuminates the ground at a grazing angle, is very much less.

The extraordinary pace of development in airborne electronics over the past two and a half decades is due largely to the advent of two important technologies. The first was the invention of the transistor by US scientists in the late 1940s, the second, the widescale change from analogue to digital techniques in the 1970s.

TRANSITION FROM THERMIONIC VALVES TO SOLID STATE

Transistor technology entered aviation during the middle 1950s and revolutionised aircraft design. Within a short time it had displaced thermionic valve (vacuum tube in US terminology) methods for virtually all airborne applications. Whereas the latter brought with it fragility and susceptibility to damage, great bulk and weight, large power demands and onerous cooling problems, the new devices were robust, small and light, used much less energy and created fewer demands on cooling facilities. The early devices were made from germanium but silicon superseded this material on account of its superior high-temperature reliability. Because of its advantages over valve equipment, transistorised systems came to be used for a multitude of new functions. Continued development led to progressive reductions in size and weight and the appearance of 'chips' containing many hundreds or thousands of transistors. All of these advantages were embodied in the description 'solid-state'.

Digital systems While these new circuits offered so many advantages, they remained analogue in nature, just like their valve forebears. Analogue systems suffer from drift and from inherent and difficult-to-eradicate errors, tend to be heavy, and are not amenable to rapid modification to suit changed requirements. But the advent of microelectronics (see Chapter Three) paved the way for airborne computing because great capacity and speed could now be accommodated within an acceptably sized package. The first computer designed specifically for airborne use was the Autonetics Verdan (versatile digital analyser) designed for the North American A-5 Vigilante, a US Navy carrier bomber that became operational in the early 1960s. Britain's near-contemporary TSR 2 tactical strike and reconnaissance aircraft (Fig. 3.1) also used Verdan during its brief, flight-test only career. Verdan comprised several quite large boxes and, as a central information-handling device, processed signals from every other major electronic system. At the same time progress was being made in the other ingredient necessary to exploit the full advantages of digital information processing – data buses. The signals handled by Verdan were tailored to individual equipment characteristics and processed independently. But with the data-bus technique signals were distributed from a central computer around the aircraft on a single cable pair, rather like a domestic ring-main electricity-supply system. This has now become the standard technique for the dissemination of information.

The efficiency of digital systems is measured by the cost (in money, weight or bulk) required to accomplish a given function. It has so improved even over the past decade that distributed processing (whereby a given system does its own processing) may now be preferable to central data-handling. The advantages of digital over analogue data handling and processing can be summarised as better reliability, lower weight and volume, more relaxed power and cooling requirements, greater flexibility for change or modification, and the vastly greater potential for

Transition from thermionic valves to solid state

self-test. The introduction of microprocessors in the 1970s made it possible for the benefits of digital technology to extend even to the smallest, fly-for-fun aeroplanes.

In that decade a revolution was also taking place in the presentation of information to pilot and crew. The increasing complexity of high performance commercial and military aircraft and the need for their efficient operation began to result in such a flood of data that crews were becoming overworked, to the detriment of safety and effectiveness. The answer lay in filtering information, perhaps adding some form of automatic priority for different types of information, and presenting it on television-like cathode-ray tubes (see Chapter Four). Military aircraft were the first to benefit, the early displays being black and white. Apart from being able to process the information into the form best suited to the mission, the other great advantage lay in the huge savings now possible in valuable panel space. No longer would a brake-pressure gauge take up permanently as much as nine square inches for the sake of just a few seconds of reference at the beginning and end of each flight. In future brake-pressure readings would form just one of many modes, the CRT screen at other times being used for navigation, target characterisation, or perhaps to display a hydraulic circuit diagram to help diagnosis of a fault in flight.

Commercial acceptance of the 'glass cockpit', as these multi-function CRT flight-decks came to be called, arrived with the launch by Boeing in 1978 of its twin-engined 767 airliner, closely followed by its smaller relation, the Boeing 757 (see Fig. 1.8). In these two airliners, virtually all flight-deck information is presented to the crew on six rectangular-face screens: two for each of the two

Fig. 1.8 Boeing 757/767 prototypes *Boeing Airplane Company*

The Evolution of Avionics

Fig. 1.9 Boeing 757/767 Century 21 flight deck *Boeing Airplane Company*

pilots showing navigation and attitude information, and two between them displaying engine or equipment data (Fig. 1.9). These so-called 'new generation' airliners have substantially digital electronics, and are the first of their kind. Their arrival heralded the demise of yet another crew member, the flight engineer.

The authority of airborne electronics took a leap forward in the early 1970s when the US Air Force ordered two types of experimental fighter aircraft, not with any idea of production but just to see how all of the then-developing technologies could be brought together into a viable combat aircraft. One of the LWF (Light Weight Fighter) prototypes was the General Dynamics YF-16, which broke totally new ground by the way in which it married aerodynamics and electronics. Designers had long recognised that aircraft could be made smaller and more efficient if somehow they could be permitted to incorporate a degree of instability. The more instability that could be allowed, the greater the benefits in weight, size, manoeuvrability and thus cost. The problem is that, as aircraft are made progressively less stable, the more difficult they are to handle. The handling difficulty becomes too great for the pilot to manage long before the real benefits begin to emerge. The answer is to incorporate an advanced autopilot (more accurately, flight control system) that could insulate the pilot from the dangerous 'raw' handling characteristics, replacing them with computer-designed synthetic ones (see Chapter Five). The stumbling block, a largely psychological one, was that such a system would have to operate with full authority from take-off to touchdown. The failure rate would have to be extremely small (of the order of one complete failure in ten million flying hours) because, by definition, such a failure would result in the loss of the aircraft and, almost certainly, its pilot. The essential new ingredient was a new technology known as fly-by-wire, or FBW. Instead of pilot commands being transmitted to the aircraft controls by reassuringly solid mechanical rods and tubes, they would now be sent in the form of electrical signals. At the same time, flight control computers would 'shape' the commands, automatically comparing them with the attitude of the aircraft to generate the desired flight-path.

Although previous aircraft had some degree of FBW (the remaining control being provided by a safety system of mechanical linkages), the YF-16 was the first aircraft to go all the way and dispense entirely with such safety features. By 1975 the costly Vietnam war was showing the need for a small but effective light fighter, and the YF-16 was adopted, in production form, as the F-16. Today it has become the best-selling Western fighter. Its original flight-control system, analogue in nature, is already being replaced by a digital system. Whilst the F-16 is fairly conventional in appearance, more advanced fighters being developed in Europe and the Middle East exploit the benefits of FBW to a much greater degree by the adoption of close-coupled, tail-first configurations. Meanwhile the first airliner with an all-FBW flight-control system is in service and shows great promise: the Airbus A320 had a greater initial order-book than any commercial aircraft before it a year before first deliveries to the airlines.

All of these digital systems call for a huge effort on the part of their designers to write complex instructions for the computer that are efficient in terms of speed and use of capacity, but yet contain no hidden errors to surface, catastrophically, in flight. The task of developing good software (as these instructions are called) is a difficult one that has been badly under-estimated by many top-class companies. For example, the flight-management system for the Boeing 767 was originally estimated to call for 70 K (i.e. just over 70,000) words of memory and instruction. During development the actual requirement was found to be nearly three times that figure.

The requirement for interchangeability of equipment from different manufacturers called for industry-wide specifications in the 1950s and 1960s on such things as box size, performance, accuracies and the characteristics of inputs and outputs. These specifications were embodied in MIL-spec (for the military) and ARINC (Aeronautical Radio Incorporated) documents and are now followed by virtually all airframe and electronics companies (see Chapter Two).

For the future, and particularly for military aircraft which have to carry many more systems than commercial transports but in a more confined space, the emphasis is on system integration. It may be wasteful to have separate power supplies for each system, for example. Likewise, despite the advantages of distributed processing already mentioned, a central computing facility may be judged more efficient. Already this trend is noticeable, with individual systems beginning to lose their identities. Thus a nav/attack suite formerly comprising a radar, Doppler, inertial platform, UHF and VHF radio, central computer, weapons management system, radar warning and countermeasures dispensing system, and head-up and head-down displays. These have been thought of and developed as a suite of sensors to measure range, position, attitude and receive and transmit signals external to the aircraft; an overall computing facility; and a set of displays. It is significant that, already, some manufacturers are specialising in sections and no longer producing complete systems. Several manufacturers are, for instance, supplying Doppler sensors to measure groundspeed; with a standardised output format, they can be used in conjunction with other manufacturers' processing and display equipment.

A very important advantage of digital systems is that some of the processing power can be allocated to diagnostic, or self-test, functions (see Chapter Three). This is an area that is rapidly growing in both military and commercial aircraft. Earlier analogue systems were able to incorporate a limited amount of self-test circuitry, but digital technology is required to really exploit this powerful new tool. With self-test, the pilot can run through an automatic test procedure before take-off that will signal any faults and provide extra confidence that the mission will be a success. In flight, the system automatically runs through the test schedule perhaps a hundred times a second, advising the pilot or crew of a failure and perhaps reconfiguring a system so that the fault can by bypassed. The need for rapid and automatic reconfiguration is notably urgent in the case of a FBW flight-control system, in which a faulty channel has to be switched out before it can jeopardise the safety of the aircraft; a typical pilot reaction, taking several seconds, could be far too slow. Another great advantage of the self-check facility is that it can isolate a fault to perhaps one box or one electronic card within a box, so that the task of the maintenance engineer is greatly eased.

The move towards greater automation is being expedited by the huge, industry-wide effort to produce a new generation of chips with substantially greater performance; the first of these 'superchips' are now being incorporated into standard items of equipment for flight trials. Their advantage lies in their extremely high processing speed, and so initial applications will be where this characteristic is most needed, notably in keeping tabs on rapidly changing radar targets and to detect the approach of hostile missiles and compute suitable evasive manoeuvres. They will be far too expensive, initially at any rate, for applications in which speed is merely a convenience.

CHAPTER 2 SYSTEM DESIGN CONSIDERATIONS

SIMON J. K. WALKER MBA

Mr Walker is an aviation consultant in a specialist company with offices in Washington, London and Hong Kong.

Qualified as a commercial pilot he has worked for corporate flight departments and airlines in the UK and was a Regional Technical Sales Manager for British Aerospace Civil Division. He followed this post with one as Marketing Manager for the Sperry Flight Systems Group of Honeywell.

Mr Walker has an honours degree in Economics and a Master's Degree in Business Administration.

THE ETHOS OF CIVIL AIRCRAFT SAFETY

Enshrined in the purpose of regulatory bodies such as the Civil Aviation Authority and the Federal Aviation Administration lies the question of aircraft safety. This underlying requirement is applied to all facets of the civil aircraft industry, from performance testing of prototype aircraft and systems to operator procedures. The same regulations and approach to equipment design and testing are applied to system and component manufacturers as to the manufacturers of the total aircraft.

This section will examine the role of the regulatory bodies in determining the standards for approval of avionics equipment. These include compliance with detailed test procedures as laid out in the Radio Technical Commission for Aeronautics (RTCA) document DO-160B or the equivalent EUROCAE document ED 14B as well as the demonstration of compliance with fail-safe procedures for specific elements of the avionics equipment. Whilst there are some variations between different regulatory bodies, the principles applied are broadly similar, with the different countries generally applying alternative values to the importance of certain procedures or items of equipment.

The procedures and approval systems to be described refer to those adopted by the European Authorities, specifically the UK Civil Aviation Authority, which provided invaluable assistance in writing this section.

Equipment approval The regulatory authorities are generally charged with the legal responsibility of ensuring, as far as is practicable, the safe operation of aircraft that fall within their jurisdiction. For the Civil Aviation Authority, these responsibilities are stated in the UK Air Navigation Order, covering design, construction and workmanship of any equipment carried aboard an aircraft.

The control systems for approval of Airframe Parts and of Equipment, excluding radios, are defined in the British Civil Airworthiness Requirements (BCAR) document A3-3. The cornerstone of the approval system is the Declaration of Design and Performance (DDP). This is a declaration of performance to a specific design standard of equipment and is produced by a signatory (organisation) approved by the Civil Aviation Authority. The Authority does not countersign DDPs.

The system thus relies on the method of approving organisations in terms of their design, manufacturing and quality control expertise, followed by an on-going involvement by the regulatory body in determining continued airworthiness of components and the competence of the approved organisation.

As in other fields that have different international bodies with similar aims, there is a continuing programme designed to ease cross-border approvals. In Europe this has manifested itself in the creation of the Joint Airworthiness Requirements (JAR), formed from the national requirements of Belgium, France, Italy, the Netherlands, Sweden, the United Kingdom and West Germany. In addition, the provisions of a Government to Government Bilateral Agreement allow for the approval, by the CAA, of materials parts or appliances to the Federal Aviation Administration's Technical Standards Orders (TSO).

Design objectives The design objectives for the manufacturer and the design and performance requirements of the regulatory authority are necessarily different in emphasis although broadly similar in intent. For the former, the objectives from a system point of view will include: the provision of required performance; acceptable levels of availability and failure conditions; ease of use and maintenance. For the regulator, the design must meet certain results in terms of safety, whilst leaving the manufacturer free to design the system.

These requirements mean that the system, when operating without fault, must perform its intended function under all operating conditions expected in service. In addition, systems must be designed so that there is an inverse relationship between the probability of the occurrence of a fault and the severity of its effect.

These issues will be discussed in more detail later. The requirements for availability of function can be achieved by the provision of multiple systems and standby services, whilst the integrity of the system is ensured by the provision of appropriate monitoring devices, capable of detecting failures, and features that counteract the failure effect.

The performance standards of avionics equipment can be defined in relation to a number of documents. For flight instruments and radios, specifications are issued by the Civil Aviation Authority, the Federal Aviation Administration, ICAO, EUROCAE, RTCA and the British Standards Institution.

Specifications for aircraft systems will be issued by the aircraft manufacturer, following consultation with the engine, avionics and systems manufacturers.

Limitations as to use of any of the equipment carried may form part of the Approved Flight Manual for the aircraft, as a component of the aircraft's Certificate of Airworthiness.

System safety For investigation purposes, avionics systems can be grouped in one of two categories. The first includes those systems, the failure or loss of which may indirectly affect safety. The second, those systems, the loss or failure of which has a direct effect on safety.

A set of airworthiness objectives is required which can classify the possible effects of failure conditions within aircraft systems according to the degree of hazard inherent in each effect. The failure condition may result from a single failure or a combination of failures.

The Civil Aviation Authority's airworthiness objectives are summarised below.

Effect	Permitted Probability Range*	Probability Classification
Minor	Greater than 1×10^{-3}	Frequent
	1×10^{-3} to 1×10^{-5}	Reasonably probable
Major	1×10^{-5} to 1×10^{-7}	Remote
Hazardous	1×10^{-7} to 1×10^{-9}	Extremely remote
Catastrophic	Less than 1×10^{-9}	Extremely improbable

*Per hour of flight

The Federal Aviation Administration adopts a similar approach, although the boundaries between each of the elements vary.

The development of complex, inter-related systems means that the system as a whole needs to be considered in order to meet the requirements of the Joint Airworthiness Requirements (JAR) 25.1309. This results in a method of safety assessment to identify and analyse critical failure modes in order to demonstrate compliance with the airworthiness objectives.

The system safety assessment

The objectives of the system safety assessment take the form of a statement listing the possible effects that the system can produce, together with a probability classification for each effect. In some cases, only part of a system may require the application of this safety system technique.

Once any potentially hazardous or catastrophic effects have been identified, it is necessary to identify the conditions which will produce these effects. Examples of hazardous effects include hazardously misleading information on the attitude displays, or a hardover condition beyond limits on the autopilot. Catastrophic effects would include the loss of all displays on the attitude displays, or the loss of glide-slope guidance, without warning, during a Cat III autoland.

An analysis must be performed to identify all failure conditions beyond those assumed to have only minor effects. Critical functions may well require further analysis using numerical and statistical techniques based on evidence relevant to the components used.

The increasing use of digital systems has increased the difficulty in this area of failure analysis for two main reasons. First, the shared use of large areas of circuitry and central processing may result in the widespread failure of multiple functions. Second, as software does not fail in the conventional analogue sense of the word, its failure modes cannot be analysed using conventional techniques.

Extensive use of monitors and a dedicated software design process can result in the necessary assurance that the quality of these systems meets the safety objectives of the system.

Environmental factors also play a large role in system design. These disciplines, covering climatic conditions, shock, vibration, electromagnetic compatibility, internal and external electromagnetic interference and lightning strikes, for example, require comprehensive and stringent tests to be performed on equipment. Under laboratory conditions the tests stress the equipment over a range of environments to ensure that it will perform under any likely environment in actual service.

The test standards vary with the use to which the equipment will be put. The Federal Aviation Administration, for example, differentiates between pressurised and non-pressurised aircraft, the height and speeds that the aircraft will operate and size of aircraft, among other variables.

System Design Considerations

Performance analysis A method of performance analysis involves the establishment of a statistical distribution for each critical performance parameter, and then determination of the probability that the performance of the system will be such as to create a hazard.

Simulations are normally performed to collect data, followed by flight testing to support the results of the simulation. The outcome of this performance analysis should allow for the establishment of system limitations, flight crew procedures and maintenance issues as well as minimum equipment lists for flight dispatch.

The architecture of a system must be designed to ensure that failures do not affect both control and monitoring functions. This is accomplished by segregation of vital components so that a single external failure source does not result in multiple system failures. Physical and environmental causes can be eliminated by the use of separate locations for duplicated equipments. Similarly, electrical power supplies, increasingly utilising digital data-buses, must demonstrate that interrupted supply to one bus does not affect the continued operation of systems supplied from another bus.

The certification of many modern systems requires the verification of software code. Although not certified themselves, software modifications will be subject to the same control as other modifications. The principles of design and control procedures for software are addressed in the EUROCAE/RTCA documents ED-12A/DO-178A 'Software Consideration in Airborne Equipment and System Certification'.

Software integrity can be improved by a number of methods, including extensive testing and analysis of system design. One method that is used with increasing frequency is that of software 'dissimilarity'. By this method, two types of processing are performed by the system, and the results compared prior to activation of the signal sent. Alternatively, it is feasible to use the same software and perform a comparative check by the use of dissimilar processors. The symbol generator on the McDonnell Douglas MD80 family (EFIS equipped) utilises this second approach for certain critical areas.

Certification task A summary of the certification task, some of which was described above is as follows:

(a) investigate and approve the equipment;
(b) investigate and approve the manner of installation;
(c) agree the system design;
(d) agree the software design and standards;
(e) demonstrate compliance;
(f) agree limitations and procedures;
(g) agree maintenance requirements;
(h) agree minimum equipment for dispatch;
(i) monitor in service.

ARINC SPECIFICATIONS

Background Throughout this book reference is made to ARINC characteristics or specifications. These represent the activities of a group established to ensure compatibility between avionics equipment suppliers to the commercial air transport world. The following foreword appears with each release of an ARINC specification, and is reproduced in full, so as to adequately acknowledge the extent of that corporation's work.

Aeronautical Radio Inc., is a corporation in which the United States airlines are the principal stockholders. Other stockholders include a variety of other air transport companies, aircraft manufacturers and foreign flag carriers.

Activities of ARINC include the operation of an extensive system of domestic and overseas aeronautical land radio stations, the fulfilment of systems requirements to accomplish ground and air compatibility, the allocation and assignment of frequencies to meet these needs, the coordination incident to standard airborne communications and electronics systems and the exchange of technical information. ARINC sponsors the Airlines Electronic Engineering Committee (AEEC), composed of airline technical personnel. The AEEC formulates standards for electronic equipment and systems for the airlines. The establishment of equipment characteristics is a principal function of this Committee.

An ARINC equipment characteristic is finalized after investigation and coordination with the airlines who have a requirement or anticipate a requirement, with other aircraft operators, with the military services having similar requirements, and with the equipment manufacturers. It is released as an ARINC equipment characteristic only when the interested parties are in general agreement. Such a release does not commit any airline or ARINC to purchase equipment so described nor does it establish or indicate recognition of the existence of an operational requirement for such equipment, nor does it constitute endorsement of any manufacturer's product designed or built to meet the characteristic. An ARINC characteristic has a twofold purpose, which is:

(1) To indicate to the prospective manufacturers of airline electronic equipment the considered opinion of the airline technical people, coordinated on an industry basis, concerning requisites of new equipment, and
(2) To channel new equipment designs in a direction which can result in the maximum possible standardization of those physical and electrical characteristics which influence interchangeability of equipment without seriously hampering engineering initiative.

As is implied from the statement of activities, an ARINC specification or characteristic is defined by reference to the end user's needs within the constraints of the equipment manufacturer to provide it at a cost that is acceptable to both. There is no regulatory pressure to produce particular design or size specifications.

The AEEC is concerned with standards of equipment and airborne systems, in particular where they affect operational maintenance and engineering functions within the industry. In this respect, the references to ARINC that were made previously, are, in many ways, too limiting. Avionics equipment is often regarded as the only area of involvement. However, the detailed requirements of supplier and purchaser go far beyond. Two main areas of interest are the size of box for each item of equipment plus its associated racking and the cooling required for the equipment.

From the aircraft manufacturer's viewpoint, as the overall co-ordinator of design, these detailed specifications are a crucial element in determining size of the avionics bay as well as space and air supply around instruments. This in turn will affect the economics of a particular aircraft and also have a bearing on emergency power supply, as a function of the aircraft's primary role – long or short distance, number of engines/generators for example.

Air transport racking (ATR) It is clearly of interest to the airlines and, to a lesser extent, the military community, to have as high a degree of commonality, and hence competition among suppliers as possible. As such, the assembly arrangements, connections and overall dimensions of airborne avionics components must be standardised in the interests of airline capital commitment whilst the minimum stowage space for the equipment must be achieved. In this way maximum customer choice is facilitated.

The extensive experience of manufacturers in recent years and the continuing quest for more powerful and smaller chips and computers have enabled aircraft builders to allocate less space for avionics bays, whilst preserving capabilities.

The main advantages of standard box sizes are in the area of navigation and communication equipment, specifically transmitter and receiver components. Here the airlines have a competitive choice between the major suppliers, with decisions based on their own assessment of the desirability of a particular supplier. Similarly, aircraft manufacturers will specify a maximum size of box for specific components. An example would be the Symbol Generator element of an Electronic Flight Instrument System, where the SG must not exceed $\frac{1}{2}$ ATR – a measure of box size.

Historically, the origins of standardisation of box sizes can be traced to the late 1930s, when United Airlines and ARINC established an industry standard racking system which was called the Air Transport Radio (ATR) unit case. ARINC specification number 11 identified three sizes of ATR cases: full ATR, $\frac{1}{2}$ ATR and $1\frac{1}{2}$ ATR, all with the same height and length. At the same time, common types of connector pins were established at the rear of the boxes for wiring connections. The use of common sizes of unit cases and standardised racking systems has also resulted in reduction of damage to the connector pins due to misalignment of the connectors.

The United States military authorities adopted the basic features of the ARINC specification, but amended it to allow for variation in height and length as well. This was referred to as the JAN Specification C-172.

Similarly in Britain, ideas on standardisation were adopted, although these differed considerably from the United States system. The British system allowed for greater freedom in the choice of widths, with variations of one inch between each size. Various lengths were catered for, as well as offering much taller boxes than the US alternatives.

This polarisation of views meant that, for practical purposes, American manufacturers would produce equipment for American aircraft manufacturers using the ATR system, whilst the British manufacturers would do precisely the same, but for different end users. This undoubtedly contributed to the relative lack of success of some British designs in the 1950s and 1960s.

Current thinking among airline maintenance personnel is directed towards a box design standardised to the point where the only variable is width. However, this must not preclude the availability of particular suppliers or a reduction in reliability for this reason alone. At the present moment, the ATR design is fixed in height at $7\frac{5}{8}$ inches. In addition, the 'short' and 'long' alternatives in case length are only applied to particular sizes of box. For example, the $\frac{1}{2}$ ATR is the only box size that has the 'short' option.

ARINC Specification 404 was established to clarify the remaining variables in box size in order that there could be true interchangeability between suppliers. These last areas concerned dimensions on the boxes for the position of connectors and cooling ducts, all measured in a particular way.

The outcome of that specification was that several new case sizes were established, such as the short and long $\frac{1}{4}$ ATR, short and long $\frac{3}{4}$ ATR and the short and long $\frac{3}{8}$ ATR. New Cannon DPA connectors were specified, as well as the option of specifying connector pin indices that would preclude the fitting of the wrong box in a particular rack position.

An approach that is gathering momentum is derived from a European standard, now being adopted by the United States avionics manufacturers. This specifies box sizes in terms of MCU. An 8 MCU box is virtually equivalent to the ATR size, with the range of options being from 1 through 8 MCU, all with the same height and length, with Cannon DPX-type connectors dimensionally specified. The trend now is to reduce the space required for specific avionic requirements, often combining previously separate devices. One example of this is the ADIRS for the Airbus A320. This combines the inertial reference system (IRS) with the air data computer (ADC), all in the space previously required for the inertial system. This has had the effect of reducing the total space required in the avionics bay by two 4 MCU air data computers.

It may be that this trend towards standardisation, probably now at a stage where further development is unproductive, is beginning to be of less importance, for two reasons. The first relates to the reducing size of components as discussed earlier, although there are still many merits in adopting certain standard features. The second is that aircraft are becoming more and more identified with a single supplier from the early certification phase of the aircraft's development. For example, a significant proportion of the ATR 42 and 72 from the avionics point of view is identified with Honeywell-Sperry products. To an even greater extent, the McDonnell Douglas MD 11 wide-body transport has Honeywell-Sperry as systems integrator. Radio navigation and communication options will remain available, but will be of reducing importance in the total package.

REDUNDANCY

The development of smaller-sized avionics equipment plus the increased use of digital data-buses has led to dramatic improvements in the levels of redundancy that a modern airliner offers. An increasingly competitive environment, with high frequency, short turn-round operations characterising much of the air transport industry, has led to a requirement for ever-improving dispatch reliability. This can be achieved either by improving the reliability of individual components, or by increasing the levels of redundancy of the components on board the aircraft. This will result in discrete parts of the system being capable of backing up failed units.

An increasingly common method of improving the redundancy level of an aircraft system is to provide a component with a higher level of capability than is normally used. An example of this would be the use of a common symbol generator aboard an aircraft, with a fifth, identical unit driving a map display when systems operation is normal. This map display, for example, the multi-function display (MFD) on a BAe 125 business jet, is not a required component for normal flight. If a primary symbol generator or display unit were to fail, this additional unit could provide back-up, allowing continued operation of the aircraft until such time as the operator was at a suitable repair station.

For components that are a vital part of the operating system, a spare unit can be carried aboard the aircraft. This can be either a 'hot' spare or 'cold' spare – the former is connected to the data bus, ready to be operational in the event of component failure, the latter method would require engineering assistance to bring it on line, but would usually be a swift operation, minimising aircraft downtime.

As implied earlier, this trend is more prevalent in situations where the engineering infra-structure is less developed. This can be due to a number of factors, such as remoteness of the airport being served, inexperience of local engineering staff, where the operator is reluctant to allow any maintenance beyond turn-round checks to be performed, or if the financial environment in which the airline is operating precludes large overheads in the form of additional engineering staff.

The outcome of these problems requires a measure of collaboration between supplier and end user so that the problems can be identified and suitable steps taken to minimise the inconvenience to airline and passengers.

RELIABILITY

Equipment reliability is an area that has received considerable attention over the years. Improvements in capability of system design have been impressive, but, without reliable components, the benefits are of little consequence. A previous section discussed the concept of redundancy in systems; here the question of reducing failure repetitions is addressed.

Two measures of equipment reliability are generally used, both related, but dependent on factors that differ. The most often used is that of mean time between failures (MTBF), a statistical analysis of component failure, either of the whole device as far as the airline is concerned, or at part level for the manufacturer or maintenance organisation. The second term is mean time between unscheduled removals (MTBUR). This refers to the number of times that a component is removed from the aircraft on the ground of suspected failure, irrespective of whether it has been subsequently proved to have failed.

Equipment guarantees are generally written around MTBF figures, although many airlines attempt to have the MTBUR statistic used, as there are clearly more removals under this measure. From the manufacturer's viewpoint, there is little or no control of an airline's maintenance policy; without adequate safeguards against the removal of components for expediency, the costs of warranties would escalate to an unacceptable level.

Reliability statistics and experience form an important element in the consideration of life-cycle costs for equipment. This refers to the total cost of owning and operating the aircraft, or its associated subsystems. Initial costs, capital and finance charges, overhaul costs, routine maintenance, unscheduled maintenance and repair, and labour costs in removing components are all part of the equation of life-cycle costs. Developments in avionics have generally led to an increase in ownership costs when compared with the earlier designs – electro-mechanical instruments compared with EFIS being a prime example. However, the improved reliability and reduced overhaul costs often mean that the true cost to an airline over the planned use of the aircraft in its fleet will be lower for the developed systems. Electronic components are currently demonstrating improvements in reliability in excess of twice that of the earlier electro-mechanical devices, whilst ring laser gyro technology offers reliability figures many times that of the devices replaced. In both cases, this is a result of removing the weakest link in the system, usually a mechanical device.

Because of these improvements, warranty periods offered by the manufacturers have been steadily increasing. From an early start of six months, a figure of two years is now the industry standard, with three and sometimes more years offered on some equipment. Competitive pressure and airline demands have assisted this change, but the manufacturers have responded with better products.

BUILT-IN TEST EQUIPMENT (BITE)

Built-in test systems are an integral component of modern avionics design. They are designed to provide maintenance assistance to confirm pilot-generated fault reports (PIREPS) and to improve the accuracy of identification of a failed component and its subsequent test after rectification and re-installation. The BITE is designed to provide a continuous, integrated monitoring system, both in flight and on the ground, available to maintenance personnel whenever power is applied to the aircraft.

Most modern aircraft use cockpit displays and avionics bay read-outs to provide access to the BITE generated data which is retained in the non-volatile memory of the aircraft computer system. This facilitates post-flight confirmation as well as storage of flight segment data which is useful for further analysis after return to the main engineering base. Digital technology has undoubtedly increased the potential of BITE concepts to give greater diagnostic capability.

The design objectives of an integrated BITE system include: a desire to minimise on-aircraft maintenance time; a reduction in the removal rates of unconfirmed failed units; facilitation of identification of failed units and their associated interfaces. These goals have become more important as aircraft systems become more complex, with an associated reduction in line maintenance technicians' ability to carry out rapid repairs. Also, the trend away from airline employed engineering staff remote from the main base, particularly in North America, has reduced the line station capability still further. Lastly, the increased cost of line replaceable units (LRUs) has made the provisioning of line stations too expensive for many airlines, particularly if the faults reported and the remedial action taken do not result in the fault being cleared. BITE can help provide the base engineers with an early warning of which systems have failed, once the line engineer has identified the fault using the system's own diagnostic capability.

To explain the capabilities of the BITE concept, a system in use on the digital flight control system of a modern jet will be used as an example.

BITE system design conforms to certain guidelines established by the designer, in association with airline and aircraft manufacturer representatives. These will generally cover the scope of the detection capability. Clearly, it is important that a significant percentage of faults is correctly identified and isolated. Most systems aim for a figure of 95 per cent of all steady state faults being so identified. Once reported by the crew, it is important, within the terms of reference of the BITE concept, to isolate the fault expeditiously. This will require simple operating techniques for the average line-maintenance technician, with faults displayed in alphanumeric symbols rather than codes that require reference to a maintenance manual.

As far as possible, all diagnostic checking should be performable by a single maintenance engineer. Some checks, particularly those that will involve assessing external control surface movement, will require more than one engineer, especially if specific ground support test equipment is required. The BITE system should also be capable of performing a full functional test of all components, particularly after the replacement of an LRU, in order that the aircraft can be cleared for continued operation.

Maintenance assistance functions are provided at various levels to simplify the requirements of everyday flight operations. The major areas are ground maintenance BITE and continuous monitoring during power-up procedures (including periodic checks, and maintenance monitoring).

Ground maintenance BITE is designed to provide an on-ground readout of the data stored in the non-volatile memory plus the ability to perform isolation and verification checks and functional tests whilst the aircraft is on the ground. Continuous monitoring, including during power-up procedures, monitors the health of the system, on a continuous basis, as implied. The maintenance monitoring function covers isolation of any system malfunction and the subsequent storage of that information in the non-volatile memory.

As such, the cycle of events can be considered in four phases – two in flight and two on the ground. Initially, the system is operating as designed, and is verified as such. A malfunction occurs, stored in memory and reported by the operating crew. The ground phases then involve the fault isolation, down to individual LRU level, followed by removal and replacement of the faulty unit. The last phase involves verification of correct function.

The differing requirements of maintenance engineers in particular phases of a system's operation dictate differing levels of capability for the BITE system. Much of the discussion above has studied the requirements of the line maintenance technician, involved in rapid fault isolation in order that the aircraft can continue on schedule. However, there are other levels possible, for example, when an aircraft does not have the demands of an immediate schedule, such as on an overnight stop at the main base. Also, some investigative capability by specialist engineers into specific, often intermittent, malfunctions can be provided.

The first capability level will have the ability to test current status of all systems, confirm in-flight faults for both LRUs and their interfaces and to verify any corrective action. This implies a limit to their capability of addressing the question of Go or No-Go decisions.

Overnight maintenance can provide the facilities of test rigs and functional test capabilities. Readout measurements from the sensors, both static and dynamic are possible, all without the need to remove the faulty unit from the aircraft.

Investigative engineering capability allows for readout of individual test results. Memory interrogation techniques allow access to the programming inherent in the system as well as triggering fault history of intermittent failures. Typically, the unit will be off the aircraft and on a test bench, with outputs from the BITE in some form of high-level computer language, contrasting vividly with the 'plain English' type of readout provided to the line maintenance technician.

Aircraft such as the Airbus A320 and McDonnell Douglas MD11 have had extensive BITE systems designed into them. To provide effective troubleshooting, some form of centralised fault display system is an integral part of the cockpit. Engineering data will be available via a multi-function control and display unit, so called because it acts in other roles at various phases of the flight. The access will be provided by the use of keys built into the unit, and by comprehensive menu-driven instructions that appear on the screen in response to commands and prompts.

BITE techniques are extensively used during the development phases of aircraft to assist the flight test programme in assessing the effects of flight procedures on avionics equipment. In service use the concept helps reduce delays by accurately identifying failed components to a level whereby remedial action can be taken by line maintenance personnel, improving dispatch reliability, with the obvious effect on passenger acceptability and hence, ultimately, the profitability of the airline. BITE reduces the unnecessary removal of unconfirmed, but suspect components, thus reducing cost as repair organisations will still make a charge for labour, even if no fault is found. It also has the result of reducing the necessary spares inventory as an additional benefit of reducing faulty diagnosis.

It can be expected that, as digital systems become more common on aircraft, more extensive use of BITE will be introduced, combining all subsystems, to provide continuous monitoring, helping improve reliability as well as providing early warning of malfunction, with a consequent improvement in aircraft safety.

Whilst of significant benefit to engineering personnel, particularly in a commercial airline, the use of BITE requires additional computing power and memory which adds weight and complexity and, inevitably, affects levels of overall reliability. It is essential that a higher level of reliability is achieved in the test equipment than in the systems under test. On military aircraft the computer capacity required may be as high as 15 per cent of the total requirement.

AUTOMATIC TEST EQUIPMENT (ATE)

The increasing complexity of modern avionics equipment with the new techniques of printed circuit boards, surface mount technology and computer generated data has led to the development of substantial test fixture requirements. These are generally produced by the avionics manufacturers themselves, for use in their own repair facilities as well as for sale to those organisations large enough to justify the cost of the equipment as well as the training and spares inventory required.

ARINC characteristics now specify the requirements for each of the devices produced to provide the facilities for test equipment interface, suitably protected when in normal aircraft use.

Automatic test equipment is designed to perform a number of roles. One of the simplest is to confirm a fault that is believed to exist, either generated by a pilot report or following normal maintenance procedures. These test fixtures can be considered as 'filtering' devices, often designed to prevent unwarranted flagging of a unit as faulty. Many airlines will choose to acquire one of these devices as a means of controlling the warranty procedure, and to avoid sending expensive components abroad.

The second type of fixture involves the diagnosis of faults, leading to their rectification by reference to maintenance manuals and techniques, followed by verification of correct function prior to re-installation in the aircraft. Modern design for these units involves the modular concept, whereby facilities can be added as the requirements develop, keeping costs under control, and allowing for purchases to take place in different budget periods. One such device is the Honeywell STS 1000, which is designed to provide test facilities for the whole range of Honeywell products on, for example, the MD11, plus most digital equipment for other types of aircraft, such as the EFIS on the Fokker F50, EFIS and AFCS on the ATR42/72.

Increasing reliability of components, the multitude of parts plus the problems of initial and recurrent training, with the problems of experience build-up, make the use of full test equipment fixtures less viable to most airlines. The size of fleet required, added to the technical skills available to the engineering department mean that the trend is more towards the provision of strategically placed overhaul and repair stations, often run by the manufacturers to service the needs of the region's airlines. Specialist maintenance organisations are springing up to offer services to airlines that feel unhappy about the cost structures of some manufacturers and the inherent control that exists as one company sells and maintains the equipment.

System Design Considerations

BIBLIOGRAPHY

British Civil Airworthiness Requirements (BCAR)
    Chapter A3–3 Airframe Parts and Equipment
    Chapter A3–4 Radio Apparatus
    Chapter A3–5 Aircraft Radio Installations
    Chapter A6–5 Specifications – Instruments, Equipment and Accessories
    Chapter D1–3 Safety Assessment of Systems

RTCA/EUROCAE
    DO 160/ED14B   Environmental Standards
    DO 178A/ED12A Software Considerations

Joint Airworthiness Requirements
    Sub-Part C – Structure   JAR 25.581
    Sub-Part F – Equipment   JAR 25.1301
                                     JAR 25.1303
                                     JAR 25.1309
                                     JAR 25.1321–1337

ARINC documentation.
RTCA document DO-160.
FAA: Certification Procedures for Products and Parts Maintenance Assist Functions paper – G. F. Ellis and H. E. Hofferber Engineering Bulletins and Training Data for the Honeywell and Sperry Corporations.

CHAPTER 3 DIGITAL TECHNOLOGY

IAN MOIR BSc, CEng, MIEE, MRAeS

Senior Multiplexing Systems Engineer Smiths Industries. Mr Moir undertook his graduate training as an Engineering Officer in the Royal Air Force being awarded a BSc (Eng) 2nd Class Honours Degree in Aeronautical Engineering. Service in the Far East was followed by responsibility, at RAF Brize Norton, for first line servicing of VC 10 and Belfast aircraft.

In 1970, in the rank of Flight Lieutenant, he took a post-graduate Course at RAF College, Cranwell, having already been awarded an Air Training Corps Flying Scholarship and the de Havilland Flying Trophy for the best pilot in RAF Officer Cadet Entry (1965).

In 1972 he joined the RAF Project Team with responsibility for the digital avionics system of the MRCA (Tornado) at British Aircraft Corporation (later BAe) at Warton.

In the rank of Squadron Leader Mr Moir served at the Government Communication Headquarters, Cheltenham, was Officer Commanding, Electronic Engineering Squadron, RAF Leuchars and the Ministry of Defence (Air) Directorate of Tornado Engineering and Supply.

In 1980 he joined Smiths Industries, becoming involved in systems integration using MIL-STD-1533B data bus hardware on the BAe Experimental Aircraft Programme. From 1984 he has been Group Leader of the Utilities Systems Management Group.

Mr Moir has presented many Technical Papers at home and abroad and written a number of Technical Articles. In 1979 he was awarded the RAF Strike Command Smallwood Trophy for a submission on Electronic Warfare. He is a Member of the Royal Aeronautical Society, a Chartered Engineer and a Member of the Institution of Electrical Engineers.

For many years the application of electronics to airborne systems was limited to analogue devices and systems with signal levels and voltages generally being related in some linear or predictive way. This type of system was usually prone to heat soak, drift and other non-linearities. The principles of digital computing had been understood for a number of years before the techniques were applied to aircraft. The development of thermionic valves enabled digital computing to be accomplished but at the expense of vast amounts of hardware. During the Second World War a code-breaking machine called Colossus employed thermionic valves on a large scale. The machine was physically enormous and quite impracticable for use in any airborne application.

The first aircraft developed in the UK intended to use digital techniques on any meaningful scale was the ill-fated TSR 2 (Fig. 3.1) which was cancelled in 1965. The technology employed in the TSR 2 was largely based upon solid state transistors, then in comparative infancy. It was not until the development of the Anglo-French Jaguar and Hawker Siddeley Nimrod in the 1960s that weapon

Fig. 3.1 British Aircraft Corporation TSR 2 tactical/strike/reconnaissance bomber was a proposed replacement for the Canberra medium bomber. The first of nine development aircraft was first flown by W/Cdr Roland Beamont in 1964. TSR 2 was the first European aircraft to embody digital technology. It had an advanced attack radar with automatic terrain-following modes, an automatic flight-control system, head-up display, mixed Doppler/inertial navigation system and sideways looking reconnaissance radar. The project was cancelled for political reasons later in 1964 *British Aerospace, Warton*

systems began seriously to embody digital computing, but on a meagre scale compared with the 1980s.

From that point, the application of digital techniques rapidly spread throughout many of the systems on the aircraft and it is now virtually impossible to find subsystems which do not utilise digital technology in one form or another. Transistors have given way to Integrated Circuits (ICs) where a number of gates could be integrated and packaged in one chip or IC. The integration of 10 to 100 transistors in a single chip is termed Small Scale Integration (SSI). This trend of integrating more devices in a single chip has accelerated rapidly over the past two decades to the point where Very High Performance Integrated Circuits (VHPIC) are embodying in the order of 10^5 to 10^6 transistors per chip. In the US similar developments are called Very High Speed Integrated Circuits [VHSIC]. The increasing scale of integration and speed of integration are largely due to the shrinking geometry of the transistors and inter-connecting tracks which comprise the IC. Present geometries are of the order of 1 micron resolution (1 micron is 10^{-6} or one millionth of a metre or 0.04 thousandths of an inch) which is why

Fig. 3.2 Integrated circuit development *Royal Aeronautical Society*

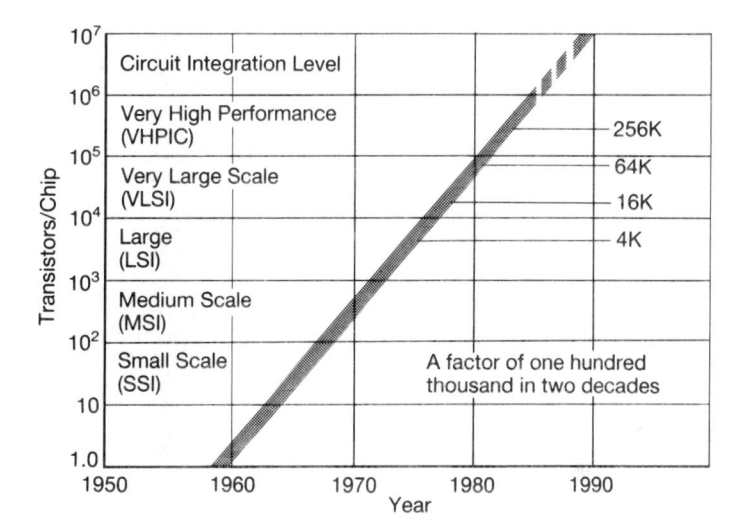

so many may be embodied upon an IC chip perhaps 0.3 inch square. See Fig. 3.2 for a comparison of IC development.[1]

The enormous increase in component densities which has been achieved has not only reduced the space required for the equipment but also increased the speed of operation. The measure of how quickly a device can switch on or off is referred to as gate delay. The gate delay of a typical thermionic valve is of the order of 1000 nanoseconds (1 nanosecond is 10^{-9} or one thousandth of a millionth of a second). Transistors are around 10 times faster at 100 nanoseconds; Silicon chips faster again at around 1 nanosecond. Gallium Arsenide chips are being developed to be at least ten times faster than silicon chips.

Another of the areas of revolution relates to power consumption. ICs consume minute amounts of power. Consumption is related to the technology type and the speed of operation. The quicker the speed of operation then the greater the power required and vice-versa. A comparison of some of the main types of modern technology together with their respective gate delay and power dissipation is given in Fig. 3.3.

There are four main areas where the development of digital technology and related disciplines have led to their application in modern avionic systems. These areas which will be examined in the remainder of this chapter are: Microprocessors, Memory devices, Data buses, Software development methodologies.

MICROPROCESSORS

The advent of cheap and effective processors, termed microprocessors, had a marked effect on the application of digital computing to aircraft avionic systems. In 1971 the INTEL Corporation introduced a 4-bit microprocessor called the INTEL 4004 as the first single chip microprocessor. In fact 4 or 5 additional support chips were required to form a viable processor system. Shortly afterwards, INTEL introduced the 8-bit INTEL 8008. By 1974 these had been developed to the INTEL 4040 and 8080 processors. A portion of the design team from INTEL split off to form the ZILOG Corporation which introduced the 8-bit Z80 microprocessor. Many other manufacturers produced their own microprocessors of comparable performance and the industry soon became very competitive, though not always with the best outcome for the user.

Fig. 3.3 Comparison of performance and power dissipation for various semi-conductor technologies

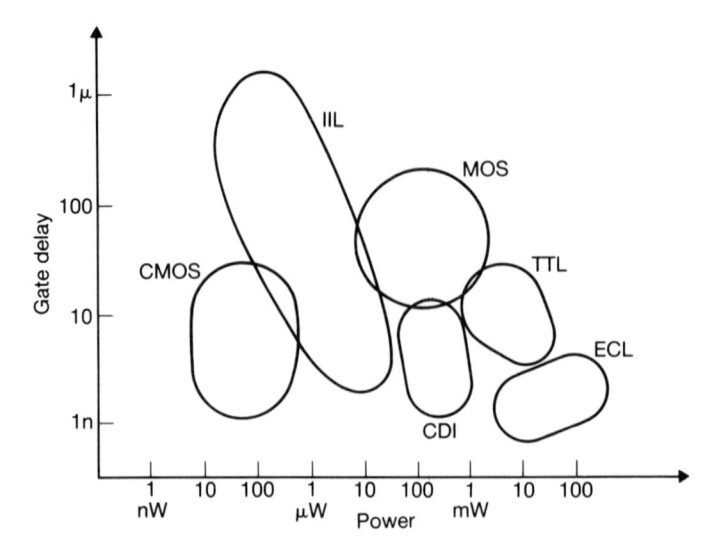

Semiconductor Technologies:

TTL	Transistor–transistor logic
ECL	Emitter-coupled logic
IIL	Integrated-injection logic
CDI	Collector-diffused isolation logic
MOS	Metal-oxide-semiconductor transistor logic
CMOS	Complementary MOS

For aircraft applications the 8-bit family of microprocessors was favoured as, with a few specialised exceptions, its accuracy was more meaningful when applied to aircraft systems. For a number of years throughout the mid-1970s a range of 8-bit microprocessors were used in many equipments being designed at the time. The availability of competing devices led to a proliferation of microprocessor implementations and accompanying software languages. The ultimate customer – and in particular the US Air Force – began to inherit an inventory of equipment with a wide range of microprocessor implementations and software support. This very soon reached the stage where the logistics, and particularly the modification of equipment in service, became a nightmare. It was to lead to standardisation initiatives – both hardware and software – which will be described later.

By 1977 first generation 16-bit microprocessors became available in volume production. The most generally used of these microprocessors were the Ferranti F100, INTEL 8086, Motorola 68000, Texas Instruments 9900 and ZILOG Z8000. At one time the Ferranti F100 was the preferred microprocessor for use in UK Ministry of Defence applications. Most equipments designed in the UK within the last five years are likely to use one of these microprocessors or a variant thereof. However the number of options available within each manufacturer's product range, as well as a general lack of compatibility between the programming languages (assembly language) used by the various microprocessors, still lead to incompatibilities. However the more widespread availability of High-level Operating Languages (HOLs) and the associated software development toolsets is beginning to offer some hope of rationalisation, as will be outlined later.

In the mid to late 1980s, further developments, particularly in the INTEL, Motorola, and ZILOG ranges, led to 32-bit microprocessors becoming available. While there are computational tasks in avionic systems which require 32-bit accuracy, in most cases 16-bit capabilities suffice.

Following the adverse experiences of its initial ownership of microprocessor-based systems, the US Air Force has pressed strong standardisation initiatives. The microprocessor was standardised upon the MIL-STD-1750A with a standardised Instruction Set Architecture (ISA); this is the widely used standard processor now specified for 16-bit operation and in time may also find 32-bit use. As well as the US Air Force, the US Navy and US Army are beginning to accept the standard. Furthermore, manufacturers in Europe are having to take increasing account of MIL-STD-1750A. Among manufacturers in general there are reservations about 1750A and it is sometimes criticised as an overkill, in many cases resulting in overspecification and overdesign. There is little doubt that at the moment it is expensive, with certain 1750A chip sets being almost ten times the cost of the contemporary 'commercial' microprocessor. Time alone will tell if this situation will continue.

MEMORY DEVICES

Many computer or microprocessor implementations may possess an inherent small amount of memory in the design. However additional memory will need to be added in order that the processor may discharge its allotted tasks. Memory is available in various forms, not all of which are suitable for incorporation in avionic systems. One form of memory used is for the location of program storage. That is memory where program algorithms, routines, data banks and look-up tables, etc., are located. Program storage will usually be implemented using some form of non-volatile memory: the program is read-only and cannot be written to during normal operation though clearly it has to be programmable. Furthermore, the memory will be such that loss of power – due to an equipment power-down or power-interrupt – does not cause loss of the program. It is desirable that these program storage areas are re-programmable when the aircraft is on the ground; in some cases re-programming may necessitate removing the equipment from the aircraft.

The program working area is that area of memory where read-from and write-to operations occur during the normal execution of the processor software. This area of working memory is sometimes referred to as 'scratch pad' memory.

The memory types used today in aircraft avionic systems to satisfy these basic requirements are largely based upon three technologies: core memory, semiconductor memory and bubble memory.

Core memory consists of a 3-dimensional array of minute ferrite cores which may exist in one of two magnetised states and which may be used to represent the presence or otherwise of a bit of information. In early implementations, four wires would pass through the centre of each core: two selection wires associated with X and Y grids for identification, a sense wire and an inhibit wire. Later developments use a three-wire system where the sense/inhibit functions are combined.

The advantages of core memory are that it is extremely rugged and is capable of surviving the most demanding environnents. It is however heavy compared with other memory types; clearly the ferrite cores are not light. It has historically also been more demanding in terms of power consumption and has a slower access time than semiconductor memory. It does have the additional advantage that the operational program can be changed by means of a Program Loading Unit (PLU) at the aircraft without having to remove the unit in question. For this reason core has been the preferred memory type for program storage of main mission/weapon aiming computers such as those used on the Jaguar, Tornado and the Sea Harrier HUDWAC.

Semiconductor memory is presently widely used in two forms with a third finding increased use. Semiconductor memory is implemented in ICs or chips as described earlier.

Electrically Programmable Read-Only Memory (EPROM) is the semiconductor equivalent to core; that is a read-only memory used for program storage. The disadvantage of EPROM is that for re-programming it has to be removed from the equipment which in turn necessitates removing the equipment from the aircraft. The memory card containing the EPROM is removed and the memory devices are erased by irradiation with an ultra-violet (UV) light source. The memory devices may then be re-programmed, re-validated and the unit checked and re-installed in the aircraft. This is a cumbersome and time-consuming process and EPROM has tended to be used for those applications where the operational programs are not liable to frequent modification.

Random Access Memory (RAM) is the read-write memory which is used as the program working memory. The disadvantage with RAM is that once power is lost so too are program working variables. It is common to support RAM such that it will survive at least limited power-interrupts without the loss of memory. This is accomplished by protecting the memory contents either by battery or capacitive back-up.

The third type of semiconductor memory is Electrically Erasable Programmable Read Only Memory (EEPROM or E^2PROM). This is a read-write memory similar to core and can therefore be used for program storage, it can also combine the advantages of EPROM with the *in-situ* re-programming capabilities of core. The disadvantages suffered by E^2PROM until relatively recently were the low packaging density (i.e. fewer memory devices for a given area or size of IC) compared to EPROM and RAM, device package pin and programming voltage incompatibilities, and slower read-write times. In general, semiconductor memories are faster than core. The advent of more user-friendly E^2PROM together with increasing packaging densities of EPROM and RAM mean that semiconductor memories are presently in favour for most avionics applications.

The final memory technology to recently find some application in avionic systems is bubble memory. This consists of a field of magnetic bubbles on a thin magnetic material fabricated to exhibit uni-axial magnetic anisotropy; the easy axis of magnetisation is perpendicular to the plane of the thin film. The application of a correct bias field enables the bubbles to store binary (digital) data. The positions of the bubbles are controlled by magnetically soft nickel-iron elements which may be directionally magnetised by a rotating in-plane magnetic field. The packaging of bubble memory is complicated as the magnetic environment required for correct operation needs to be carefully controlled. The rewards are that the storage capacity of bubble memory is far ahead of that for comparable semiconductor memories. Access time is slower, but still faster than magnetic tapes, drums and discs with which this type of memory is most comparable. Bubble memories are more rugged than tape and disc memory techniques allow. Present bubble memories are available to withstand temperatures of $-55°$ C to $+85°$ C which is rather limited for some (certainly military) airborne applications. A typical example of the application of bubble memory is the Flight Management Control System (FMCS) described in Chapter Six. In this example bubble memory is used for the data bank memory which holds all navigation and aircraft data.

Future advanced applications within avionic systems will increase the requirement for mass memories to store large quantities of mission data. Mass memories may be satisfied by further developments of semiconductor and bubble memory technologies. However it appears more likely that mass memory requirements of 100 megawords and upwards will demand a new technology such as laser disc to be developed.

DATA BUSES

The availability of reliable digital semiconductor technology has enabled the inter-communication task between different equipments to be significantly improved. Previously, large amounts of aircraft wiring were required to connect each signal with all the other equipments. As systems became more complex and more integrated so this problem was aggravated. Digital data transmission techniques use links which send streams of digital data between equipments. These data links may only comprise two or four wires and therefore the inter-connecting wiring is very much reduced. Recognition of the advantages offered by digital data transmission has led to standardisation in both civil and military fields. The most widely used digital data transmission standards are described later: ARINC 429 for civil and MIL-STD-1553B for military systems. First, a few basic system principles will be outlined.

Common types of digital data transmission link are:

- Single source/Single sink – This is a dedicated link from one unit transmitting data to another receiving the data. Where data is transmitted from one unit to another and not back again, it is called half-duplex. If the link is able to transmit data both ways it is termed full-duplex. The serial data transmission link used on the Tornado and Sea Harrier is single source, single sink, half-duplex.
- Single source/Multiple sink – This describes a situation where one transmitting equipment sends data to a number of receiving equipments (sinks). ARINC 429 as used on large civil transports, such as the Boeing 757 and 767 is an example of a single source, multiple sink transmission system.
- Multiple source/Multiple sink – In this system multiple transmitting sources may transmit data to multiple receivers. MIL-STD-1553B and the embryonic DATAC (ARINC 629) are examples of this type of system.

As summarised above, the commonly used ARINC 429 is a single source, multiple sink system; it also acts as half-duplex, therefore different links need to be installed for two-way traffic to be established. If a number of equipments – say four – are inter-connected by ARINC 429 and a further equipment is added then the additional data links required are as shown in Fig. 3.4. Therefore, where the number to be inter-connected is large, or where an additional link is required for modification reasons, the number of ARINC 429 links needed is large. Modern airliners such as, again, the Boeing 757 and 767 may have as many as 150 data links to inter-connect the aircraft avionic systems.

The characteristics of ARINC 429 were agreed among the airlines in 1977/78 and the data bus is widely used on the Boeing 757, 767 and Airbus A 300/A, 310/A, and 320 aircraft. An ARINC 429 data bus comprises a screened, twisted-wire pair with the screen usually earthed at both ends and at all intermediate breaks. The transmitting element is embedded in the transmitting equipment and may interface with up to 20 receiving terminals in the receiving equipments. Information may be transmitted at either a high rate of 100 kilobits per second or at

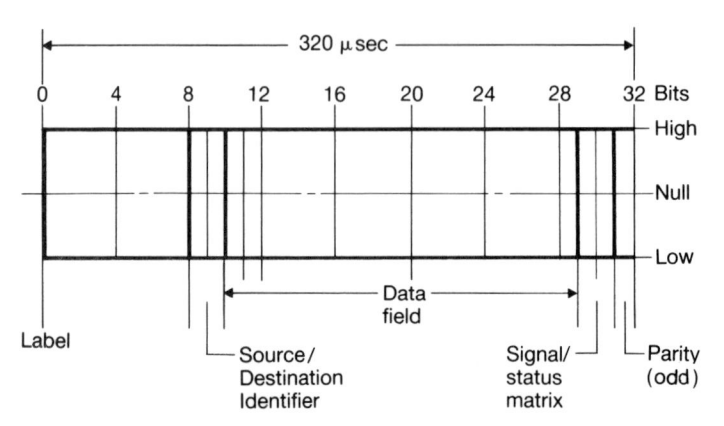

Fig. 3.4 ARINC 429 – addition of equipment

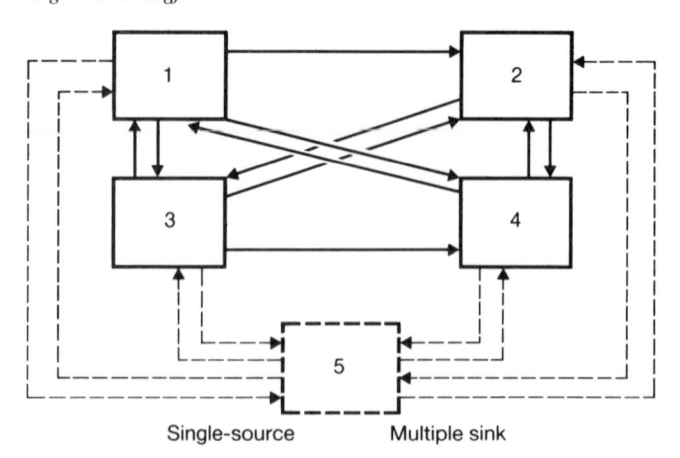

Single-source   Multiple sink

a low rate of 12 to 14 kilobits per second. The modulation technique is bipolar RTZ (Return To Zero) which has three signal levels: high, null or low. A logic state 1 is represented by a high state returning to zero; a logic state 0 is represented by a low state returning to zero. Refer to Fig. 3.5 for an example. Information is transmitted down the data bus as 32-bit words as shown in Fig. 3.6. The standard has many fixed labels and formats so that a particular type of equipment always transmits data in a particular way. This has the advantage that all manufacturers of a particular equipment know what data to expect; where necessary, additions to the data format may be made. ARINC 429 is now widely accepted and understood by the civil aircraft industry and is relatively inexpensive to implement. If there is a disadvantage it is that the single source, single sink half-duplex nature of the system is not the most efficient.

Fig. 3.5 ARINC 429 – Data RTZ modulation

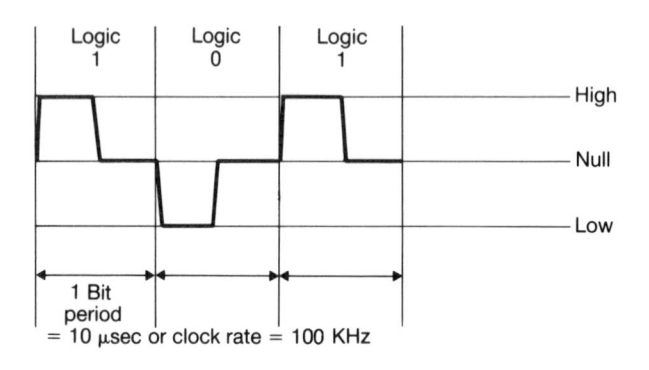

1 Bit period = 10 μsec or clock rate = 100 KHz

Fig. 3.6 ARINC 429 – Data format

Fig. 3.7 MIL-STD-1553B addition of equipment

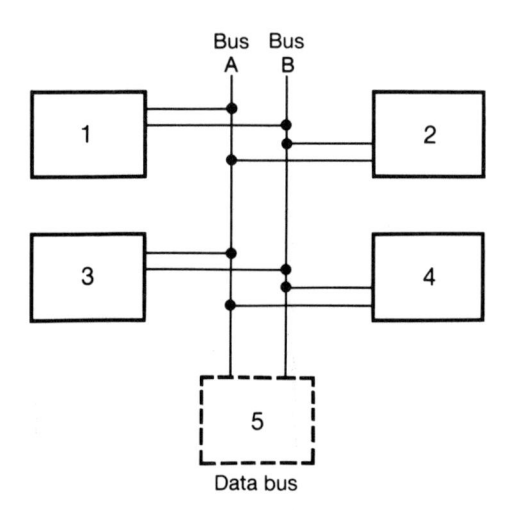

In Fig. 3.4 an example of expanding a system inter-connected by four ARINC 429 data buses was given as a fifth element was added. The same example of adding a fifth element using a MIL-STD-1553B data bus is shown in Fig. 3.7. It is clear that adding an additional element is far easier to implement than was the case for ARINC 429, though not without modification to the controlling element (bus controller) software. This ease of modification is of great benefit for military avionic systems which are often updated to maintain effective operational capability.

MIL-STD-1553B has evolved since the original publication of MIL-STD-1553 in 1973. The standard has developed through a 1553A standard issued in April 1975 to the present 1553B standard which was issued in September 1978.

The application of 1553 has been apparent for some time in US military aircraft systems. The F-16, B-1, F-18 and AV-8B have all used variants. More recently, the UK industry has used the 1553B standard for the integration of aircraft and helicopter avionic systems and also for land based and marine systems.

The basic layout of a MIL-STD-1533B data bus is given in Fig. 3.8. The data bus consists of a twin wire twisted pair along which DATA and $\overline{\text{DATA}}$ are passed though future applications may utilise a fibre-optic transmission medium. The standard allows for dual redundant or even quadruplex redundant terminals although for most applications the dual redundant implementation is sufficient.

Control of the bus is effected by a Bus Controller (BC) which is connected to a number of Remote Terminals (up to a maximum of 31) via the data bus. Remote Terminals (RTs) may be processors in their own right or may interface with a number of sub-systems (up to a maximum of 30) with the data bus. Data is transmitted at 1MHz using a self-clocked Manchester bi-phase digital format. The transmission of data in true and complement form down a screened twisted pair or fibre-optic link, together with a message error detection capability offers a high integrity digital data link which is highly resistant to message corruption. Words are formatted as Data Words, Command Words or Status Words as shown in Fig. 3.9. Data words encompass a 16-bit digital word whereas the command and status words are compartmented to include various address, sub-address and control functions.

Control of the data transactions on the data bus and the issue of the appropriate command and status words are carried out by the bus controller; in a practical system there is likely to be more than one bus controller for redundancy reasons.

Fig. 3.8 MIL-STD-1553B basic layout

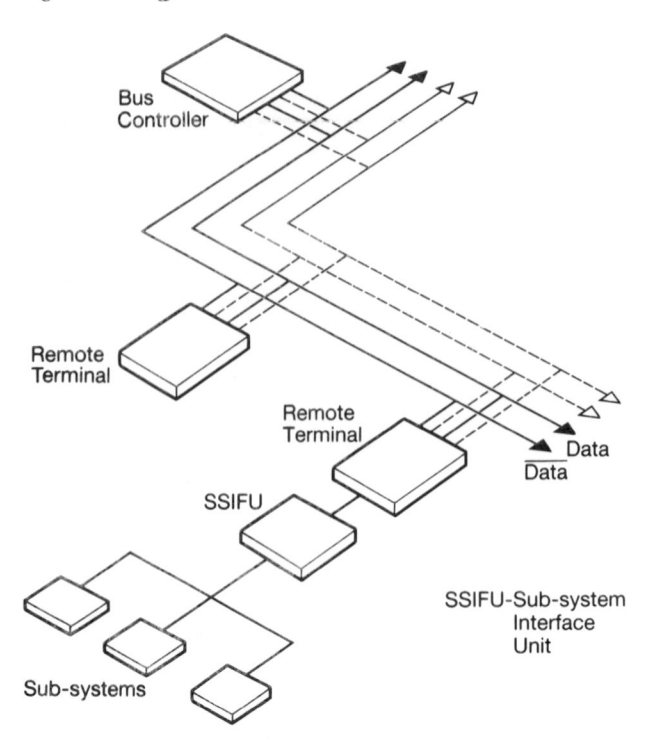

Fig. 3.9 MIL-STD-1553B word formats

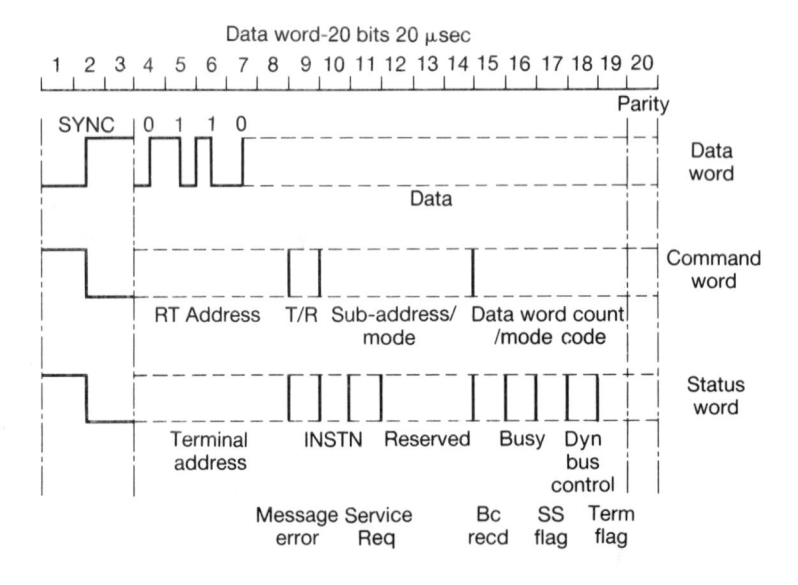

In a simple transfer of data from an RT to the BC, the BC sends a transmit command to the appropriate RT, which replies after a short interval gap with a status word, followed immediately by one or more data words up to a maximum of 32 words. In the case shown the transaction of one word from an RT to the BC will take approximately 70 μseconds. For the transfer of data between RTs the BC sends a receive command to one RT followed by a transmit command to the other RT. The delivering RT sends a data word (or data words up to a maximum of 32 words) preceded by a status word, to the recipient RT which in turn sends a status word to the BC, thereby concluding the transaction. In the example shown a time elapse of around 120 μseconds is required to transfer one data word from one RT to another which is rather expensive in overheads. If however a maximum of 32 data words had been transferred the overheads would be the same, though now representing a much lower percentage of the overall message time. Refer to Fig. 3.10.

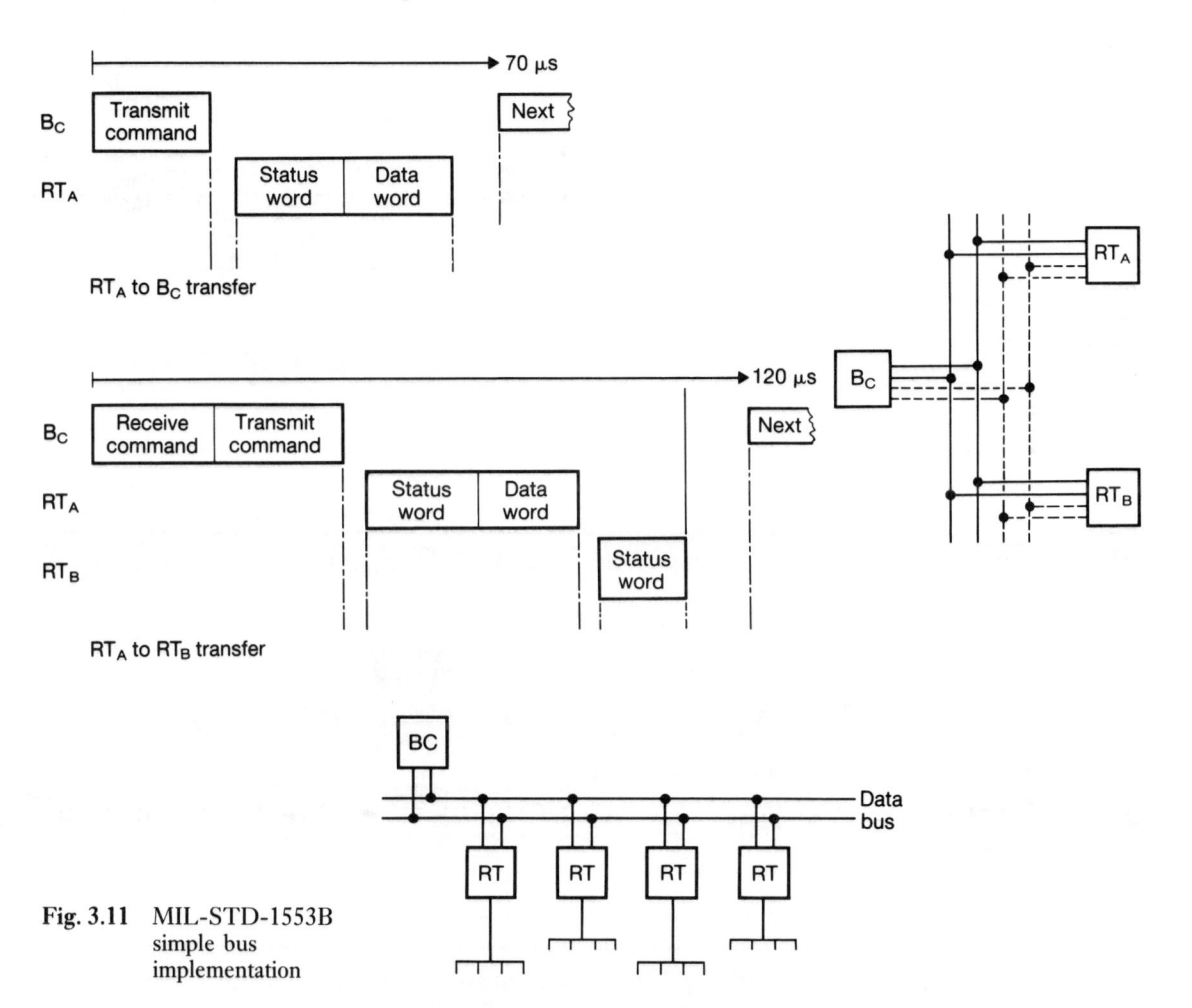

Fig. 3.10 MIL-STD-1553B typical data transactions

Fig. 3.11 MIL-STD-1553B simple bus implementation

A typical MIL-STD-1553B system is shown in Fig. 3.11 the BC controlling a number of RTs on the data bus: the parallel lines are the standard representation of a dual redundant bus implementation. A more complex system is shown in Fig. 3.12, where two data buses are linked by means of a Bus Interface Unit (BIFU). In a typical application of 1553 to an aircraft system BiFUs may link the main system avionic or mission bus to one or more subsidiary buses. In this situation the BIFU acts as an RT on the main system bus while also being the BC for the subsidiary data bus. In the example shown the RTs or processors on the subsidiary bus have the ability to interface with analogue, discrete, digital and optical data. By this means the benefits of 1553 may be extended to large and complex systems which embrace a wide variety of signal parameters and functions.

A prime advantage of MIL-STD-1533B is that it is, by definition, a standard. This stimulates the widespread availability of components and expertise throughout the life of a system due to wide acceptance of 1553 across all or most military

Fig. 3.12 MIL-STD-1553B multiple bus implementation

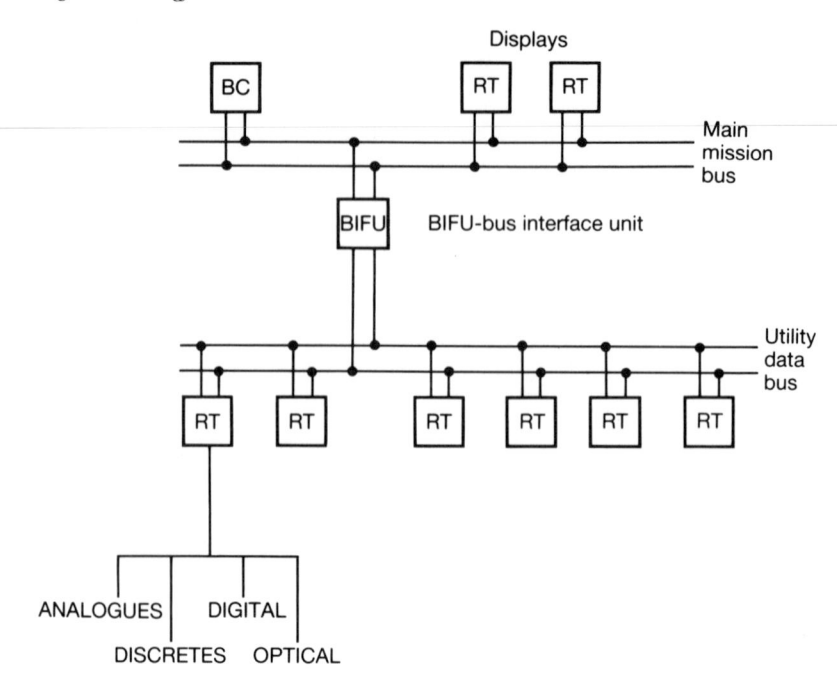

projects. A further advantage is the acknowledgement of the standard, not just in the US, but in European, Middle and Far Eastern applications. The inherent advantages of MIL-STD-1533B are:

- weight savings resulting in cost savings and performance improvements;
- high integrity data transmission system;
- built-in redundancy with consequent fault tolerance;
- battle damage resistance;
- flexibility;
- life cycle cost savings – easier to modify, maintain and test;
- reliability – available in the latest LSI technology.

In the civil field the shortcomings of ARINC 429 have been long understood. For more than five years the Boeing Company have been developing a civil equivalent of MIL-STD-1553B for use in civil applications. In the civil sphere, 1553B is viewed with reservation because of the need for a central control element – the bus controller – though it has already been stated that practical military systems have more than one controller per bus. The alternative that Boeing have been at the forefront in proposing is known as Data Autonomous Transmission And Communication (DATAC). This is presently being debated as a possible new ARINC standard (ARINC 629), embodying improvements upon ARINC 429. Like 1553, DATAC terminals would be embedded in functional equipments being inter-connected using the data bus.

In the DATAC system, the control function is distributed among all the participating terminals present on the bus. These terminals autonomously determine the transmission of data on the bus by means of a protocol termed Carrier-Sense/Multiple Access-Clash Avoidance (CS/MA-CA). In simple terms the protocol guarantees access to any participating terminal, but prevents any terminal from having access to the bus before all other terminals have had an opportunity to transmit data. Data transmission by DATAC comprises strings of data words accompanied by label and address data. The data word length is 16 bits as for 1553. However DATAC offers more flexibility in terms of the number of terminals; 120 terminals may be supported in lieu of the maximum of 31 offered by 1553. The data transmission word string is also greater; a maximum number of 256 words may be transferred as opposed to the maximum of 32 for 1553. The proposed implementation for DATAC is serial Manchester bi-phase, 1 MHz data transmission over twisted wire, screened pairs as for 1553. Therefore the practical limitations for both systems are the same, namely a bus length limitation of around 100 metres.

Boeing are actively considering the adoption of a DATAC type of transmission system for systems integration on their next generation of civil transport aircraft of which the Boeing 757 will be a prime example.

All of the data buses so far described – ARINC 429, MIL-STD-1553B and the proposed DATAC system – are presently implemented using twisted, screened wire pairs. There are limitations to this implementation. The constraint of a physical bus length of around 100 metres is applied to the 1 MHz systems. The use of electrical wire transmission media may be adversely affected by external effects such as electro-magnetic interference (EMI) or lightning strikes. For these reasons and for the need to achieve higher data rates of transmission, the use of fibre-optic transmission techniques is under active consideration.

Fibre-optic transmission techniques will permit data rates of 50 MHz and even as high as 100 MHz, some 50 to 100 times faster than those of existing avionic systems. In the Society of Automotive Engineers (SAE) data bus forum, discussions have been under way during which two primary candidates are being considered. These topologies or architectures are the linear bus and ring networks. See Fig. 3.13 for a simple comparison of these architectures.

Fig. 3.13 Fibre-optic bus candidate architectures

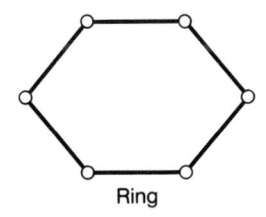

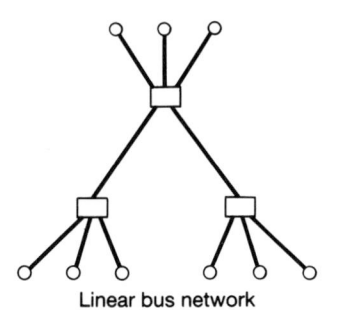

The ring, as the name suggests, consists of a series of point-to-point links which connect the participating terminals into a ring. Signal re-generation occurs at each stage and therefore the overall reliability of the system is less than if a passive inter-connect system were employed. The design of this type of system needs to take account of how such failures may be tolerated.

The linear bus network, using a token passing protocol, appears to be the more favoured architecture of the two, certainly by potential users in the military field such as the US Air Force. Inter-connection is by means of 'STAR' couplers as it is not possible to configure a 'TEE' junction in a fibre-optic bus without incurring unacceptable losses.

The main shortcomings which have been experienced with fibre-optics have been the signal attenuation at connectors and the repeatability of such connectors. Only recently have durable connectors with acceptable loss rates become reality. Therefore although fibre-optics offer significant potential for future applications it will be some time before such systems are widely employed.

SOFTWARE DEVELOPMENT METHODOLOGIES

The application of electronics, and in particular digital technology, has had an enormous impact upon the performance and capabilities of avionic systems. As part of the digital technology revolution, the availability of digital computers and microprocessors has been the key in giving inherent intelligence to these systems. The advent of microprocessors in the past fifteen years has greatly accelerated this trend, to the point where some of the second generation 16-bit microprocessors are now able to rival, in performance and capabilities, the minicomputers of five to ten years ago. Furthermore these microprocessors are cheap and compact, requiring little electronic real-estate to support them. The memory types needed to support the microprocessor have also developed apace and the data bus technology required to inter-connect centres of intelligence is mature. Everything therefore appears set fair for an unimpeded technology revolution: but this is not entirely true as there is still the matter of software development.

The basic problem with any processor or computer, as home computing enthusiasts will be all too aware, is that a computer is little other than a very fast working idiot. The computer or processor is only as good as the software that is driving it. If the software is deficient, so then will the final system be deficient in performance. Unfortunately, as the first generation of airborne digital computers entered service the implications of software upon in-service support were not immediately recognised. In fact the need to support equipment hardware and software became critical to the performance of the equipment in service.

Earlier in this chapter the adverse experiences of the US Air Force were mentioned. The US Air Force were not alone in the problems they experienced, they merely had the problem on a grander scale because of their greater inventory of equipment.

Computer programs are commonly expressed at three levels. At one extreme is the language which the processor hardware best operates; that is machine code. At the other end of the spectrum is that which the computer programmer best understands: the High-level Operating Language or HOL. In between the two is an assembly or low level language. See Fig. 3.14 for a simple explanation of the relationship between these languages.

The HOL is converted to assembly language by means of a compiler program. The assembly language is converted to machine code by means of an assembler program. Compiler and assembler programs may reside in the target processor (that processor for which the program is ultimately intended) or in a host processor (a processor used for convenience to develop the program). Host processors may be modified or adapted to emulate the target processor. Alternatively a cross-compiler or cross-assembler program may be necessary to 'translate' the program. It is usual for the assembly language and the assembler

Fig. 3.14 Relationship of computer languages

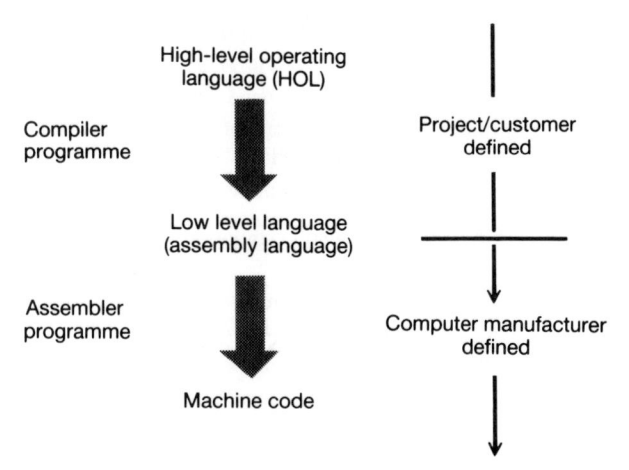

program to be supplied by computer or processor manufacturer. The assembly language for a given processor will closely resemble the machine code instruction set for the processor for which it was designed. It is therefore clear that different processors will have an accompanying wide range of (usually incompatible) assembly programs.

When an equipment is specified to fulfil a certain task or function, the operational tasks will dictate to a large degree the final technical solution or implementation. At the heart of the technical solution will be the choice of processor and this choice is largely one of 'horses-for-courses'. Some machines are more suitable for arithmetic or mathematical rather than logical manipulation. The designer may have a preference for one processor as opposed to another as he may already have had design experience with the former. The customer or the manufacturer may dictate the processor to be used. The processor which the designer wishes to use may not be mature or be available in sufficient quantities; or the software support and design tools may be deficient. The list of factors is almost endless; however it can be seen that there are a number of constraints which affect processor selection and the associated software languages and support.

It was to avoid these problems that the US Air Force started a standardisation initiative with MIL-STD-1750 (now 1750A). This aims to standardise the Instruction Set Architecture (ISA) so that a range of processors supplied by different manufacturers will execute the same instruction set. As mentioned earlier, this has not met with universal acclaim from the industry. The prime objections are that the standardised ISA is cumbersome and is far more than is required for many applications. The second is that of cost; MIL-STD-1750A chip sets are presently much more expensive than the commercially derived equivalents. Nevertheless, the incentive to standardise across all three US armed services as well as the US allies in NATO is a very powerful factor. MIL-STD-1750A implemented in VLSI (VHSIC) technology is being specified at the heart of a multi-million dollar development termed PAVE PILLAR which will comprise the avionic and mission systems for the US Air Force Advanced Tactical Fighter (ATF). There is very strong pressure from the US Congress to apply this technology to the US Navy Advanced Tactical Aircraft (ATA) and the US Army LHX programs in the interests of standardisation and risk reduction. Whether this will succeed remains to seen, however it is indicative of the standardisation pressures which exist in the US.

Digital Technology

A key area in the development of modern computer development is the choice of the high-level operating language or HOL. Examples of high-level languages are FORTRAN, COBOL, CORAL, PASCAL, JOVIAL (US) and ADA (ADA is a registered Trade Name of the US Department of Defense, ADA Joint Program Office). The use of a HOL is an aid to easing software development and minimising software development time. This is true for a variety of reasons.

- High-level language statements are more intelligible and meaningful to the programmer.
- Each high-level statement usually corresponds to several or many machine code instructions.
- The ease of use means that the programmer is less likely to make errors.
- The high-level language allows a better program structure than assembly language. This leads to a more structured, modular program; thereby aiding visibility and understanding of the program and minimising semantic errors. It is also easier to debug and modify.

It is usual for most of the available microprocessors to have the associated high-level language and software developments tools available. As for the other standardisation issues, the customer is likely to specify the high-level language in which he wishes the program to be written. The scope for waiver of these requirements is becoming progressively less. In practice assembly language is still used where the software function is intimately involved with hardware device function. Examples are processor initialization or Built-In-Test (BIT), or where time-critical software functions are being programmed. It is generally recognised and accepted that software produced using high-level languages is less efficient than that which may be produced by an optimum assembly language program. Against this must be weighed the fact that high-level programs are easier to write. In particular, if the program is large and sophisticated and is to require the services of inexperienced programmers, then the use of a high-level language will bear dividends. Furthermore, the availability of cheap high density memory devices as described earlier in this chapter removes the constraint of memory size which used to apply to former systems.

The most common high-level languages are effectively becoming standards. The UK Ministry of Defence specified CORAL as the preferred high-level language for a number of years. PASCAL has been an accepted standard in avionic systems for several years; the use of ADA is commonly mandated as the standard language today. The US Air Force specified JOVIAL for some years but are now insisting on ADA to be used in conjunction with the MIL-STD-1750A ISA. The US Department of Defense have invested a lot of money in ADA and this development is led by the ADA Joint program Office. This development includes the software development tool set and development environment called ADA Development Support Environment (ADSE) which aims to tighten control during software development projects. It has been claimed that the use of ADA will halve the software development costs compared with JOVIAL, itself already offering considerable advances over former development languages.

The choice of processor and accompanying assembler, and the selection of high-level language are important. However if the avionic system requirements are not correctly identified and understood at the outset then no amount of software will produce an acceptable solution. There is an increasing reliance upon top-down design approaches which decompose or break-down the system requirements from the highest level in a methodical and logical fashion. In this way inconsistencies in requirement or system implementation may be recognised at an early stage in development. If necessary, performance trade-offs can be conducted without an enormous amount of effort and time being expended. These top-level design approaches break the system down to smaller and more easily understood modules or elements. Therefore the activities which take place within the modules or elements, as well as any communication between them, may be better comprehended and documented. This activity precedes software design and coding, and it will assist in identifying the major software tasks which will need to be performed and help to size the overall scale of the development task.

The Experimental Aircraft Programme (EAP) aircraft developed by British Aerospace and UK industry (see Chapter Five) made extensive use of a technique called Semi-Automatic Functional Requirements Analysis (SAFRA). This methodology was applied to the avionic system as well as the Utilities Management System; the latter is described in detail in Chapter Six. An important feature of this methodology was Controlled Requirements Expression (CORE). This enabled all the information and data strings to be identified and checked for consistency. System processing tasks were identified at an early stage in development, in turn allowing the software designers adequate time to design modular software, and code and test in compliance with strict programme timescales. The whole emphasis has been that of allocating more time to correctly identifying and specifying systems and less for coding, testing and debugging software.

This chapter has given a brief overview of some of the digital technologies which have made existing avionic systems possible, or which will be implemented in systems presently being designed. Advances in technology in the avionic systems field occur so often that it is impossible to forecast where the next development of note will occur. It appears likely that the increasingly sophisticated technology becoming available will be applied in these systems and will find increasing application in high integrity systems. The introduction of high technology in these areas will doubtless be approached with caution by the aircraft manufacturers; however, the benefits in performance will be difficult to resist. If there are areas where increased emphasis is to be applied, it is believed to be in the increased integration of system functions, increased standardisation and a greater reliance upon the timely and efficient production of the large software programmes involved. (See Chapter Eleven.)

It will also be apparent that substantial weight savings will be achieved by replacing multi-cable looms by twin-wire data buses, with consequent reduction in installation times, particularly in such locations as the rear of the instrument panels.

REFERENCE

1. **Stamper, J.** 'Information Technology in Aerospace Engineering', Part 2, *Aerospace*, March 1986.

CHAPTER 4 FLIGHT DECKS AND COCKPITS

LESLIE F. E. COOMBS. I Eng. MPhil.AMRAeS, FRSA

Mr Coombs has been researching and writing on aviation subjects in general and on ergonomics and avionics in particular during his 50 years in civil and military aviation.

He has contributed to many technical journals and, for the past 19 years, has been editor of 'Aerospace and Defence Review', a specialist technical publication on avionic systems.

As an ergonomist Mr Coombs lectures on the human factors aspect of flight deck design and, in the 1950s, as technical Secretary of the British Airline Pilots Association, helped to integrate pilot opinion in the design of new aircraft cockpits. As early as 1955 he advocated the side stick control concept and direct data links between aircraft and air traffic control.

The modern flight deck and cockpit have developed to their present standards through the application of avionic systems. Without avionics on the flight deck and in the cockpit modern civil and military flying could not be carried on with safety and at an acceptable cost.

Understanding the role of avionics in the control positions of modern aircraft is facilitated if we consider the flight deck and the cockpit as a whole. By reviewing the development and functions of the non-avionic elements we can understand better the importance and the contribution of avionics to the design of the pilot's place and the needs of the pilot.

Therefore, before going into the details of avionics in this context, the following paragraphs cover the design of that part of an aircraft allocated to the operating crew.

THE PILOT'S PLACE

From the earliest years aircraft designers have had to provide a dedicated position for the pilot and the controls and instruments needed for safe and effective control. There have been two major lines of development, the single-engine, single-seat and two-seat aircraft and, the multi-engine aircraft with side-by-side seats for two pilots. These are generalisations because there have been, and are, numerous exceptions. However, for the purposes of this study these two principal cockpit and flight deck arrangements form the background to the subject of avionics on the flight deck and in the cockpit.

It is also important to emphasise that although civil and military aircraft have been designed to different standards so as to meet requirements which are, in general, unique to each of these major classifications, there are a large number of systems and a lot of control equipment which for all practical purposes we can consider as being common to both.

Structural design and avionics

The introduction of pressurisation, particularly for civil aircraft, resulted in a significant change in the structural design of the flight deck section of the airframe. In particular it imposed a severe limitation on the area available for the forward windows. One result of this was an increase in the risk of mid-air collision, and therefore avionic systems began to proliferate to meet the demand for more accurate track keeping. This is an example of non-avionic factors forcing the pace of avionics.

Speed and avionics

Until about 1950 the majority of civil transport aircraft cruised at about 250 knots. Within only ten years cruising speeds of over 400 knots became commonplace. These speeds were, of course, a benefit conferred by the jet engine. However, a price had to be paid and this was the requirement for accurate navigational aids to ensure that the optimum tracks and flight profiles were flown to avoid excessive fuel consumption. So, once again, avionic systems had to be advanced in technology to keep pace with the ever increasing speeds and operating heights. At the same time, as we have seen, accurate track keeping demanded more and better avionic aids to navigation.

Crew positions and avionics

An important distinction between civil transport flight decks and the cockpits of single and tandem seat air force aircraft is the concept that in a side-by-side flight deck some of the avionic systems can be shared by the two pilots. Obviously in a tandem seat military aircraft, such as a trainer, avionic systems often have to be duplicated. This also serves to introduce another difference between civil and military pilots' places. This is the safety concept for civil transport aircraft in which vital, that is essential systems are, at least, duplicated and often triplicated so that in the event of a failure of one system the crew will not be deprived completely of information or control. By contrast less multiplication, i.e. redundancy, of systems is acceptable in a military single or two seat aircraft; in the last resort the crew can eject.

Flight deck and cockpit environment

An obvious requirement in any flight deck or cockpit is an arrangement of panels and pedestals on which to mount the instruments and controls.

They have to be positioned and their sizes selected to match the ergosphere of the pilot or pilots. The ergosphere is an imaginary space surrounding the pilot. It defines the limits of his or her physical characteristics; such as reach of hands and feet and the normal range of eyesight for reading the symbols and alphanumerics displayed by the instruments and indicators.

In a side-by-side transport aircraft flight deck the ergospheres of the two pilots overlap the centre line of the flight deck and each other. As already mentioned this feature makes it possible for some of the controls and displays to be shared.

The control interface

The control interface on the flight deck and in the cockpit is an imaginary boundary across which are exchanged information and control actions between the pilot and the aircraft.

The pilot of an aircraft cannot maintain safe or efficient control without information. In the simplest aircraft, possibly with only a few instruments, the pilots depend on their natural senses and vision to effect control. The view of the earth and in particular the line of the horizon provide an essential reference for keeping control of the aircraft.

The history of aviation is marked by a number of milestones in technology. Each represents a new idea or technique which has enabled aircraft to be flown faster,

Flight Decks and Cockpits

higher and over greater distances than the previous generation. Some mark the introduction of systems and equipment for improving the safety and regularity of flights. Others mark the development of systems which have provided an important upward step in aircraft effectiveness: lower fuel consumption; lower operating costs; and greater versatility of use.

Clearly, avionics has been one of the most important contributors to the growth of aviation and has advanced in parallel with the development of more powerful and efficient engines and stronger and lighter structures. But nowhere has this been more important than at the control interface between pilots and their aircraft.

Sensors, processors and displays

A theme which occurs frequently in this book is 'sensors, processors and displays'. This chapter is concerned primarily with displays. However, the control interface, as described, forms a two-way interchange of information and actions between the pilot and the aircraft. Therefore any survey of the avionics which is of direct concern to the pilot – which means the majority, must include the systems and equipment which the pilot uses to effect or implement his or her responses to the information presented by the instruments and displays. Input devices, such as levers, knobs, switches, buttons and alphanumeric keypads are part of the interface and its ergonomics.

Figure 4.1 shows the 'Electronic Instrument System' (EIS) architecture of a civil aircraft. As can be seen, the system and sub-systems are made up of sensors,

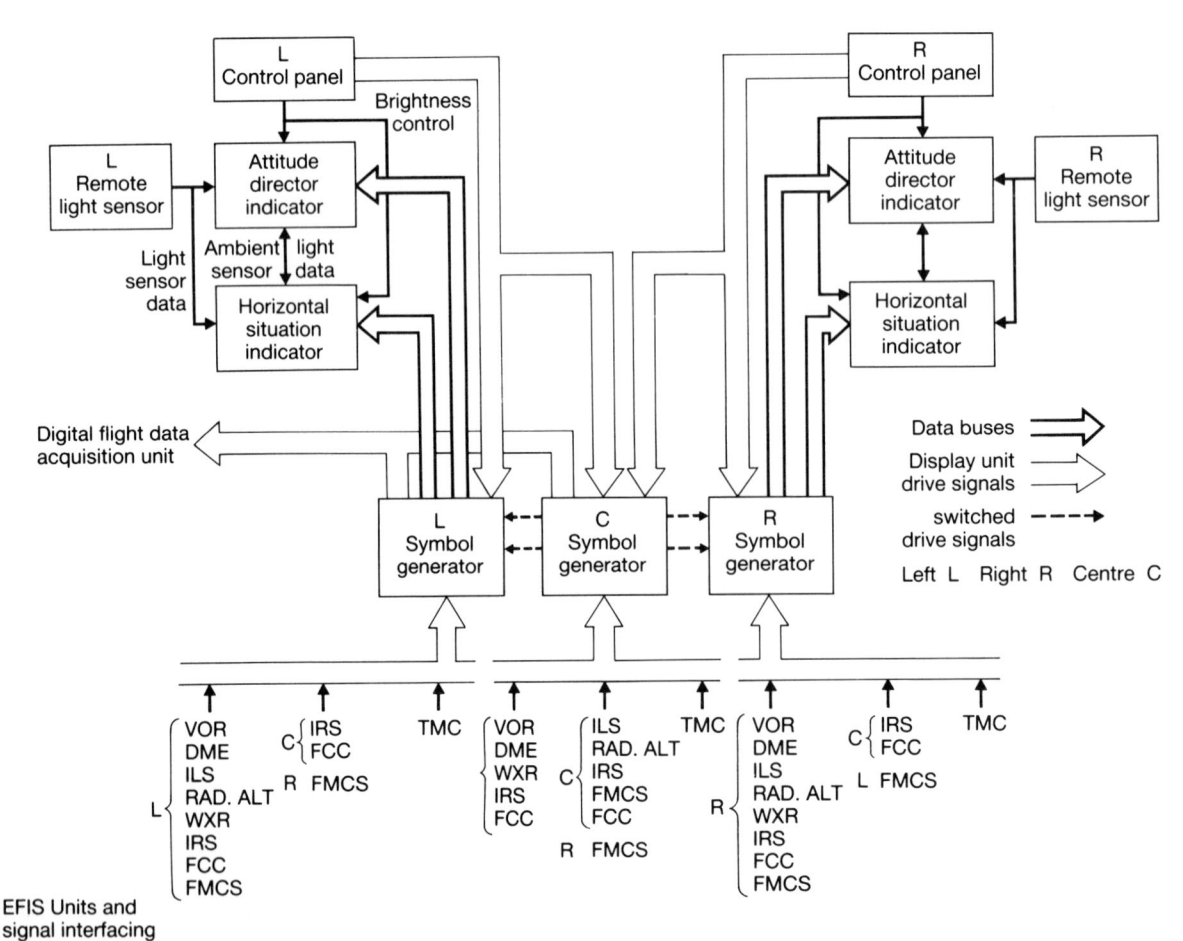

Fig. 4.1 EFIS units and signal interfacing
*(Micro-electronics in Aircraft System,
E. H. J. Pallet)*

EFIS Units and signal interfacing

50

processors and displays. Each discrete unit or element is linked to the others by one or more data buses. An interesting feature of the layout is the large number of input sensors. At one time each sensor would have had a unique link to an appropriate processor or flight deck instrument. With an EIS, as shown, there is interoperability and interchange of data between the system elements. Redundancy, to ensure that the crew is not deprived completely of information in the event of a failure, is part of the architecture adopted.

If the pilot needs to know, and can make adjustments to an avionic system then data signals must link it to displays, annunciators and controls within sight, sound and reach of the pilot. All avionic systems can be classified as primary, secondary, tertiary or maintenance. However, these classifications are not immutable. They can vary with changing circumstances.

Primary displays and controls are positioned as close as possible to the pilot's forward line of sight and within the envelope which defines the limits of comfortable reach.

Ergonomics   As is to be expected, any book on avionics, particularly in those chapters dealing with controls and displays, will describe in detail the technologies used to achieve movement, colour, symbols, shapes and alphanumerics. Methods by which the pilot can insert commands and responses, such as control levers, switches and keypads, are also important subjects.

Ergonomics is the study of man in his working environment and therefore includes the study of both man and machine. It is intended to achieve safe and efficient use of both and to prevent the human operator making mistakes.

The history of ergonomics in aviation is characterised by a continuous process of looking at why something went wrong and then finding ways to prevent it happening again. Ergonomics has also tried to anticipate problems by extensive research into human factors and particularly in the context of this chapter, into better ways of presenting visual data to the pilot.

Until electronic displays became practicable there was little that could be done to improve the quality of information provided by electro-mechanical instruments. These had been developed to the limits of the technology. Any further advances could only be achieved at the cost of complexity which in turn affected reliability.

Many of the advances made in electro-mechanical instruments followed incidents and accidents in which it was clear that the pilot had either been led into making a mistake or had received insufficient data. These advances, such as the abandonment of the three-pointer altimeter, only took place after a significant number of accidents had occurred. The three-pointer altimeter, whose pointers might be read by the pilot as indicating that the aircraft was ten thousand feet higher than it actually was, remained as a standard for over 20 years. Only after many people had been killed was legislation introduced to have it replaced by something better. In the days before solid-state techniques, such as light emitting diodes, (LEDs), the solution had to be electro-mechanical. The problem was solved by adding a drum counter display of three digits to the conventional pointer-on-dial altimeter so as to provide an unambiguous read-out of altitude.

The significance of this particular item of instrument history to avionics on the flight deck is the limitation on better instrument design imposed by the available technology. Today's technology provides an embarrassment of riches: a wide range of colours for differentiating types of data, symbols and graphics and alphanumerics provide virtually unlimited choice so as to present the pilot with everything essential for controlling the aircraft. But the most significant limitation imposed by pre-avionic displays was inflexibility. A pointer moved across a dial, for example, but its shape, colour and range of movement could not be varied by the pilot to match a particular operating mode, whereas one of the most distinctive characteristics of electronic displays is their flexibility. This applies not only to things the pilot can select but the ability to change the characteristics and range of a display to match changes in aircraft performance and handling following modifications.

INSTRUMENTS AND DISPLAYS

The modern aircraft instrument panel usually consists of a number of solid-state cathode ray tube (CRT) colour displays. Typically each of these has a 6 × 5 inch screen. Other displays use LEDs or liquid crystal.

The solid-state or, as it is sometimes called, 'glass cockpit' is rapidly becoming the standard for both civil and military aircraft. However, before describing in any detail the application and use of these modern displays, it is important to look at the history of instrument development. This will give a better understanding of the function of the information displays on the flight deck and in the cockpit.

In the beginning there were few if any instruments. In 1903 when the Wright brothers achieved the first controllable, man-carrying, powered flight with a heavier-than-air machine their primary instruments were their ears, the sensation of movement imparted to their senses and the attitude and movement of their frail flying machine in relation to the earth's surface. As their early flights rarely achieved more than a hundred feet above the ground there was no need for an altimeter even if one had been available. Similarly speeds were not very high and the sound of the engine was of greater importance than worrying about how fast or how high.

With the rapid progress in aircraft design during the ten years from 1903 instruments were gradually introduced as aircraft became capable of flying significant distances and climbing to heights above which it was difficult to assess altitude by eye alone.

The tremendous upsurge in both technology and numbers and in the types of aircraft between 1914 and 1918 was matched by the development of instruments specifically for aircraft use. Hitherto they were often adaptations of automobile or marine equipment: the aneroid barometer of the meteorologist was one of the first altimeters because its use had been anticipated by the Victorian balloonists.

Certification requirements

The certification requirements (see Chapter Two) serve to introduce the minimum number of instruments and displayed parameters whether for electro-mechanical or fully electronic systems. As with the pointer-on-dial, electro-mechanical instruments, the characteristics and performance of electronic displays and instruments for civil aircraft must conform to national and international standards.

Although there are variations among the different civil aviation certificating authorities there is general agreement on certain minimum standards. Using the Joint Airworthiness Requirements (JARs) of Europe as an example these specify that a primary flight display, such as an attitude director, must be clearly visible to the pilot when he is looking forward along the flight path. This applies, of course, to both electro-mechanical and electronic displays. JARs also require that when a 'stand-by' instrument is fitted it must be positioned so that it can be seen by both pilots.

The JARs also specify the 'basic tee' configuration for airspeed, attitude, altitude and navigation displays and the use of red for warning, amber for caution and green for safe operation.

They also specify that instrument systems must be arranged so that those of the principal pilot's position (P1) are independent of the instruments at other crew positions. This is to ensure that any failure or combination of failures, unless extremely improbable, will not deprive the pilot of the speed, attitude and heading information needed to maintain control of the aircraft.

JARs also specify the minimum number of displayed parameters for a civil passenger aircraft. For each pilot's position they are: airspeed, Mach number, altimeter, vertical speed, rate of turn and slip/skid, bank and pitch angles, heading and aural speed warning. In addition and visible to both pilots: outside air temperature, clock and a stand-by direction indicator.

Engine instrumentation for a turbo-jet aircraft includes for each engine: fuel pressure warning, oil quantity, oil pressure warning, oil temperature, fire warning, gas temperature – usually EGT (exhaust gas temperature), fuel flow rate, shaft speed, and ice protection. In addition there has to be a separate indication of the fuel quantity in each tank of the aircraft.

Instrument evolution   The modern electronic instrument system provides a useful example at the end of a long line of development from the direct reading compass, through the radio bearing indicators to the 'map' or horizontal navigation situation display. By following the evolutionary steps which led to today's solid-state electronic displays we can better understand the way in which, step by step over the years, avionics has provided the only solution to the problem of improving both the quality and quantity of data available to the pilot.

We can start with the magnetic compass. This was one of the earliest instruments and was derived from the marine compass. Until the mid-1930s the magnetic compass was just one of six or eight flight instruments. The introduction of the gyroscope mechanism for flight instruments provided the heading indicator from which the pilot could quickly see any change of heading. As with all gyroscopic devices the heading indicator began to 'wander' after a time so that the pilot had to reset it by reference to the magnetic compass. However for many decades, particularly in the smaller types of aircraft as well as in aircraft of the Royal Air Force, the magnetic compass and the heading indicator remained as essential instruments. However, there was an alternative. This was the remote reading compass developed primarily in the US and Germany. The advantage of this was the easy-to-read compass rose dial with a moving pointer operated remotely from a magnetic compass system located in a part of the aircraft where the adverse effects of stray magnetic fields were at a minimum.

In parallel with the evolution of the remote reading compass the radio bearing indicator was developed. Within the somewhat wide definition of avionics, both this and the remote reading compass could be classed as 'avionic'. The direction-finding antenna was arranged to align itself automatically with a selected radio beacon and to indicate the relative direction of the beacon in degrees on either side of the aircraft's nose.

The next step was to combine the compass indicator with the radio bearing indicator. This produced the radio magnetic indicator (RMI). With this instrument the pilot can read both the magnetic heading of the aircraft as well as the bearing of the radio beacon and, without ambiguity, instantly relate one to the other.

A further development in which a number of electro-mechanical and electronic elements were combined produced the horizontal situation indicator (HSI), (Fig.

Flight Decks and Cockpits

Fig. 4.2 Typical electro-mechanical horizontal situation indicator (HSI) *Smiths Industries*

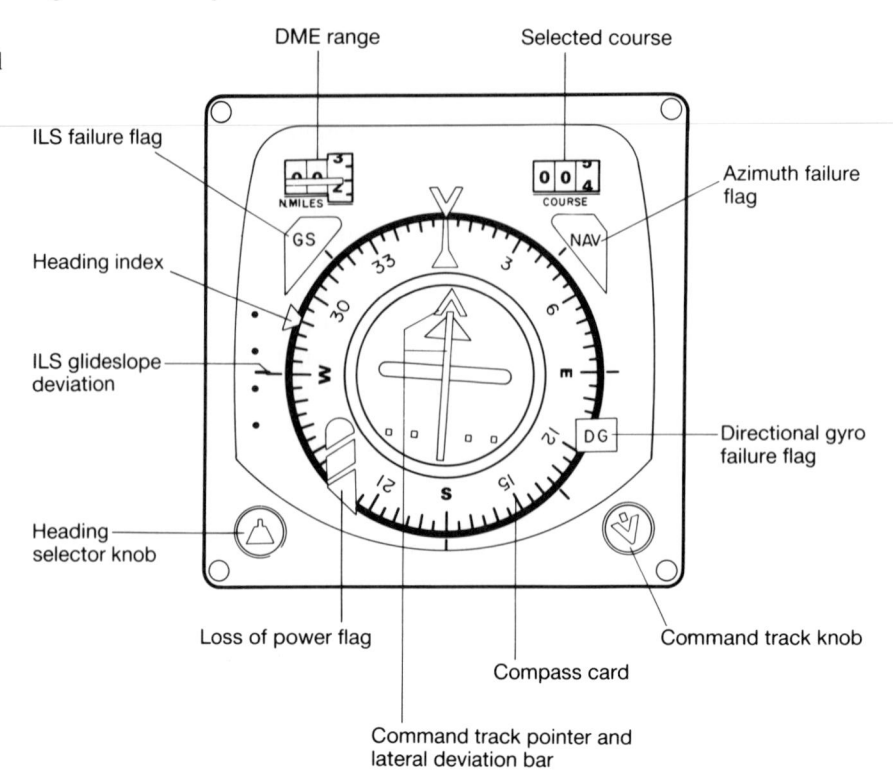

Labels: DME range; Selected course; ILS failure flag; Azimuth failure flag; Heading index; ILS glideslope deviation; Directional gyro failure flag; Heading selector knob; Loss of power flag; Command track knob; Compass card; Command track pointer and lateral deviation bar

4.2). This, together with the attitude director (ADI), became the first 'pictorial' instruments because they went some way toward providing the pilot with an easily interpretable picture of where the aircraft was at that moment, its attitude and where it was likely to be at some future time.

Both the HSI and the ADI have been developed to the limits of electro-mechanical technology. Even with the help of some solid-state electronics to improve accuracy, reliability and versatility of information little more could be done to improve the 'picture'. This was produced by pointers, scales, spheres and rotating discs and symbols. The pilot could neither add to nor reduce the amount of information provided; a facility only available with modern electronic displays.

It is important before dealing with electronic displays in detail to emphasise the large measure of standardisation of instrument positions which was achieved internationally for civil aircraft.

As instruments began to proliferate on the flight deck in the 1930s not much thought was given to where they should be located in relation to the pilot's centre line and eye point. Nor was much consideration given to the relative positions of the different instruments in front of him. For example it was not unusual to find that a clock, some RPM indicators, compass repeater and altimeter were grouped on the centre of the panel. The airspeed indicator and other important instruments might have been 'scattered' elsewhere.

In the mid 1930s the Royal Air Force decided that there would be a Basic Six panel of instruments which would be a standard for all future aircraft types. This was in contrast to the practices of the other countries. In the United States, for example, instrument panels of all types of aircraft continued to have a 'mixed' arrangement.

The importance of this review of instrument panel layout to the subject of avionics on the flight deck and in the cockpit, is shown by the fact that today the relative positions of the different symbols and alphanumeric groups on an EIS display reflect, to a large extent, the standard arrangements of electro-mechanical instruments arrived at by about 1950 for the majority of aircraft types in the Western world.

AVIONICS THE ONLY ANSWER

Many of the uses of avionics are as alternatives to electro-mechanical technologies. However, as already mentioned, avionics can do things which were never possible with other technologies which would have needed excessive development times and complex, not easy to set-up and run systems, to achieve the abilities of an avionic system.

One of the most important functions of avionics is the presentation of data to a pilot when human senses, such as direct vision, have been cut off because of poor or zero visibility at night or in rain, snow and fog.

In the late 1920s companies such as Sperry in the US, Reid & Sigrist in the UK and Lorenz in Germany applied themselves to instruments and systems which would overcome the pilot's lack of sight of the horizon, the earth's surface and in particular the landing area. During the 1930s a number of important systems were introduced: the Sperry artificial horizon indicator, the Reid & Sigrist turn and bank indicator and the Lorenz landing guidance beam system. There were others, but these three will serve as examples of the effort put into methods which enabled a pilot to maintain stable flight without visual reference to the earth, to make precisely banked and timed turns and to complete a landing in poor visibility.

The designers of the mechanical, gyro-based instruments achieved remarkable progress without the help of avionics

Visual displays of information

Functions:

- monitoring the progress of flight;
- decision-making in relation to some future time or aircraft position;
- monitoring the correct functioning of all parts and systems of the aircraft.

These functions can be displayed separately or can be combined.

The number, type and amount of data displayed on all or on individual visual displays can be varied with modern electronic instrument systems. It can vary in accordance with the progress the flight and with the status of the various systems. The amount of detailed information can be varied in extent either by the system itself, in order to match the current operating mode, or by the aircraft crew.

Therefore, unlike pointer-on-dial, electro-mechanical or partly electronic instruments, there is no dedicated, unvarying display format other than specific formats for specific operating modes. Essentially, but as an extreme example, the CRT displays on a transport aircraft flight deck might be, at a particular moment, blank. This situation occurs when the aircraft is stationary, on the ground and 'dead', i.e. no electrical or electronic systems working. Between that condition and the operating mode which requires the maximum amount of information there are a great number of variations available to the crew.

Electro-mechanical instruments

Before describing the types and variations to be found among today's extensive range of solid-state electronic displays and instruments, another reference is made to the electro-mechanical flight instruments which still make up the greater part

Flight Decks and Cockpits

of the world's aircraft instrument panels. This is also necessary because, as we have seen, the evolutionary rather than revolutionary progress of aviation technology has required the new to work alongside the old. And often the new technologies, such as solid-state displays, have first been introduced as part of hybrid instruments in order to overcome the limitations of electro-mechanical techniques.

Flight instruments, that is those used to control heading, speed, vertical speed and attitude, were the subject of numerous patents directed at giving higher accuracy and versatility. Ease of reading on the part of the pilot was not given much attention until after the Second World War. Floodlighting of the instrument panel and integrally lit instruments gradually replaced individual instrument lamps and the use of fluorescent paint. These went some way to meeting the problem of seeing the instrument readings at night without conflict with the pilot's view outside the cockpit.

In the meantime a greater use was made of electro-mechanical techniques to improve accuracy and to provide instruments which combined more than one parameter or function in one display.

Hybrid instruments   The attitude director indicator (ADI) (Fig. 4.3) is a typical example of the hybrid electro-mechanical/electronic instrument.

Fig. 4.3 Attitude director indicator (ADI) – electro-mechanical with some electronics *Sperry Flight Systems*

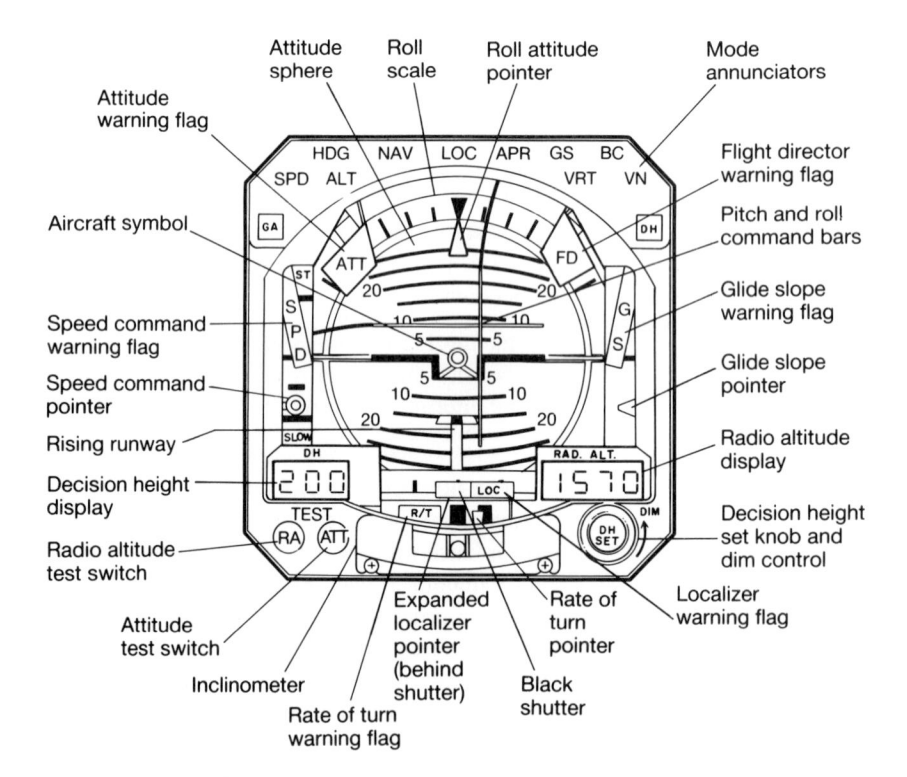

The modern ADI originated with the Sperry artificial horizon of 1929 which enabled James Doolittle to complete the world's first take-off, circuit and landing without reference to the view outside the completely enclosed cockpit. Since that important event a succession of improvements has been made to the sensitivity and quality of the display. By the 1960s the ADI had been developed into a multi-parameter display with colour used to differentiate between parts of the display. Much of the advance in instrument technology was achieved by using miniature servo motors for driving the mechanical system including the sphere painted in the sky and ground hemisphere colours.

The latest versions of the ADI incorporate a microprocessor and micro-servo. In addition the pilot is presented with LED numeral readouts of decision height and radio altitude. This is therefore an important step towards the electronic instrument system.

Electronic displays Recently developed avionic technologies have revolutionised the ways in which information is presented to the crew of an aircraft – and continue to do so. As aircraft systems become more and more complex emphasis has to be on making the most effective use of the human component of the man–machine interface: a component which does not change.

The passive role of the human operator in which he or she responds to changing data on a display screen by operating a set of controls or a keypad is being superseded by the 'interactive' role for the human.

The advent of powerful and not too expensive digital computers allows us to replace dedicated displays, such as an altimeter, by multi-function, colour, control and display units. Interactive elements of a display allow the human component of the interface to communicate with the systems by touching the screen or by using direct voice input (DVI) and, therefore, without necessarily having to press, turn or move other controls.

None of the avionic techniques described in the following pages makes the pilot less important. The object of the new control and display techniques is to make the most of human potential by eliminating the need to perform or monitor routine tests. At the same time they can interface directly with the digital systems which are now in the majority, and it is those digital systems which have made possible solid-state, colour, displays. The result is a thorough rationalisation of the aircraft man–machine interface.

Electronic display requirements To pilot an aircraft with safety and efficiency it is necessary to have an understanding of the principles of aerodynamics. However, a detailed knowledge of the functioning of an electronic display, such as the CRT, is not necessary either to use it or service it. Provided the maintenance crew know how to change a failed unit and how to complete the routine test procedures then the fact that the CRT (Fig. 4.4) consists of such elements as different circuits, three electron guns, directing coils, shadow mask and phosphor coating, made up of thousands of red-green-blue pixels, is of little importance. After all such display units are usually classified as line-replaceable. That is, if they fail, either in flight or during routine test, they are 'pulled' at the first opportunity on the ground and a replacement unit is substituted.

Therefore it is enough to understand that, within limitations, an electronic display can provide a wide range of symbols, graphics and alphanumeric characters in a range of contrasting colours. The concern of the pilot and the designer can be limited to the quality of the information on the screen: is it unambiguous? Can it be seen in all ambient conditions and over a wide viewing angle? Does it avoid imposing adverse physical effects on the user? As an electronic display may be the primary interface on the flight deck or in the cockpit, both the designer and the pilot have to consider the ability of the system to overcome failures by having sufficient built-in redundancy.

Of less concern to the pilot but of major concern to the designer are the physical dimensions, power supply, interchangeability, built-in-test and safety and

Flight Decks and Cockpits

Fig. 4.4 Shadow mask colour cathode ray tube for electronic displays *Smiths Industries*

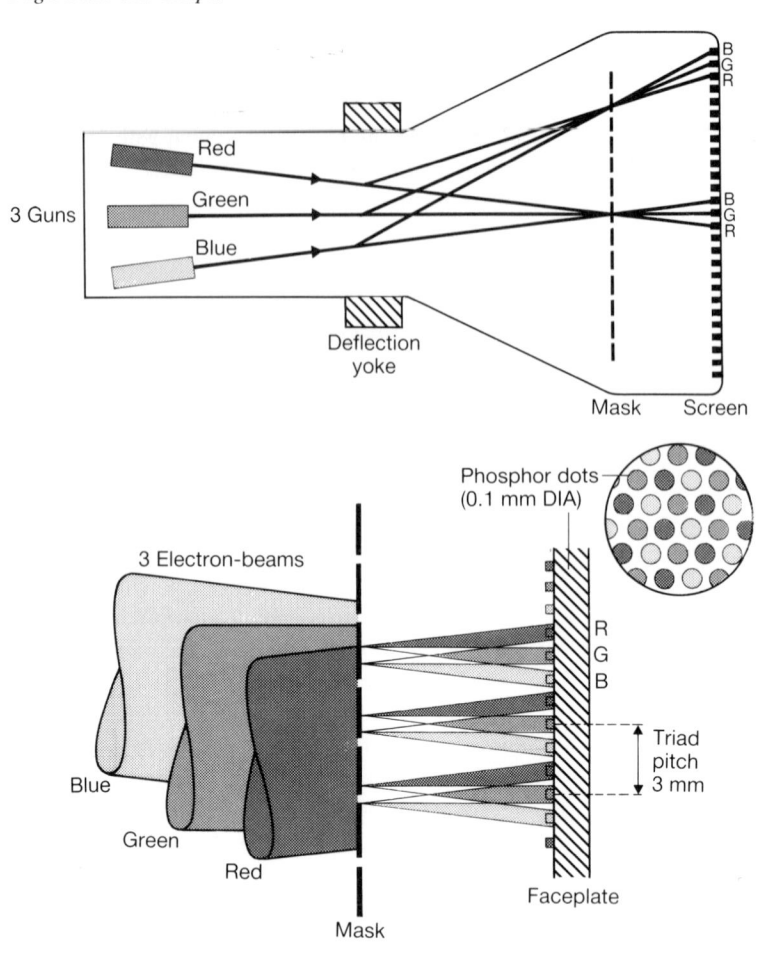

cost of ownership. The last includes such contingencies as the failure of a display unit at a remote airfield on an airline's route or on an STOVL fighter operating from the pad of a naval auxiliary vessel. Is the cost of keeping spare units at a remote location justified?

Display technologies compared

Figure 4.5 compares the principal display technologies which are currently available as production equipment. It is important to emphasise the development gap which often exists between the prototype and the production standard. The latter can only be realised after an acceptable number of units have been proved in flight evaluation.

The colour CRT display, now so familiar for primary flight information and for systems displays, can be used as the ergonomic standard when assessing the performance of the other display techniques.

The viewing angle referred to is an obviously desirable quality of any display in a cockpit or flight deck. Modern avionic displays can provide far wider viewing angles than 'conventional' pointer-on-dial instruments. The sort of lateral angle which is written into specifications is 60 degrees to each side of the normal, i.e. an arc of 120 degrees over which the pilot or pilots have to be able to read indications on a electronic display without ambiguity or loss of legibility. A required vertical viewing angle might be plus or minus 35 degrees.

Fig. 4.5 Display technologies compared

Display technologies compared

Cathode ray tube (CRT)

Plus features:
  Well proved technology
  High resolution and contrast
  Wide range of colours
  High brightness
  Wide viewing angle
  Low cost
  Potential for 'thin' display units

Minus features:
  Heavy and bulky
  High voltage
  High power consumption
  High temperature
  Sensitive to external electromagnetic effects
  Affected by bright light
  Vulnerable to shock, vibration and catastrophic failure

Light emitting diodes (LEDs)

Plus features:
  High display brightness
  Wide viewing angle
  Well proved technology
  Extensive application
  Low voltage operation

Minus features:
  High power demand
  Low resolution
  Not yet suitable for large arrays

Liquid crystal display (LCDs)

Plus features:
  Low voltage and power needed
  High resolution
  High contrast in high ambient light
  Non catastrophic failure.

Minus features:
  Slow response at low temperature
  Has to be backlit in low ambient light
  Narrow viewing angle
  Limited display size
  High cost and circuit complexity for full colour

Electroluminescence (EL)

Plus features:
  Rugged, lightweight and reliable
  Long life
  High contrast and good resolution
  High brightness
  Wide viewing angle
  Extensive application
  Low power consumption

Minus features:
  Brightness reduces with increasing area
  Complex circuits needed to drive display
  Full colour range yet to be perfected

Plasma

Plus features:
  Rugged technique
  High reliability
  Suitable for large displays
  High resolution and contrast
  Wide viewing angle

Minus features:
  Expensive to produce
  'Wash out' in high ambient lighting
  Affected by low pressure
  Restricted control of brightness
  Complex circuits needed
  Limited application in industry and transport

Resolution is an important display characteristic. This is the quality of providing the viewer with the ability to discriminate between the discrete elements which make up the display so that there is no confusion between the different symbols and alphanumeric characters. It also applies to the characteristics of replication and the ability to display, without confusion, complex shapes and relationships. For example the finite minimum size of an LED affects the degree of resolution, which is not always equal to that provided by other display technologies.

The majority of displays include symbols and alphanumeric characters and the very nature of this technology provides a virtually unlimited range of shape, size, proportions and colours. So what are typical values?

Line definition is an important characteristic of a CRT. A typical specification calls for a line width with the primary colours, at 50 per cent brightness, to be 0.3 to 0.5 mm. Finer lines may need to be displayed but when symbol strokes are drawn at critical angles relative to the geometry of the shadow mask then interference patterns may appear.

An important specification item is the rate at which the display is refreshed or renewed. At 50 Hz and less the eye becomes aware of a distracting flicker. Therefore the majority of CRTs used for EIS applications have a refresh rate of 70 Hz.

Another critical factor is the need to limit variations in brightness between symbols and parts of the display to less than 20 per cent.

Increasingly there is a demand for 'video' enhancement of primary flight displays so that Low Light TV (LLTV), thermal imaging and Forward Looking Infra-Red (FLIR) sensor data, particularly for military aircraft, can be combined with computer-generated alphanumerics and symbology. A CRT can either use the cursive, stroke-written technique or the raster, as with a domestic TV, or combine the two. This is why a 1050 line colour raster TV display of the real world combined with computer generated symbology is one of the standards at which other display technologies must aim.

A CRT has high contrast as do the other displays. But the important criterion here is the ability to maintain contrast in high ambient lighting conditions. For example when flying at 30,000 feet and higher, direct sunlight is so intense that it can 'wash out' a display. Colour CRTs can 'wash out' in high ambient light. The LCD can also prove to be inadequate in these conditions because it has to be backlit for use in low ambient light, such as at night, and this compromises contrast performance in direct sunlight. Similarly the plasma display has poor readability in direct sunlight.

LED displays are inherently rugged and reliable and their first cost has been kept down by large quantity production. They provide a wide viewing angle and high brightness without the need for high voltages. Large arrays of LEDs are being developed but there remains the problem of speed of response which is too low for dynamic displays.

As opposed to its ergonomic advantages and use as a datum for comparison, the CRT has a number of disadvantages which directly affect cost of ownership. CRTs are comparatively heavy and bulky, require a high voltage supply and high power, they sometimes need forced air cooling, are vulnerable to shock and vibration and incur high maintenance costs. By contrast an electroluminescent display combines robustness with light weight and is a 'flat panel' shape; the power consumption is much less than that of a CRT and it is inherently reliable. It also requires little or no cooling. Furthermore the flat panel CRTs being developed for the 1990s are less bulky and are lighter than the conventional tubes, which can weigh as much as 8 kg. The thin tube CRT has been a design goal for many years: not just for aviation use but for all types of vehicle and for commercial and domestic applications. However, as for all alternatives to the long established CRT, an important factor is the provision of graphics equal to if not better than those of the shadow mask 'deep' CRT.

The thin tube display (Fig. 4.6) is an example of a completely new approach to the generation of symbols and graphics. The designers of the thin tube CRT claim that it is no longer necessary to design a flight deck around the volume requirements of the conventional 'deep' fore and aft tube.

Liquid-crystal flat-panel displays of primary flight symbology and alphanumerics have been developed. Rockwell-Collins, for example, plans to replace its CRT displays with flat panel LEDs.

Once again using the CRT as the datum: the vulnerability to shock and damage and the real probability of catastrophic failure weighs against the installation of

Fig. 4.6 MEL thin tube CRT *MEL/Philips*

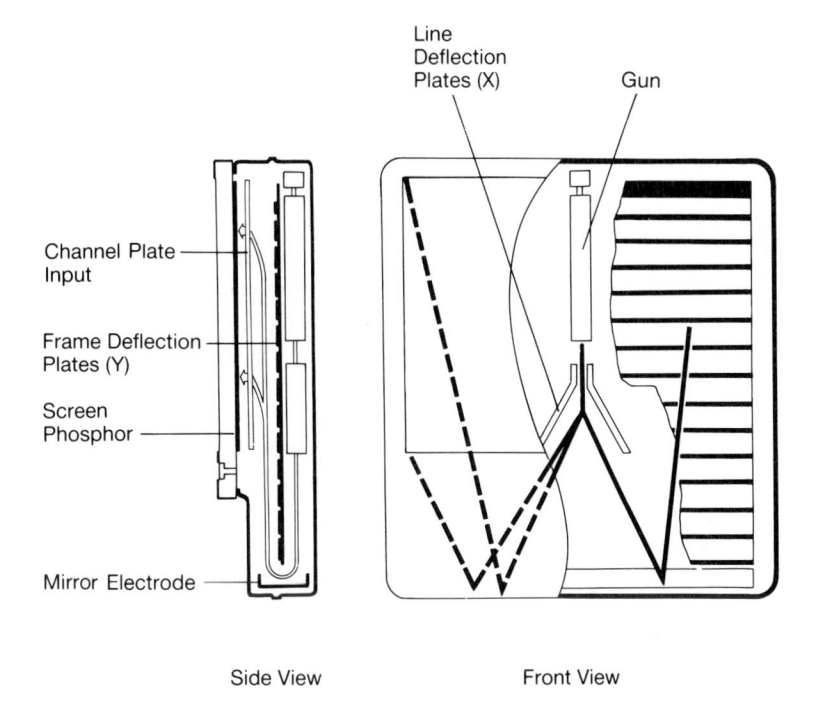

just one CRT in a civil transport aircraft. This is why an EIS has at least two CRTs and usually six so as to provide redundancy. The CRT of a military aircraft's HUD is not duplicated because of the different rules concerning safety and reliability which apply.

'Graceful failure' or 'fail soft' are attributes of systems which have been designed so that in the event of the failure of an element or component the level of redundancy allowed for in the design will ensure that there will not be a complete loss of information.

A liquid crystal display can be designed so that it fails gracefully and not catastrophically whereas certain failures in a CRT can be catastrophic. One or more of the in-parallel elements of an LED array can fail without affecting the others. An electroluminescent display and a matrix of LEDs can also be arranged to fail soft, as also can a plasma display.

The comparison table (Fig. 4.5) can be used to compare many different criteria, some of which have been emphasised above. One of the most significant conclusions to be drawn from the table, in the light of the technology of 1988, is the failure of the gas plasma technique which at one time appeared to have an assured future in aviation. It is important to consider the limitations of the plasma display in order to emphasise the advantages of the other techniques. For example among the factors which need to be overcome or improved are the following: it requires an expensive production process; the symbol generating circuits are very complex; the lack of interest for non-aviation applications does not encourage its development; and its ergonomics, such as readability in high ambient light levels and the limited range of brightness, have to be improved.

As with all avionics, first cost and cost of ownership of displays are important considerations. First cost is a function of the specification which can range from a simple display to a complex multi-function unit, such as an EIS. Simply this is a matter of 'you get what you pay for'. Cost of ownership, on the other hand, is a far more complex factor. It includes such system characteristics as robustness and reliability, and cost of repairs and spares.

In-service experience with a particular display technology is an important factor when making a selection. This is important not only from the point of view of the experience gained from the use of a particular type of display, from which its reputation can be established, but from the consideration of cost. The greater and more widespread the use of a specific display technology the lower the first cost. This, when associated with the cost of ownership, is an important factor when choosing from the different techniques, particularly for civil aircraft. However, an advanced display technology may be selected for military aircraft even when it has no well-proved extensive application. The CRT-based HUD was introduced into military service in the 1960s because the operational need outweighed the lack of previous application and experience.

Standing high in importance above the ergonomic characteristics and cost of ownership are the requirements of the certificating authorities whose primary concern, with civil transports, is safety. Air forces are concerned more with survivability of equipment but, essentially, this is still founded on the principles of safety.

The present trend of display technology development is towards the full colour CRT, with raster (TV) overlaid by cursive (stroke-written) and liquid crystal displays for large area primary displays.

Data highways The linking of all the principal flight deck display and control units by a digital data bus or highway is one of the key concepts in modern avionics (see also Chapter Three). The present civil aviation standard for data transmission between systems is ARINC 429 (see Chapter Two).

ARINC 429 has served well for many years but its data-handling capacity is now insufficient for the more advanced flight deck avionics about to enter airline service. Therefore ARINC 629 has been introduced to meet the needs of a wide range of systems including fibre optic data links.

DISPLAY AND CONTROL INPUT TECHNOLOGIES

Having reviewed some of the technologies used to produce information displays on the flight deck and in the cockpit we can now consider the different methods used for introduction of demands and answers into the various systems which provide the visual part of the aircraft control interface.

A display can include many different types of data: some is informative, some a response to a question, some an alert or warning. Informative displayed data is likely to be presented whenever the pilot makes some change to the aircraft's systems. When the aircraft is under manual control the primary flight displays will respond in step with the pilot's control inputs. The most obvious display change is that of the attitude director indicator and of the horizontal situation indicator but other displays will also change to a greater or lesser degree. When the pilot initiates changes to the aircraft's configuration by selecting aerodynamic devices, such as flaps, one or more sets of data and symbology will change.

Here we are relating major system changes to the different displays. The effects these changes have on the operation of an aircraft's avionics are described in Chapters Five and Six. This chapter is concerned more with pilot inputs other than those applied through the primary controls. The FMS described in Chapter Six (Fig. 6.6) has a control interface consisting of a matrix of keys below the CRT display and a number of 'soft' keys around the edges of the CRT.

A range of input devices is available for effecting inputs to an electronic display. For examples there are knobs, switches, keypads, touch screen, voice and, importantly, the data buses and wiring looms which provide both inputs and output to each display.

A typical ergonomic standard for control knobs will specify not only ease of use, by a gloved hand in military aircraft, a shape which cannot be confused with other knobs if that could lead to an error, and a finish in a matt paint or self colour such as grey. Switches may be designed to a similar set of ergonomic requirements. More than likely both knobs and switches will be in a different colour from that of the panel on which they are mounted.

These may seem somewhat obvious standards of design, but unfortunately the path of aircraft cockpit design is littered with exceptions to good design.

Any lettering and numerals which have to form permanent markings on display panels and on the input and selector devices must be clearly legible over a viewing arc of 60 degrees.

Keypads need special attention to the size and shape of characters because the pilot is sometimes presented with an array, on an FMS CDU, of 30 or more keys. Some of the keys may have a dual function so two characters have to be displayed.

Keys, switches and knobs   The majority of avionic interface equipment includes some form of tactile input device or devices for the use of the operator. The design of switches and knobs usually conforms with the best available ergonomic standards which should take into account the needs of the pilot, particularly when subjected to high G loads during combat and when wearing heavy gloves.

The design of key pads requires special attention to ensure that the advantage of their compactness is not negated by any difficulties the pilot might have in selecting without error one key from a number of similar keys.

A typical key size is 0.5 inch, square set at a pitch of 0.75 inch. Along with finger guards between the rows and columns of keys these dimensions have been found suitable for use in combat aircraft cockpits. Because of the 'gloved hand' requirement the keys are usually designed to a particular operating force level so that the pilot has some feedback through his fingers to indicate that the key has been actuated. The civil flight deck requirement is not so demanding because of the infrequent application of high G and the use of bare finger tips.

Touch screens   All input controls used by a flight deck crew, other than verbal and eye and head pointing, can be classified as tactile. Of these the touch screen CRT is a notable example of combining an electronically generated display with the input controls.

The advantage of a touch screen is the simplification of display and control relationships: a particular group of alphanumerics or even a symbol can be immediately 'interrogated' or responded to, i.e. YES or NO, by the touch of a finger.

There are four principal touch technologies. These are: scanning infra red (IR); resistive screen overlay; capacitance overlay; and surface acoustic wave. The last, however, has not been developed to the same extent as the other techniques. Each of these has one thing in common: they form part of the CRT screen.

One 'touch' technique uses the interruption of an infra-red field across the front of the display screen. When the pilot's finger penetrates the field, a 'touch' is detected.

With the resistive overlay type of CRT there is a glass substrate over the display screen. By touching the surface, which alters the resistance, the system is activated.

A glass substrate is also used with the capacitance overlay. When the user touches the screen the change in capacitance is measured and the location of the touch is defined.

With another technique surface acoustic waves are transmitted through a glass substrate. When the pilot touches the substrate to activate the system the change in energy is measured and the location of the touch is detected.

Comparisons   There are advantages and disadvantages to each touch input method. Each can be assessed on the basis of: resolution, speed of response, image quality, environmental resistance, and reliability.

Resolution.   This is an important factor in any system of communication. The cockpit and flight deck interfaces are no exception. Whatever method is used to display and select data, including the touch screen, the resolution must enable the users to discriminate without ambiguity or error between the different elements or 'bits' of data.

One of the design choices is between 'coarse' and 'fine' resolution. As an example a 'coarse' touch screen may have a matrix of 256 × 256 discrete points. A 'fine' screen might have 4096 × 4096 touch points.

The resolution, or density points per unit of screen area, depends on the size of the display, the 'touch' technology and on the capacity of the computer and the sophistication of the software.

For example, resolution in infra-red touch systems is expressed as the amount of space between each 'touch element'. Physical resolution of an infra-red system could be 0.25 inch from the centre of one element to the centre of the next.

The physical resolution of a surface acoustic wave system is approximately 0.03 inch between reflectors, but with software techniques resolution can be increased.

Speed of response.   Resistive overlay, infra-red, and surface acoustic wave techniques provide nearly instantaneous sensor responses. Capacitance overlays are slower because of the time taken to store and measure changes in capacitance.

The importance of response time depends on the application: graphics require a fast response whereas menu selection need not have such a fast response.

Image quality.   Display systems which require a glass substrate on the screen result in some visual obstruction between the viewer and the display. Some overlays can reduce the light transmitted from the display by as much as 50 per cent.

Surface acoustic wave systems have higher transmissivity than resistive or capacitance overlays. The infra-red technology is based on a matrix of sensors which frame the screen and do not affect the light transmitted. Infra-red light is invisible, therefore the pilot's visual path is not degraded. At the same time the colour and sharpness of the CRT video image is maintained.

Environmental resistance.   A touch system must withstand environmental conditions which would otherwise degrade its performance. Capacitance systems may be affected by moisture and static from the user's hands or may not be activated if touched by a non-conductor such as a gloved hand. Variations in the pilot's body capacitance also have to be allowed for.

Infra-red systems use solid-state components, not electro-mechanical parts which can corrode or fail. Furthermore, these can withstand temperature extremes, as well as exposure to the elements. However, special techniques are needed if the system is used in bright sunlight. Infra-red systems can meet stringent civil and military requirements and be completely sealed against corrosive atmospheres and liquids.

Reliability. In general, touch systems are rugged and reliable with few parts subjected to user operation. Since the touch action is integrated with the display, damage or loss from a failed input device, such as a switch, is minimised. Overlay systems on a CRT can operate without loss of effectiveness over a wide range of temperature and humidity.

Infra-red touch systems in particular can operate over a large range of shock, vibration, and chemical exposure.

Direct voice input (DVI) (see also Chapter Eleven) The major aviation nations are spending vast sums on developing practical and effective DVI systems. For example the Royal Aerospace Establishment in the UK, working with Marconi Defence Systems and GEC Avionics has, like other establishments, demonstrated that humans may make fewer mistakes using DVI compared with using an alphanumeric keyboard when entering commands and requests into a computer-based avionic system.

DVI provides a method of matching two parallel-processors; the digital avionics and the human brain. The latter constructs, in micro-seconds, complex concepts, but then has to translate them into serial, and therefore slower, inputs to the control interface using eye and hand coordination. Therefore this process is one of parallel operations converted to serial operations which are then operated on by the parallel processing potential of the avionics. DVI, even though it employs serial speech data, can still improve both the speed and accuracy of the pilot input processes.

The speech recognition systems of the late 1980s incorporate personalised templates. These allow the systems to respond to the particular and unique characteristics of each pilot's voice. These recognisers can accept and use contextual inputs relating to the flight plan or mission, systems and general flight information.

Speech recognition systems have to operate against the wide spectrum of background noise experienced in an aircraft cockpit or flight deck.

DVI is used to select, among other things, multi-function display formats, navigational aids and approach and landing parameters. It can also be used to input commands and to interrogate both the navigation and communication systems.

FLIGHT DECK SYSTEMS

There have been a number of evolutionary steps between the electro-mechanical instruments and the 'glass', i.e. electronic display dominated, cockpit. The use of LEDs, for example, to enhance the overall performance of primary flight instruments such as the ADI and the HSI. Perhaps the most important development in the last decade has been the integration of control and display systems along with the increasing application of CRT colour displays. Radio navigation, such as the radio magnetic indicator (RMI) and its integration with inertial navigation (IN) are steps toward the present flight management systems (FMS). These in turn have been developed to a high level of sophistication and are now integrated with the automatic flight guidance system (AFGS) of which the A320 systems are important examples.

Flight management The flight management system was one of the more important systems introduced

Flight Decks and Cockpits

in the 1980s to improve the flight deck control interface and to reduce crew workload. It is now a prominent feature of the flight deck of the AIRBUS, MD88, B757 and 767 and similar aircraft types.

As mentioned earlier flight deck design has rapidly moved ahead; particularly in the last ten years. Few instruments, few sets of controls and few systems are designed to operate as discrete elements of an aircraft. They may be designed to 'stand-alone' so as to be unaffected by the problems of other systems, such as power supply failure, but they must act in concert with the majority of other systems.

Figure 4.1 is a simplified diagram of the flight deck avionics of a transport type aircraft.

It is important to note that few of the systems included in Fig. 4.1 act only as an input or output device. The majority both provide and accept information for processing. This is why the order in which the individual systems are described can only be arbitrary.

Flight management system methodology

The flight management system is one of the first to be initiated by the crew because some of the input data, as will be described, is used during flight briefing, and therefore we can use this as the starting point.

The control and display unit of an FMS (Chapter Six, Fig. 6.6) provides the pilots with alphanumerics and symbology by which they can verify the essential details of the flight, make changes to the flight plan, i.e. manage the flight and, if need be, select simulations of alternative flight plans.

At all times when the crew is not actually controlling the aircraft by hand, i.e. operating the primary controls and the engine controls, they use the FMS control and display units as the principal interface with the aircraft.

The facility to select the next waypoint or series of waypoints, vertical flight profile and optimum airspeed and engine settings, enables the pilots to 'fly' the aircraft through the control and selector keys of the CDU.

Therefore, the CDU of the FMS is, in effect, a miniature flight deck with fingertip control. On the face of this unit, 12 × 8 inches, is all the visual display data and all the input controls. Therefore its design imposes a number of special ergonomic requirements: the ability to see without ambiguity and the ability to select and command without ambiguity.

It is a fact of aviation history that this revolution in flight deck design has been introduced into service with few ergonomic problems. Both monochrome and multicolour CRTs are used in FMS CDUs.

The description of an FMS can be simplified if we consider first the basic features of a typical scheduled passenger flight in an aircraft not fitted with an FMS. This will highlight the way in which an FMS can enhance safety, efficiency and reduce crew workload.

The flight deck crew report for duty at least one hour before departure in order to complete the flight briefing which includes a lot of 'paperwork'. The briefing room staff will usually have ready a prepared computer-generated flight plan which specifies routeings, flight altitudes and radio navigational facilities as well as emergency diversions to alternative airfields. An important part of a flight plan is the relationship selected between fuel load and revenue load. This can be varied to account for availability of destination and alternative runways as well as navigational aids, some of which may be inactive.

The 'briefing' computer can also hold data relating to individual members of the crew so that an immediate print-out is available to the captain from which he can instantly check the qualifications and experience of a member of the crew. For example: is the second pilot fully qualified to take over should the captain collapse and if the aircraft has to be diverted to an airfield in limited visibility conditions?

But when the crew leave the flight briefing they leave the computer behind if their aircraft does not have such modern systems.

Other data to which the crew must have access relates to the forecast meteorological conditions and how this may require changes to the fuel load, the routeing and the flight levels at each sector of the flight.

The greater part of the data needed by the crew both before and during a flight is contained in route charts and airway manuals. An important group of charts come under the heading of SIDs and STARs. These acronyms refer to standard instrument departures and standard terminal approach routes. They are detailed charts showing the precise headings, minimum or maximum speeds, altitudes, frequencies of navigational aids and other data needed so that aircraft depart from and arrive at airport terminal areas along defined paths, both horizontally and vertically. In other words, ATC establishes three-dimensional paths, four-dimensional if time and speed are included, to which the crew of each aircraft must, in the absence of alternative instructions from ATC, adhere.

At one time the crew of an aircraft operating a route which included a number of landings and take-offs and therefore a complex series of legs, had to carry flight bags crammed with airways and airfield data and charts as well as the aircraft's flight and performance manuals.

On reaching the aircraft the crew has to perform a considerable number of internal and external checks of the aircraft and its many systems. Before the introduction of electronic centralised aircraft monitors (ECAM) and engine indicating and crew alert system (EICAS), the crew had to exercise a very disciplined and sequenced checking procedure to ensure that the aircraft was 100 per cent fit to fly. Printed check lists could easily lead to one particular item being missed, just as an external visual check of the position and condition of equipment could miss a vital item.

By contrast with the above procedures an FMS equipped aircraft reduces considerably the amount of pre-flight 'paperwork' on the part of the crew. In other words, much of the pre-flight briefing data is already on board the aircraft stored in the FMS data base which can be called up through the CDU. As described in Chapter Six, the information in the data base is updated at intervals.

The primary function of the CDU is to act as the interface between the aircraft and the crew so that the latter can select optimised flight profiles, i.e. the best route, the minimum time or the minimum fuel consumption between selected waypoints. The CDU can be used to command completely automatic control of the aircraft or semi-automatic with varying degrees of pilot involvement including full manual control.

The CDUs, because there are always at least two, are the 'tip of the iceberg', made up of a number of different avionic systems integrated and working in concert. The 'iceberg' is represented by the architectural schematic shown in Fig. 4.1.

In other chapters reference has been made to data bus technology and to the important role this performs in linking disparate avionics.

As Fig. 4.1 clearly shows, an FMS and its CDUs are connected via the data bus system to the majority of an aircraft's information, control and sensing units. The present data bus standard is ARINC 429.

Fig. 4.7 Records in data base memory *Microelectrics in aircraft systems* (E. H. J. Pallett)

Record	How identified and defined
Radio-nav aids: VOR, DME, VORTAC, ILS, TACAN	Identifier ICAO region, latitude and longitude, frequency, magnetic declination, class (VOR, DME, etc.), company defined figure of merit,* elevation for DME, ILS category, localizer bearing
Waypoints	Each waypoint defined by its ICAO region, identifier, type (en-route, terminal), latitude and longitude
En-route airways	Identified by route identifier, sequence number, outbound magnetic course
Airports	Each identified by ICAO four-letter code, latitude and longitude, elevation, alternate airports
Runways	Each identified by ICAO identifier, number, length, heading, threshold latitude and longitude, final approach fix identifier, threshold displacement
Airport procedures	Each identified by its ICAO code, type (SID, STAR, profile descent, ILS, RNAV), runway number/transition, path and termination code
Company routes	Origin airport, destination airport, route number, via code (SID, airway, direct, STAR, profile descent, approach), via identifier (SID, name, airway identifier, etc.), cruise altitude, cost index

* The figure of merit is a number assigned to each navigational aid to indicate the maximum distance at which it can be tuned.

The operation of an FMS is described in detail in Chapter Six. Figure 4.7 shows the principal sets of data stored in the FMS computer memories. This table emphasises the comprehensive nature of an FMS. The total data held in memory represents the contents of many flight and operations manuals as well as information to be found in the manuals of air traffic control and the publications of regulatory authorities, such as ICAO.

The design of a CDU has to be related to its importance as one of the primary control and information areas on the flight deck. As already mentioned, during some flight modes the CDUs are the primary interface between crew and aircraft.

Despite the many different functions and sets of information data available for an FMS, modern control and display technology provides a multi-function interface within a comparatively small area. The comparison is, of course, with the total area which would be needed if all the systems with which an FMS interfaces had their own individual CDUs. Fortunately, the multi-function facility afforded by modern avionics enables a limited number of display and control elements to be concentrated in a small area.

Keeping to the methodology of starting with a display, there is essentially just one. This is a monochrome or colour CRT which takes up the upper half of the CDU face. On this are displayed different 'pages' of selectable data stored in the FMS memory.

Cursively generated symbols and alphanumerics are presented at, typically, 24 characters for each of the 14 lines. Different sizes of characters are used to indicate the mode or status of the information. For example the smaller characters are used to indicate default or predicted values for the different parameters. The larger characters indicate data changed or selected by the crew. Colour CRT display can obviously increase the range of importance or status available for different sets of parameters or flight modes.

Moving from the display to the selector key panel, both the two principal FMS systems of the Airbus, for example, one by Smiths Industries the other by Sperry, depending on the operating airline, have a common keypad function. The function, alphabet and numeral keys are arranged in accordance with the best available ergonomic principles. The following list will help to explain the relationship between display and controls for a number of typical flight modes.

Select or return to:	*Meaning*
PPOS	Present position
DATA	Data index pages for:
	Lateral and vertical performance
	Key waypoints
	Sensors
	Navigation
	Aircraft configuration
	History
NEXT PHASE	Display next leg of flight plan
PERF	Performance page
DIR	Direct entry of flight plan revisions
FUEL	Fuel pages
AIRPORTS	Next airport page on leg
HDG	Headings to be flown automatically
FIX	Check for update position
START	Start data pages
ENG OUT	Engine and performance pages
SEC F-PLAN	Secondary flight plan
EXEC	Promote a temporary plan to active status
MSG	Acknowledge message displayed
CLEAR	Delete scratch pad

Electronic attitude direction

Complementing the CDUs of the FMS are the electronic primary flight displays of attitude and navigation with which the majority of civil transports of the 1980s are equipped.

The electronic attitude director display (Fig. 4.8) can present over 25 separate parameters, indications and legends in the form of symbols, alphanumerics and sky and ground colours. In addition to the blue, cyan and the yellow for the sky and ground hemispheres, white, green, red and magenta are used to differentiate between functions and changes in status.

This is a limited display compared with a full primary flight display of an EIS. However when compared with the electro-mechanical ADI of Fig. 4.3 it emphasises the significant increase in the amount of data which can be conveyed to the pilot.

Electronic primary flight display

The electronic ADI leads on to the EIS which is, as its title suggests, a complete electronic system using two or more display units to convey all the information required by the pilot.

In its most usual form an EIS consists of six colour CRT displays. The two display units immediately in front of each pilot provide primary flight and navigation data and graphics. By 'primary' is meant aircraft attitude, speed (in both knots and Mach No.), vertical speed, heading, altitude (both radio and barometric) as well as limiting values and warnings. The navigation display shows aircraft plan position

Flight Decks and Cockpits

Fig. 4.8 Electronic attitude director *Rockwell Collins*

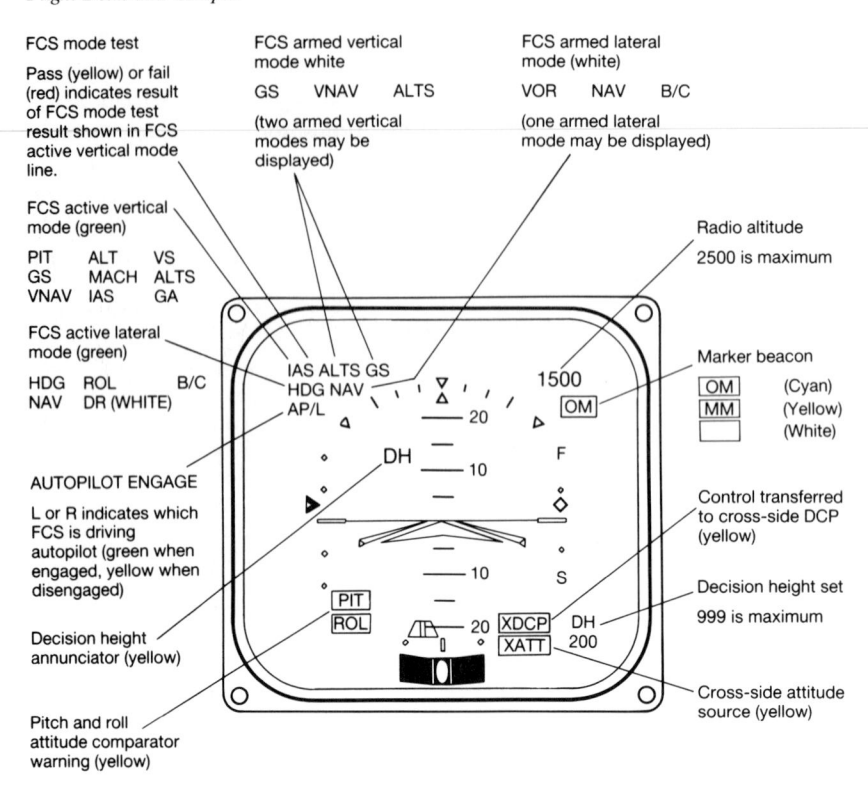

Labels on figure:

- FCS mode test — Pass (yellow) or fail (red) indicates result of FCS mode test result shown in FCS active vertical mode line.
- FCS armed vertical mode white — GS VNAV ALTS (two armed vertical modes may be displayed)
- FCS armed lateral mode (white) — VOR NAV B/C (one armed lateral mode may be displayed)
- FCS active vertical mode (green) — PIT ALT VS GS MACH ALTS VNAV IAS GA
- FCS active lateral mode (green) — HDG ROL B/C NAV DR (WHITE)
- AUTOPILOT ENGAGE — L or R indicates which FCS is driving autopilot (green when engaged, yellow when disengaged)
- Decision height annunciator (yellow)
- Pitch and roll attitude comparator warning (yellow)
- Radio altitude 2500 is maximum
- Marker beacon — OM (Cyan), MM (Yellow), (White)
- Control transferred to cross-side DCP (yellow)
- Decision height set 999 is maximum
- Cross-side attitude source (yellow)

relative to navigational aids such as radio beacons and approach and landing systems. It also displays the weather radar picture, the aircraft's track relative to waypoints, wind speed and direction and all other data needed to control the aircraft in time and space.

The remaining two display units are used for aircraft systems management such as Engine Indication and Crew Alert System.

The primary flight display (Fig. 4.9) is a typical example of a modern colour CRT EIS. The usual power supply is 115V 400Hz with a power consumption of about 50W which means that forced air cooling is not necessary.

Weather and lightning display

The weather radar display was one of the first examples of a production standard CRT on the flight deck of civil aircraft. Prior to about 1955 the CRT was usually to be found only in military aircraft as part of the radar navigation and bomb-aiming systems.

The advent of the digital computer resulted in a considerable improvement in the quality of a weather radar display by better discrimination between the different areas of precipitation and turbulence.

The digital computer has also been used to detect, measure and locate lightning discharges. This information is now combined with the visual display of precipitation and turbulence. This type of CRT display maximises the advantages of the colour CRT. Turbulence shows on the screen as a solid blue integrated with the green, yellow, red and magenta to indicate areas of increasing precipitation.

This easily interpretable display clearly shows the pilot the position and intensity of rainfall, the location and number of lightning discharges per minute and the areas to be avoided.

Fig. 4.9 Primary flight display on an electronic instrument system (EIS), colour CRT *Smiths Industries*

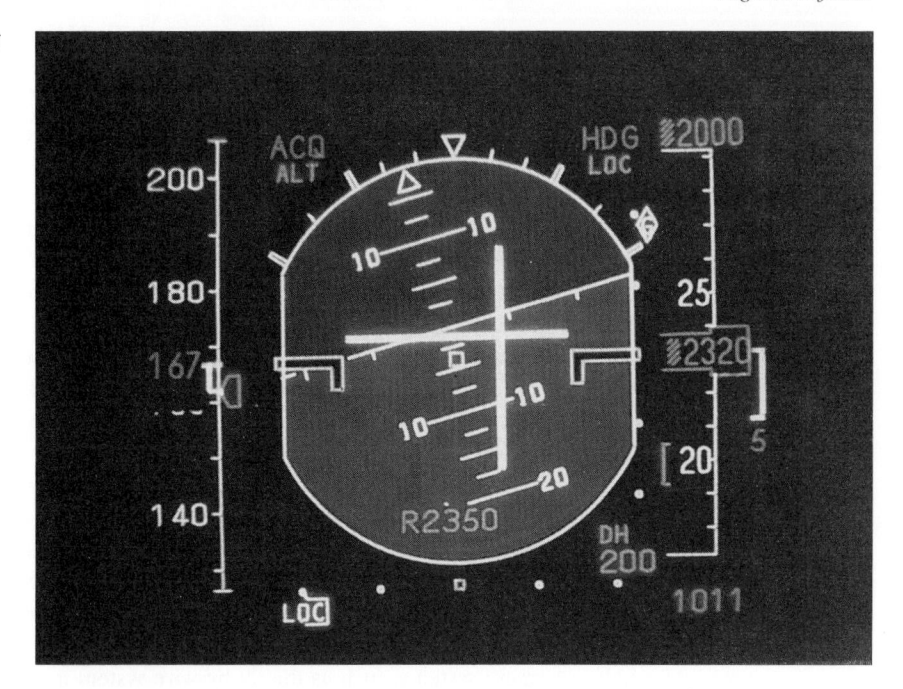

Civil head-up display From the inception in the early 1960s of the head-up display (HUD) for fighter-attack aircraft research was also directed at possible civil aircraft application.

Essentially there are no major problems in mounting a HUD on the flight deck of a transport aircraft, and from 1970 onward a number of HUD installations were evaluated. The comparatively slow advance of HUDs for civil aircraft, compared with military systems, is the result not so much of technical problems but of a lack of internationally agreed operational requirements. In other words, what is the role of the HUD and what is its potential contribution to improving the safety and efficiency of transport aircraft? Also to be considered is the cost which, like all large avionic systems, can run into six figures.

Even with an automatic landing system a point is reached during an approach and landing at which the pilot has to take over manual control of the aircraft.

The transition from monitoring the progress of the automatic system to looking forward through the flight deck windows is a critical time. Transferring the eye point from the instrument panels to the view ahead has always posed problems even with a non-automatic approach when following the commands of the attitude director (AD) during an ILS approach.

Although there are many different ideas about the way in which an approach and landing should be controlled the one common factor is the pilot's visual problem. The pilot has to assimilate rapidly the essential cues needed from the approach and runway lights and from the complex, constantly changing, perspective of the airfield and its features.

The pilot of a fighter-attack aircraft can fly by reference to alphanumeric data and symbology displayed in his forward line of sight by the HUD so that there is no need to transfer the eye point from the 'outside' scene to 'inside' information during critical phases of a flight. This 'head up' concept is obviously applicable to the transport aircraft flight deck but a HUD used only during the take-off and landing phase is a very expensive 'passenger' for the rest of the flight.

Among a number of companies which have developed head-up display systems for civil aircraft, such as Thomson CSF, Hughes, Kaiser, Sundstrand, Smiths Industries and GEC Avionics is Flight Dynamics Inc which developed its HGS (Head-up Guidance System) in 1986. This is a HUD whose computer, software and display symbology and alphanumerics, is dedicated to giving the pilot guidance during the approach and landing phase with the important feature of windshear warning and corrective action.

The objective of this system is the safe lowering of landing minima below the 50 ft decision height and runway visual range (RVR) of 700 ft (Category 111A). Just as the automatic landing systems of the 1960s were introduced to maintain scheduled operations in adverse visibility conditions so the HGS is intended to free civil operators of the handicap of airports which have a significant number of fog-bound days – Seattle, for example, and, also, provide vital guidance which will enable the pilot to react quickly to windshear and to take the most effective action.

In function and components the HGS is similar to a military HUD (Fig. 4.13); it has the same holographic combining glass set at an angle and intercepting the pilot's normal forward line of sight. The principal physical difference is the mounting of the display unit under the flight deck roof and not, as in a fighter aircraft, at the top of the primary instrument panel.

The potential use of HGS must be seen in relation to automatic flight management systems, such as the fly-by-wire system in the A320. These not only prevent an aircraft exceeding its safe flight envelope or stalling but also manage the aircraft's potential energy so as to offset the adverse effects of windshear and line squall conditions during an approach to land.

Continuous monitoring Flight data acquisition and health and usage monitoring systems, which record a number of operating parameters against real time, provide both the flight crew and the maintenance specialists with data, in digital form, which can give an immediate indication of an adverse trend or, over a finite time, 'print out' patterns and trends so that remedial action can be taken to prevent component failures.

As a flight deck or cockpit display these systems now provide an important read-out of data from which the pilot can gain assurance that all is well with vital systems, such as the transmission system of a helicopter or a warning that a component or system is showing signs of accelerated wear.

In Chapter Six reference is made to Utilities Systems Management (USM) and the way in which such systems provide data to the avionic data buses and thence to the cockpit displays. Figure 6.17 is an example of a typical fuel system display. The data handled by the USM system is used to 'drive' one of the MPCDs to show clearly and without ambiguity the amount of fuel in each tank and the open or closed positions of every valve in the complete system.

Colour for displays Mention has already been made of the facility of using different colours on an electronic display, such as a CRT, to highlight and differentiate among the various sets of data. A typical allocation is:

WHITE for scales and present information;
GREEN for present information needing contrast and of a lower priority than WHITE: also used for pointers and selection titles;
MAGENTA for commands, deviation pointers, active flight paths, and 'fly to';
CYAN for sky background colour on an ADI format, for low priority data and for selections made by the crew;

AMBER for lubber lines (datums), aircraft symbol, cautionary and transitory data;
YELLOW for ground background colour on an ADI format, caution and adverse trend data, failure 'flags', limits and alert indications and fault messages;
RED for warnings and limits.

Among the manufacturers of electronic displays, particularly for EIS and ECAM, there are slight variations of the above list.

Instrument lighting Both the flight deck and the cockpit are control interfaces which will be subjected to extreme variations in ambient light level. Not just the variation between sunlight at low altitudes and the fierce light at 30,000 feet and higher but the sudden changes as the aircraft flies in and out of cloud. Therefore modern electronic displays usually have light level sensors which are used to control automatically the display brightness in accordance with the ambient light level. The response characteristics of the sensors are matched to those of the human eye.

FLIGHT DECK EXAMPLES

At one time it was usually possible to tell a medium size business/corporate jet from a large airliner just by looking at the number and type of flight deck instruments and equipment. The former was unlikely to have such systems as these were within the domain of the big wide body and long-range types. Today and for the future, the suite of flight deck interface control and display equipment of the small jet is little different in number and type from that of, say, an Airbus A320. Both have dual AFCS, EICAS or ECAM, multi EIS displays (such as CRTs), and dual FMS. These systems are monitored and controlled through a number of different keypads and display their information on, say, six colour CRT display units.

Therefore we can select a specific aircraft type to represent general flight deck practice for all passenger jet aircraft. The A320 is a recent example which serves this purpose.

A320 Airbus flight deck The design philosophy adopted for the flight deck of the A320 provided a clearly defined set of requirements. The majority of these were achievable only by using avionic techniques to the full.

The principal feature of the flight deck and one which dominates the overall design and layout is the side-stick controller: one to the outboard side of each pilot. Once the decision had been reached to go for 'fly-by-wire' primary control using side-stick controllers in place of the conventional wheel there was no longer any need to worry about obstruction of the pilots' view of the principal instrument panels.

The primary display of information takes place on six colour CRT units. These are used for both the electronic flight instrumentation systems and the electronic centralised aircraft monitor. The pilots can 'dedicate' four of the six display units to either EIS or ECAM. In addition there are two other CRT displays. These are part of the two multi-purpose control and display units which, in conjunction with the flight control unit, command the automatic flight control and navigation systems (Figs. 4.10, 4.11).

The A320, like some of its contemporaries, has the CRT displays located in prime positions on the main panel in front of the pilots. The CRT units are also given responsibility for displaying all types of data ranging from primary, such as

Flight Decks and Cockpits

Fig. 4.10 Airbus A320 flight deck, the world's first airliner to go into production with side-stick controllers and integrated displays *Airbus Industrie*

flight and navigation, through engine and systems to lower order systems, such as cabin environment.

Of course not all the panel area of the A320 flight deck is dedicated to CRT-EIS/ECAM There are a number of 'stand-by' conventional instruments and also of great importance are the numerous levers, switches, keypads and other devices through which the crew controls and interrogates the different aircraft systems. Many of these controls or input devices have some form of illuminated legend or indication of status. In the A320 the overhead panel is designed to the 'lights out' philosophy. That is, unless a system requires attention, is failing or has failed there are no lights or other indications. This philosophy, which is now adopted by most aircraft designers, is also part of the concept of 'the need to know'. Unless it is essential for the safe and efficient conduct of the flight the pilots should not be overburdened with unnecessary information.

One of the most significant effects of ECAM and its associated systems on flight deck design is the opportunity it affords for two-pilot operation of the aircraft. Earlier generations of transport aircraft are characterised by a minimum crew of three. The flight deck controls and displays were arranged for three crew positions of which the third, the flight or systems engineer, monitored a large panel of instruments and controls. Such panels were usually arranged so that the controls

Flight deck examples

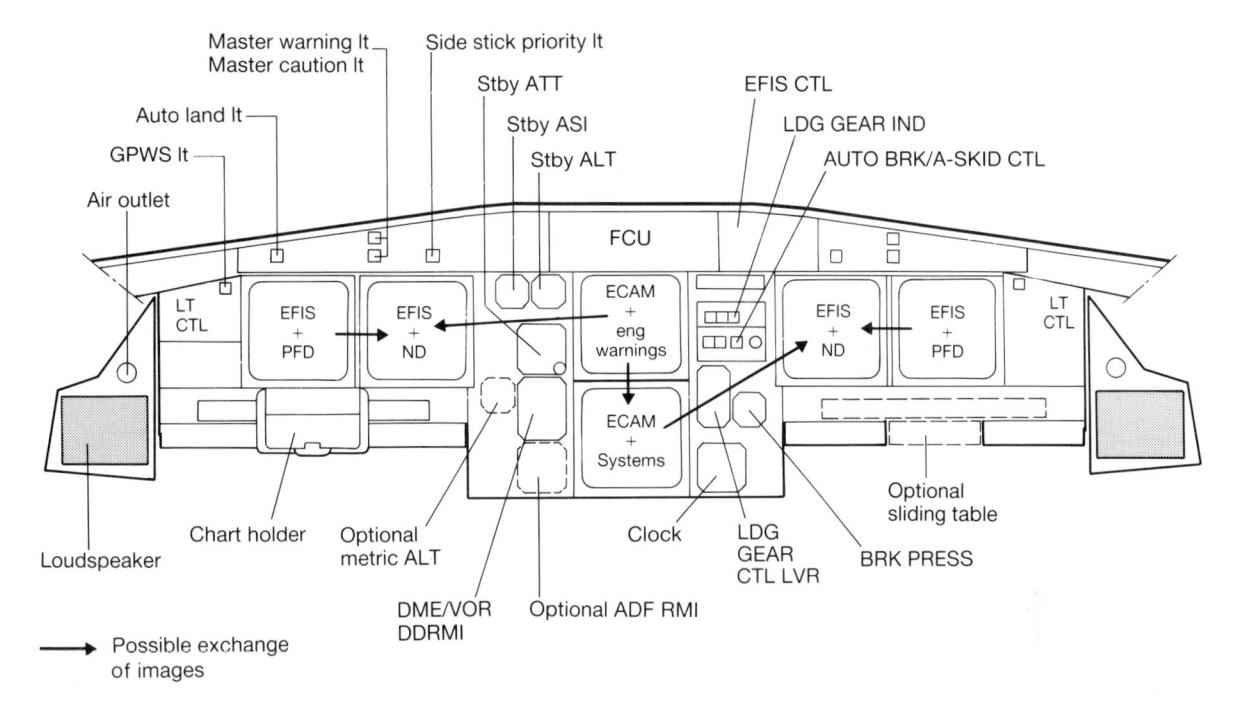

Fig. 4.11 Airbus A320, layout of main instrument panel *Airbus Industrie*

for individual systems formed a schematic diagram with controls and associated indicators close together.

The advent of the digital computer and, in turn, such systems as ECAM which were programmed to indicate the condition or status of a system only when requested or automatically display a caution or warning message, made practicable the two-pilot only flight deck. The computer now performs the task of monitoring and reporting on the condition of all systems and all their elements and components – a task once performed by a human operator (Fig. 4.12).

Fig. 4.12 Airbus A320, electronic centralised aircraft monitor (ECAM) *Airbus Industrie*

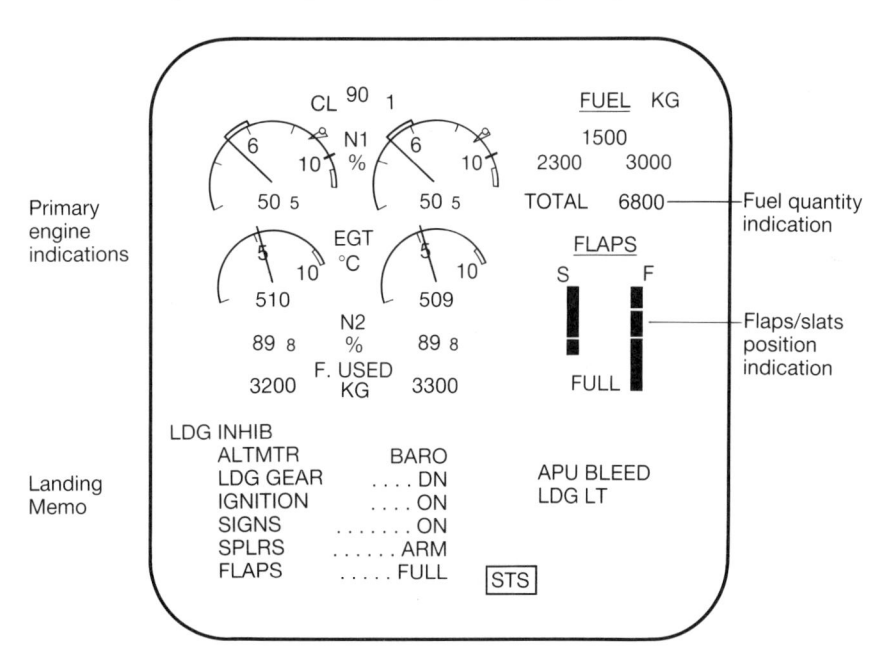

MILITARY COCKPITS

The interface equipment of the cockpits of fighter/attack and air superiority fighters uses display and control technologies which are similar or common to those in civil aircraft.

The major differences between a civil or transport aircraft flight deck and the cockpit of a military aircraft are, for the latter: controls and displays for tactical and weapon-aiming and selection systems; controls for the pilot's environment and safety, such as anti G-loc (G. induced loss of consciousness) and ejection seats; controls and selectors, such as input keys, designed for operation by the pilot's gloved hand and when subjected to high G forces; target and real world enhancement to enable a flight to continue in reduced visibility including the acquisition of and aiming at targets irrespective of the visibility or the location of waypoints and targets hidden from the pilot's direct line of sight.

Therefore there are a number of avionic systems which, in general, are peculiar to military aircraft. However, many of the avionic techniques applied, such as the use of CRT and solid-state displays, direct voice input (DVI) and some electro-optical systems, as we have seen, are also applicable to civil aircraft. Perhaps the most important differences are not in the particular avionic technologies used but in their function. In a civil aircraft the avionics are there to ensure safety, regularity and efficiency of aircraft operation; in a military aircraft the avionics have, to some extent, the same overall requirements but, more importantly, they are there to ensure that the total weapon system gets to the target and completes the sortie. It is a difference of priorities.

Head up display and helmet mounted sights

Until the successful integration of optics and electronics which resulted in the modern electronic head up display, in the early 1960s, the pilot of a fighter or fighter/attack aircraft had to make frequent and large eye movements between the view ahead and the instruments inside the cockpit. At 600 knots and close to the ground, two or three seconds are needed in which to switch the direction of eye attention, the point of regard, and re-focus from the target to the instruments. At that time the aircraft will have travelled 3000 feet. A similar eye adjustment problem occurs in air-to-air combat.

Therefore, the increased performance of jet aircraft had to be matched by electro-optical systems which allowed the pilot to fly 'head up' without taking his attention from the view ahead.

The electronic HUD, (Figs 4.13, 4.14, 4.15) has an importance far greater than its size might suggests. It is now the focus of cockpit design. The combination of a computer, a CRT, an optical projection system and the combining glass projects symbols and alphanumerics into the pilot's forward line-of-sight.

Because a HUD system is computer-based a number of different programmes can be accommodated. Depending on the selection made by the pilot, using the up-front control panel the HUD displays either: flight and navigation data, used when cruising to or from the target; or attack data, used when aiming the aircraft as weapon system; or gun and missile aiming during aerial combat. Within reason there is no limit to the different formats of symbols and alphanumerics.

The computing elements of a HUD derive their inputs from the aircraft's air data and inertial systems so as to display airspeed, altitude, heading, vertical speed, attitude, angle-of-attack, gun, missile and bomb aiming and release, and so on. All or some of these can be selected by the pilot to match the immediate phase of the sortie.

Military cockpits

Fig. 4.13 Electronic head-up display (HUD) for General Dynamics F-16 *GEC Avionics*

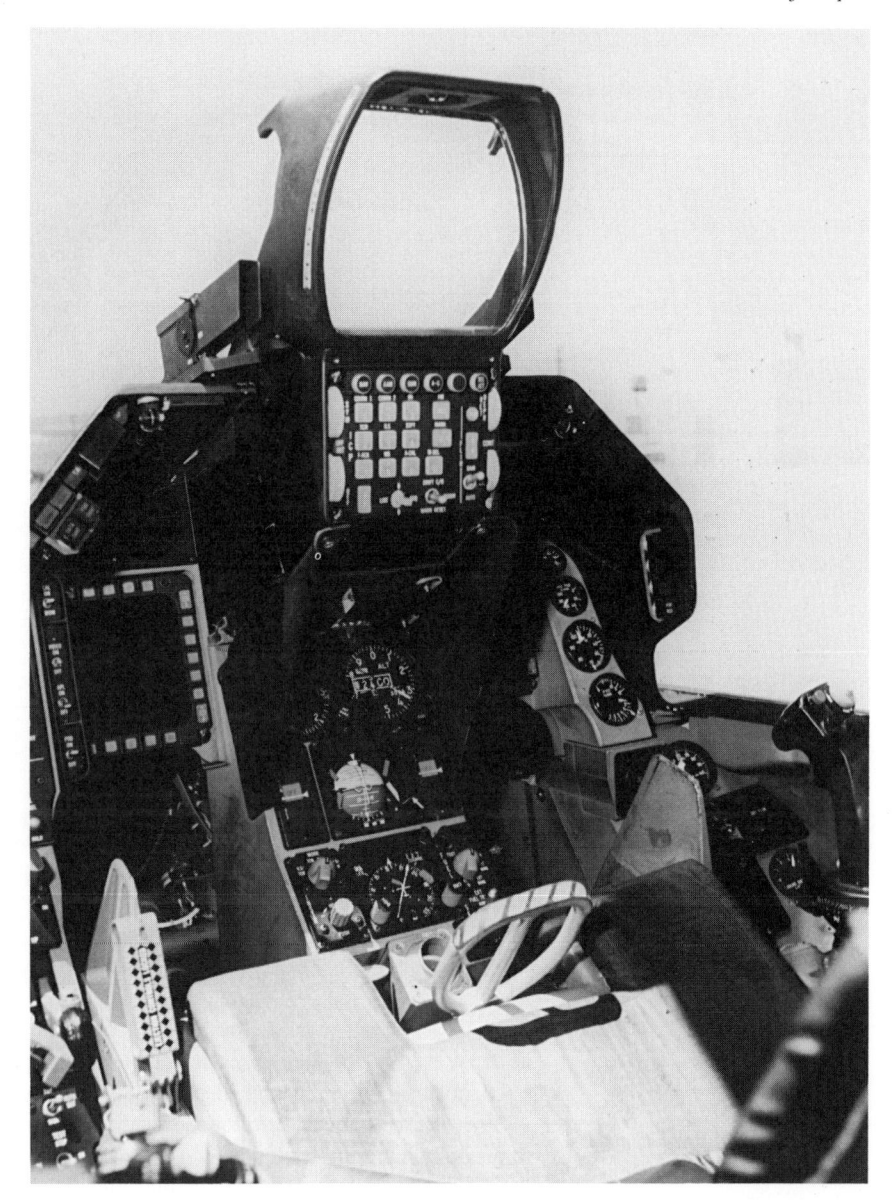

The electronically generated information is seen by the pilot superimposed on the view ahead of the real world. In addition the HUD images can be combined with a Low Light TV (LLTV) or Infra-red (IR) picture of the view ahead.

Of course, the quality of the HUD is only as good as the software programmes and the efficiency of the CRT, the optical path of the lens system and the combining glass. Holographic techniques used in the manufacture of the combining glass now provide a significant improvement in its reflective properties. A thin layer sandwiched between glass is treated by a laser to form an optical interference pattern. This layer only reduces the transmissivity of the light rays between the pilot's eyes and the view ahead by 10 per cent. The alphanumerics and symbology can be seen clearly superimposed on the pilot's view of the outside world.

Figure 4.15 is an example of a typical set of HUD information. As mentioned, different amounts and types of data can be projected into the pilot's forward line

Flight Decks and Cockpits

Combining glasses of the HUD displaying alphanumerics and symbols

Up-front control panel of the HUD

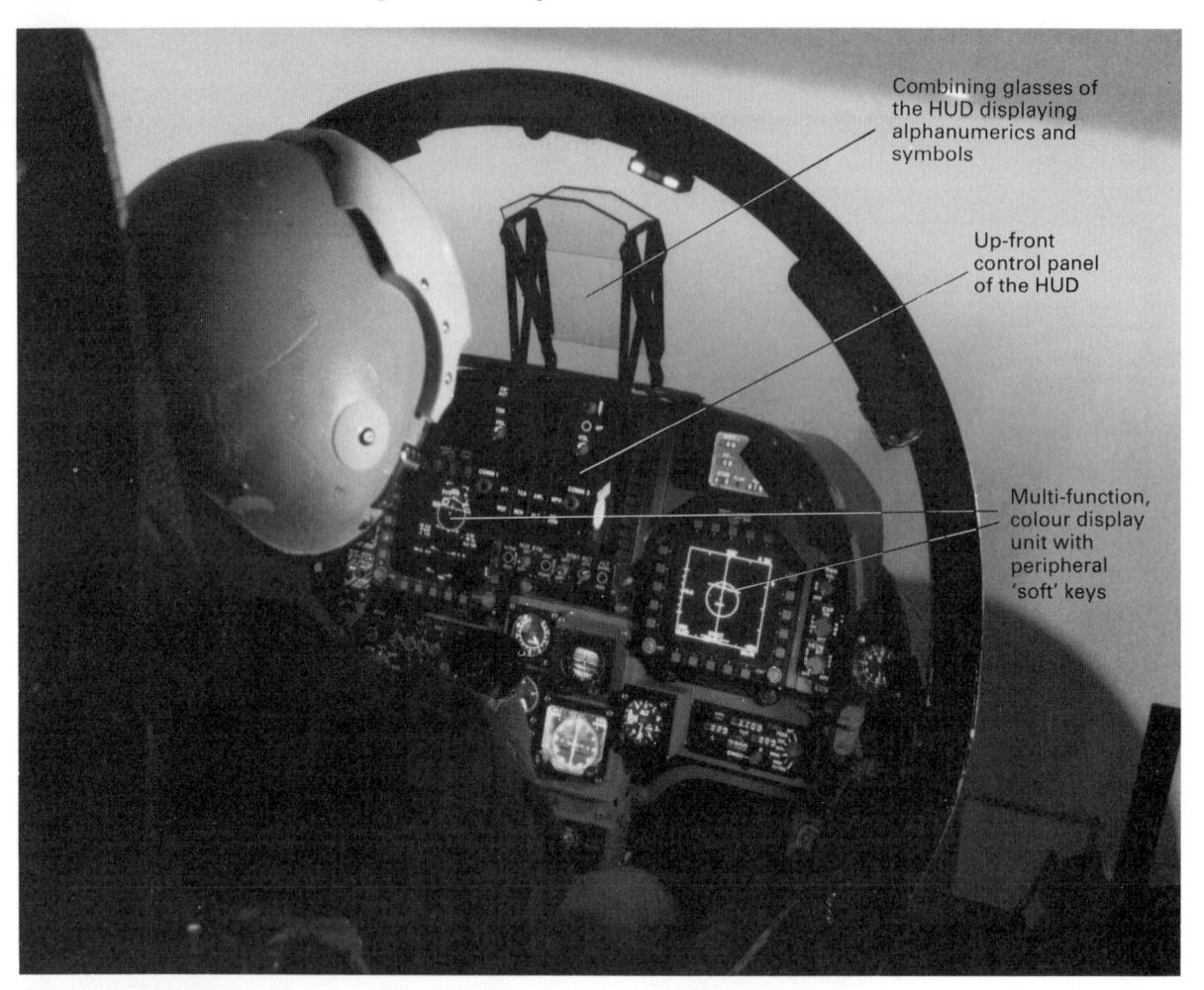

Multi-function, colour display unit with peripheral 'soft' keys

Fig. 4.14 Cockpit layout of AV-8B Harrier showing HUD *McDonnell Douglas Corpn.*

of sight. In practice the data is selectable in groups to match specific flight modes so as not to 'clutter up' the display with unnecessary information.

The electro-optical technique used for the head up display has been applied to the helmet mounted sight (Fig. 4.16) so that the pilot can see computer generated symbology and alphanumerics in his line of sight in whichever direction he is looking. This type of sight is a compact version of the fixed HUD unit and is mounted on the eyebrow part of the pilot's helmet. It is an important example of modern avionics. If the pilot moves his head so that he is looking directly at a target then a helmet position detecting system feeds data to the weapon-aiming and navigation computers. If the pilot then presses a 'Mark' key on the up-front control panel, or if using DVI, says 'Mark' the computer stores the location in its data base for future reference. Or, as the pilot selects and speaks 'Shoot' the missiles are released and directed to the target.

Multi-purpose colour displays (Fig. 4.17)

As we have seen the HUD rationalises the vital sets of information needed by the pilot for particular flight modes. The HUD cannot provide all the data needed for all phases of a flight and therefore the pilot has to use the conventional 'head down' instruments. Until recently these were usually of the pointer-on-dial and numeral counter types. These instruments had to be scanned to ensure that critical

Military cockpits

Alphanumerics and symbols displayed on the combining glass of the head-up display.

Multi-functional display with peripheral soft keys.

Control panel of the HUD

Conventional flight instruments.

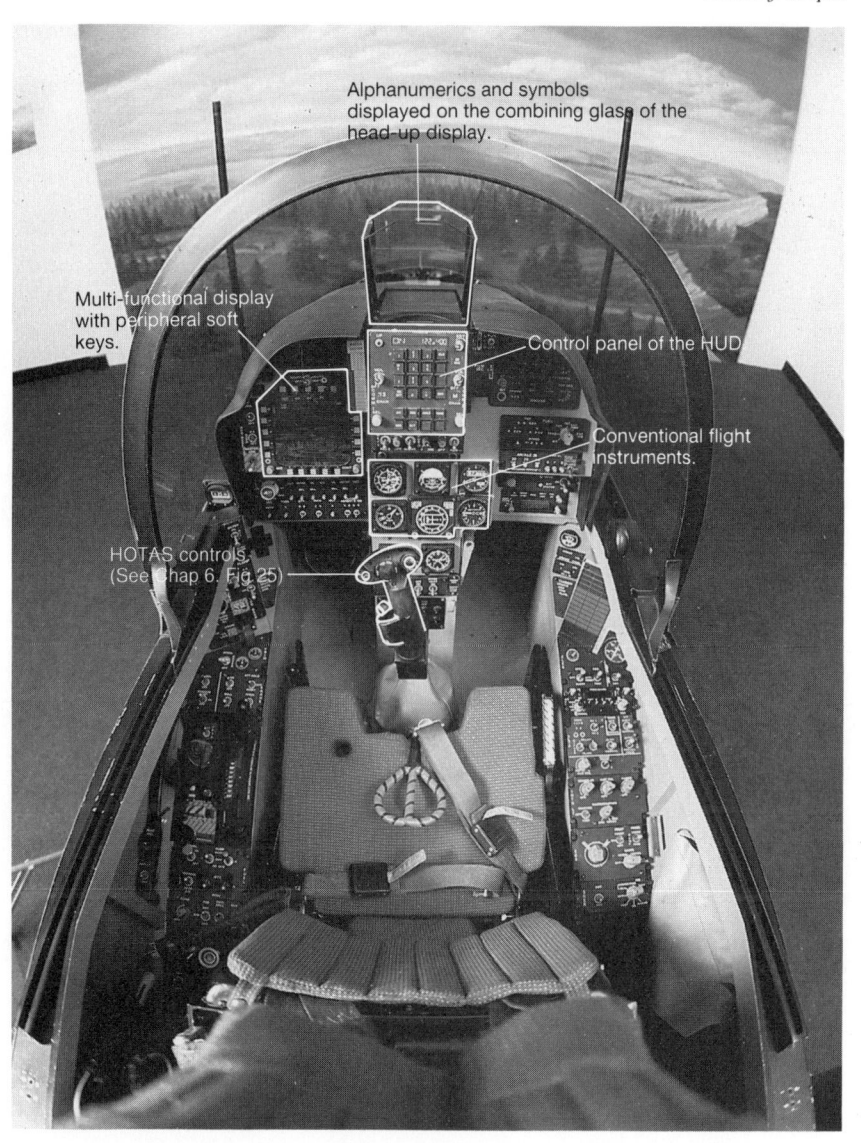

HOTAS controls (See Chap 6, Fig 25)

Fig. 4.15 Cockpit layout of AV-8B simulator showing HUD display and image of aircraft carrier superstructure *McDonnell Douglas Corpn*.

data was not missed, and furthermore they did not 'interface' effectively with the increasing number of avionic systems, such as the HUD, which from the mid 1970s became a feature of military aircraft. Many of the conventional instruments used analogue data processing and therefore required digital-to-analogue techniques as an additional interface. Therefore it was logical that increasing use should be made of the colour CRT displays, similar to the civil EIS, to improve both the quality of the interface and its efficiency.

The multi-purpose colour display (MPCD) is the companion to the HUD. It gives the pilot sets of selectable data and formats, including that of the HUD itself if need be, and so provides the primary 'head down' display in the cockpit. The ability to select different sets of data on different CRTs provides both flexibility of use and also a measure of redundancy for offsetting damage or failure.

Stealth The low level, nap of the earth, high speed approach to a target by an attack aircraft depends on avionic computer systems requiring a number of inputs. Of

Flight Decks and Cockpits

Fig. 4.16 'Falcon Eye' helmet mounted display system developed for General Dynamics F-16
GEC Avionics

these Doppler radar is one of the more important. But its use immediately alerts the ground defence systems which can detect all forms of electronic emission from an aircraft.

Therefore control and display systems have been developed which allow a covert, 'stealth-like' approach to a target at minimum altitude.

The Ferranti PENETRATE is an autonomous system which has an electronically stored profile of the terrain over which the aircraft has to fly in order to reach the target. The system combines a radar altimeter, the range of which is limited, and an inertial navigation unit with a three dimensional 'model' of the terrain held in the computer memory. The data output from these 'drives' a head up display and a head down display.

This is an example of a specialised avionics system which provides automatic terrain following with an integrated display of data on the HUD and the head

Fig. 4.17 Multi-purpose colour CRT display *Smiths Industries*

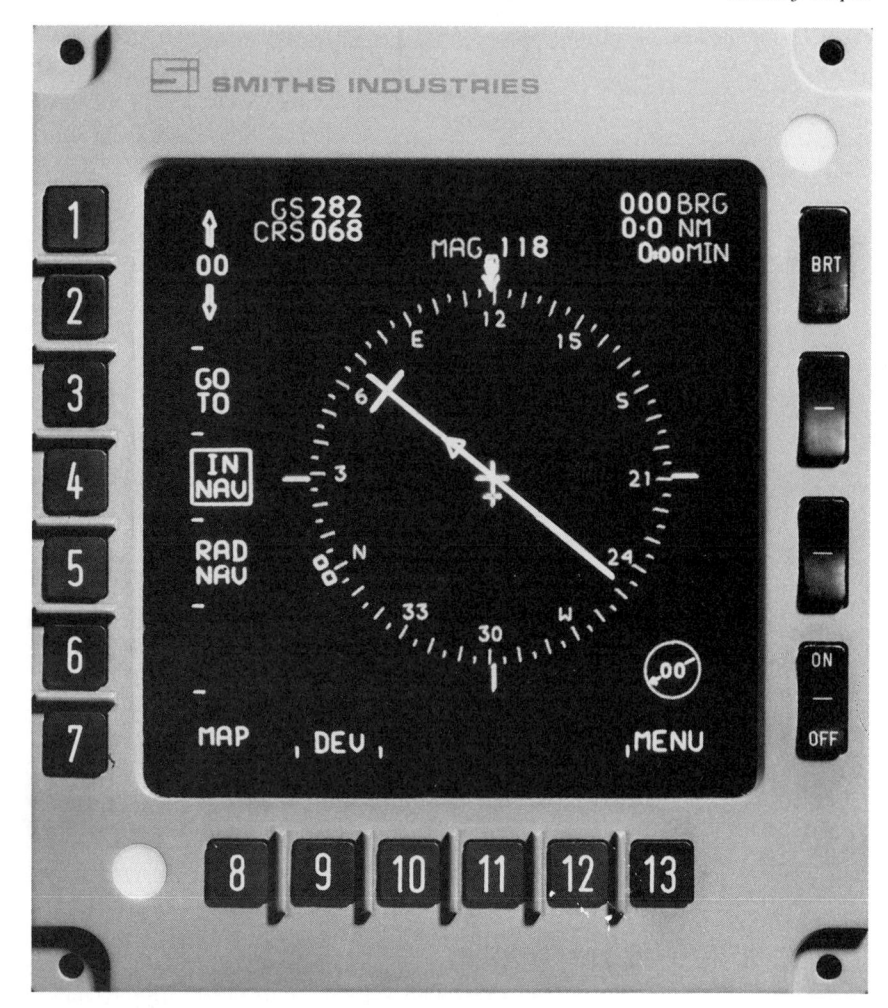

down 'map' display (Fig. 4.18). The HUD information is similar to that described earlier, with the addition of 'highway in the sky' graphics for the guidance of the pilot.

The digital 'map' display shows terrain and topographical information on a high brightness CRT. The 'map' data is derived from a data base which is inserted by means of a portable data transfer module and in this respect is the equivalent of the data insertion procedure for a civil flight management system. All the features of a conventional map are included together with selectable colours, the ability to delete unwanted information and, of course, zooming-in on a particular feature.

The PENETRATE display is a good example of the versatility of the modern colour CRT. In this system a pilot can instantly see those zones over which the aircraft must not be flown. An interesting feature is the facility for 'shading' those areas where the aircraft will be hidden by the intervening high ground so that its electronic emissions, such as IFF, radar and communication, can be used with less chance of being detected.

LLTV, thermal imaging and FLIR

Low light TV, thermal imaging and Forward Looking Infra-Red techniques, described in Chapter Nine, provide important inputs to the cockpit displays of military aircraft. These sensors enable aircraft to be flown safely at low altitude

Flight Decks and Cockpits

Fig. 4.18 'Penetrate' electronic map display *Ferranti*

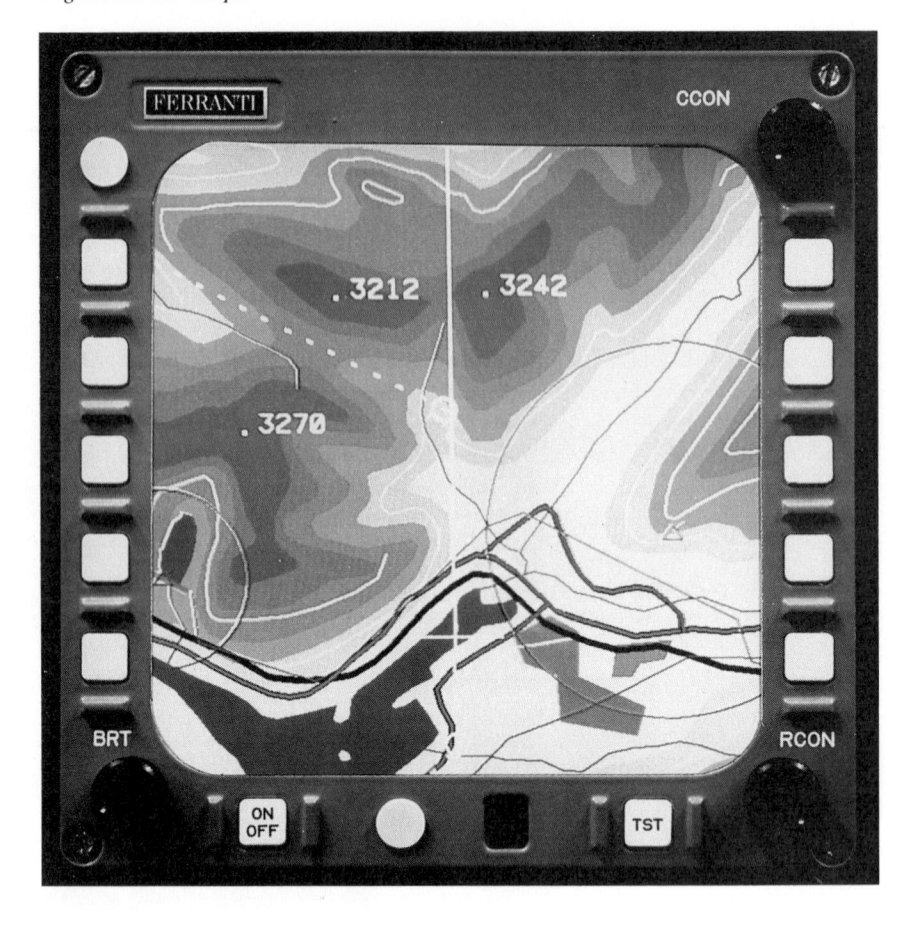

in visibility conditions of starlight or when there is an overcast quarter moon. The processed images can be displayed on the combining glass of the head-up display with the television image overlaid on the real world with a magnification of unity.

A thermal imaging sensor can be used to provide additional information presented on the head-up display.

Stand-by instruments Stand-by instruments are specified whenever there is a probability, however remote, that the primary displays might fail or be damaged. The concept is that of 'get you home' by providing an alternative and independent display of speed, heading, attitude, vertical speed and altitude. Figure 4.19 is an example of three solid-state, LED, air data displays for the B. Ae. Experimental Aircraft Programme.

The future (see also Chapter Eleven.) Without delving into the fanciful, there are a number of control and display technologies which will, with some certainty, have a significant effect on the design and equipment of the future flight decks and cockpits.

If the pilot is to remain in the control loop, even in the most critical and demanding phases of flight, and the workload is to be kept to an acceptable level, then the avionics must provide a 'real' or artificial view of the world irrespective of the visibility conditions.

The combination of electronic displays, fully automatic control and monitoring systems and non-tactile, e.g. voice input, controls, is already changing the pilot's working environment. In addition the pilot associate concept will increasingly contribute to significant changes in the relationship between the pilot and the aircraft.

A pilot associate computer is programmed with the knowledge and experience of all the pilots who operate a particular type of aircraft on specific routes or, for military aircraft, on specific operations. The pilot associate figuratively sits alongside the pilot and prompts, advises, reminds and warns. Because it has microsecond access to a vast fund of knowledge about the pilot, the route and its navigational aids, it can provide the best of a number of options in the event of an emergency.

BIBLIOGRAPHY

Airbus Industrie, *A320 Briefing for Pilots*, 1986.
Chorley, R. A. 'Seventy Years of Flight Instruments and Displays,' Royal Aeronautical Society Third Folland Memorial Lecture, 1976.
Coombs, L. F. E. *Aerospace & Defence Review*, Smiths Industries plc, 1969–1989
 'Aircraft Ergonomics', *World Aerospace Profile*, Sterling, 1986.
 'The Cockpit of the Year 2000', *MILTECH 7/86*, 1986.
Hirst, M. 'Avionics Analysed', *Air International*, Jan. 83, Mar. 83, Jul. 83, Sept. 83, Nov. 83, Feb. 84, Apr. 84.
Hurst, R. Pilot Error, Crosby Lockwood Staples, 1976.
Owen, C. A. Flight Operations, Granada, 1982.
Pallett, E. H. J. *Microelectronics In Aircraft Systems*, Pitman, 1985.
Smith, J. H. 'Colour Multi-Function Displays For Aircraft', Royal Institute of Navigation, NAV87 paper, 1987.

Fig. 4.19 Light emitting diode (LED) stand-by instruments *Smiths Industries*

CHAPTER 5 FLIGHT CONTROL SYSTEMS

IAN MOIR BSc, CEng, MIEE, MRAeS

Senior Multiplexing Systems Engineer Smiths Industries. Mr Moir undertook his graduate training as an Engineering Officer in the Royal Air Force being awarded a B Sc (Eng) 2nd Class Honours Degree in Aeronautical Engineering. Service in the Far East was followed by responsibility, at RAF Brize Norton, for first line servicing of VC 10 and Belfast aircraft.

In 1970, in the rank of Flight Lieutenant, he took a post-graduate Course at RAF College, Cranwell, having already been awarded an Air Training Corps Flying Scholarship and the de Havilland Flying Trophy for the best pilot in RAF Officer Cadet Entry (1965).

In 1972 he joined the RAF Project Team with responsibility for the digital avionics system of the MRCA (Tornado) at British Aircraft Corporation (later BAe) at Warton.

In the rank of Squadron Leader Mr Moir served at the Government Communication Headquarters, Cheltenham, was Officer Commanding, Electron Engineering Squadron, RAF Leuchars and the Ministry of Defence (Air) Directorate of Tornado Engineering and Supply.

In 1980 he joined Smiths Industries, becoming involved in systems integration using MIL-STD-1533B data bus hardware on the BAe Experimental Aircraft Programme. From 1984 he has been Group leader of the Utilities Systems Management Group.

Mr Moir has presented many Technical Papers at home and abroad and written a number of Technical Articles. In 1979 he was awarded the RAF Strike Command Smallwood Trophy for a submission on Electronic Warfare. He is a Member of the Royal Aeronautical Society, a Chartered Engineer and a Member of the Institution of Electrical Engineers.

PRINCIPLES OF FLIGHT CONTROL

As an introduction to the application of electronics to flight control systems it is necessary to have a theoretical basis in order to be able to understand some of the more sophisticated systems. The control axes of an aircraft are defined by the convention of a right-handed orthogonal (mutually perpendicular to one another) axis set. It is within the definition of such axes that the motion of the aircraft is described and, of course, this needs to be adequately done before the control of the aircraft is discussed.

The right-handed axis set used is shown in Fig. 5.1. The three axes originate at point O and are defined by the lines Ox, Oy, and Oz:

Ox defines the Longitudinal Axis – the direction in which the aircraft flies and in which the aircraft propulsive and drag forces act.

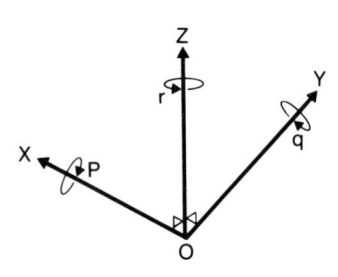

Fig. 5.1 Right-handed aircraft axis set

Oy defines the Lateral Axis – the direction in which sideslip or lateral forces act.

Oz defines the Normal Axis – the direction in which the aircraft lift and weight forces act.

An aircraft may also manoeuvre or rotate around each of these major axes:

The aircraft rolls by rotating around the Ox or longitudinal axis. The control surfaces which can cause the aircraft to roll are the ailerons on a conventional aircraft; the elevons on a delta aircraft (Vulcan); tailerons where aileron and elevator functions are combined (Tornado); fore-planes on a canard configuration (EAP), where aileron and elevator functions are combined; spoilers; and flaperons which are similar in operation to elevons. The rate of roll is commonly denoted by the annotation p.

The aircraft pitches by rotating around the Oy or lateral axis. The control surfaces which can cause the aircraft to pitch are the elevators; the elevons; the tailerons or the foreplanes. The rate of pitch is denoted by the annotation q.

The aircraft yaws by rotating around the Oz or normal axis. The control surface which causes the aircraft to yaw is the rudder. The rate of yaw is usually denoted by the annotation r.

On the simplest aircraft the pilot operates a control column and a rudder bar. Control in pitch is achieved by fore and aft movement of the control column; pushing the column forwards lowers the nose while pulling it to the rear raises the nose. Control in roll is exercised by moving the column from side-to-side; movement to the right lowers the right wing and vice-versa. Control of the aircraft in yaw is achieved by pushing the rudder bar with the feet; pushing the right pedal causes the aircraft to yaw right and vice-versa, both column and rudder bar being connected to the appropriate control surfaces by metal cables or tubes.

To give the aircraft additional control at certain stages during flight the following control surfaces may also be used:

– flaps to increase the lift generated by the wings during take-off and landing but also during manoeuvre;
– slats to increase lift during take-off and landing and during manoeuvre;
– airbrakes or speedbrakes to create additional drag when airspeed needs to be rapidly reduced;
– wing sweep. Wings locked in the forward position increase the lift generated at low speeds and improve the handling characteristics. Sweeping the wings aft improves the high-speed characteristics including gust response.

Operation of all of these secondary control surfaces affects the control characteristics of the aircraft and, where automatic flight control is being used, the system will need to take account of them. Conversely, in a highly integrated flight control system the surfaces may be automatically deployed as the system reacts to the demands of the pilot. For example, the wing high lift devices such as flaps and slats may be automatically deployed when the pilot of a fighter aircraft is engaged in combat. In another system, control surfaces may be activated for load alleviation during gusty conditions. To illustrate how some of these control surfaces may be used refer to the example of the Tornado in Fig. 5.2 and 5.3 and the Experimental Aircraft Programme (EAP) in Fig. 5.23 and 5.24.

Figure 5.4 is an example of a simple open loop control. As any pilot will know,

Fig. 5.2 Tornado flight control surfaces. The primary flight controls are the tailerons and rudder, the others being secondary controls. Nevertheless the requirements of combat manoeuvre may require the use of the secondary controls to achieve the necessary agility.

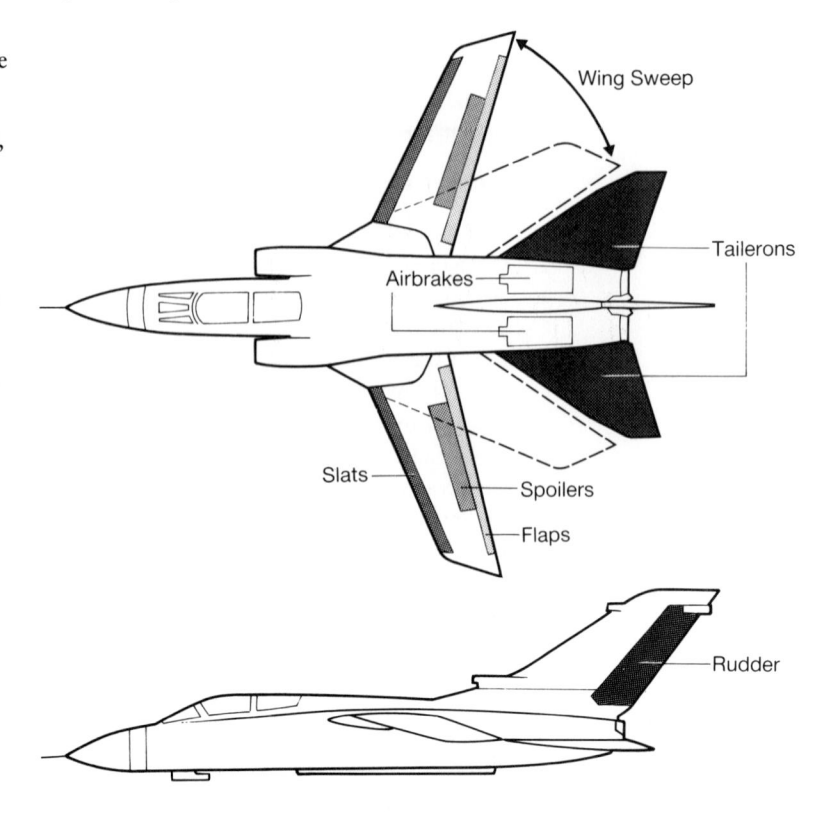

a correctly trimmed aircraft should maintain a given pitch attitude for a given set of flight conditions – say straight and level flight. However if he alters one of the conditions by perhaps increasing aircraft power to increase speed, then the pitch attitude would in all likelihood increase and require a control correction to prevent the aircraft from climbing. The wider the range of variables which the aircraft encounters during flight then the greater the corrections required. In practice, the open loop type of control system described is totally unacceptable for flight control as it would require continual intervention.

By providing continual corrections as flight conditions alter the pilot is in fact part of a control system shown in Fig. 5.5 where he provides the feedback and closes the loop. This system is known as a closed loop control system and we all exercise such control as we walk, ride a bicycle or drive a car. Therefore when a pilot flies an aircraft he is simultaneously closing a number of control loops by controlling pitch, roll and yaw attitude, as well as aircraft altitude, speed and heading. Both the commercial and military pilot are employed to do this; however as aircraft have increased in performance and complexity the workload upon him has increased to the extent that he now requires considerable help.

Because of these performance increases it has become necessary to use power actuation to overcome the extremely high forces acting on the control surfaces. As a result of this development, artificial feel systems have had to be introduced to give the pilot simulated 'feel' forces proportional to demand. Otherwise he could unwittingly overstress the aircraft during manoeuvre. Also, as aircraft speeds have increased, the need for automatic trim systems based upon air data have become necessary. Furthermore as aircraft shapes have become aerodynamically more

Principles of flight control

Fig. 5.3 Tornado F2 with wings in forward position. The slats and flaps are deployed in the slow speed mode to match the speed of the photographic Hercules. The tailerons can be seen at negative incidence to counteract the nose-down pitch   *Brisith Aerospace, Warton*

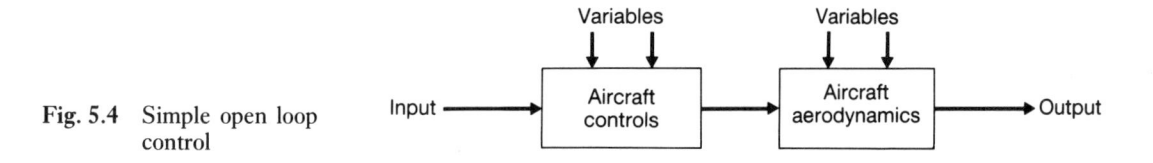

Fig. 5.4 Simple open loop control

Flight Control Systems

Fig. 5.5 Closed loop control using pilot

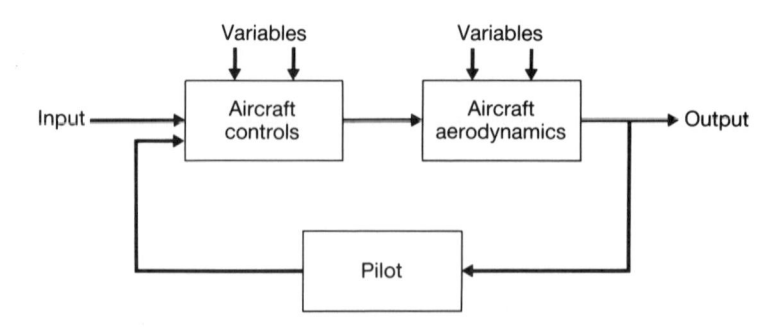

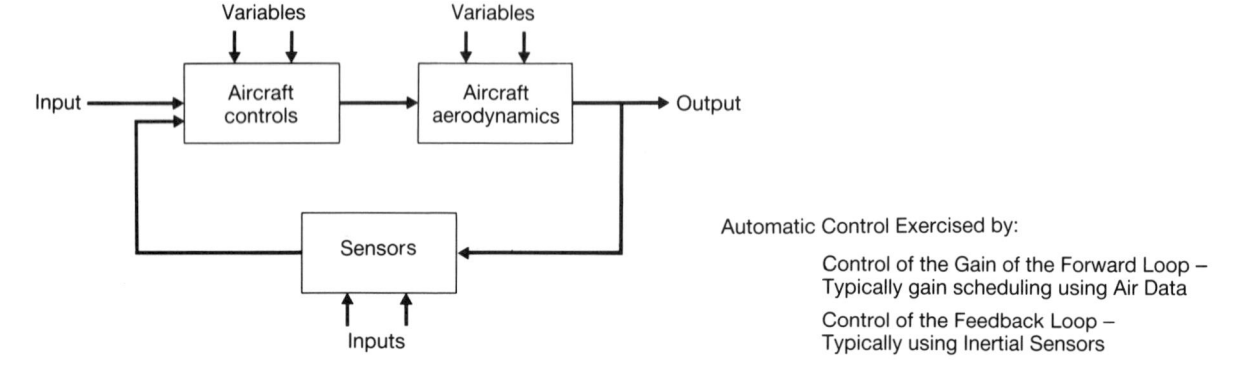

Fig. 5.6 Closed loop control using sensors

Automatic Control Exercised by:

Control of the Gain of the Forward Loop – Typically gain scheduling using Air Data

Control of the Feedback Loop – Typically using Inertial Sensors

advanced to attain specific performance goals, stability has been eroded thus requiring artificial auto-stabilization or stability augmentation. These more complex shapes and accompanying increased performance have greatly increased the number of variables which need to be used around the control loop and have necessitated loop closure to provide the pilot with automatic control. Therefore the elements of automatic flight control are typified by the diagram at Fig. 5.6. A number of examples will be given of actual flight control systems. First it is necessary to understand something of the types of sensors used by the modern flight control system. The subject of control and stability will be briefly outlined.

The types of sensor used to modify aircraft control characteristics may be split into two distinct types: air data sensors, and inertial sensors.

Air data sensors provide information relating to the atmosphere through which the aircraft is flying. The characteristics of the atmosphere change appreciably with altitude as does the performance of the aeroplane. Therefore, to minimise undesirable features and to maintain stability, air data parameters need to be measured and the necessary corrections or compensations in the control loop made. Air data measurements relate in the main to pressure and airstream direction detection; in some cases temperature too may be measured. Pressure is measured by using the aircraft pitot system, commonly by the use of a probe which extends forward into the airflow. These sensors are called pitot static probes and they measure the dynamic or pitot pressure of the incident airstream and the local static pressure. An example of an advanced form of probe with an in-built transducer unit is shown in Fig. 5.7. This probe also has the capability of sensing airstream direction. The embedded transducer unit removes the need for pitot-static tubing within the airframe. The unit shown is manufactured by Rosemount and probes similar to this are used on the B-1B and the X-29.

It can be seen that pitot pressure is sensed by the inner section which detects

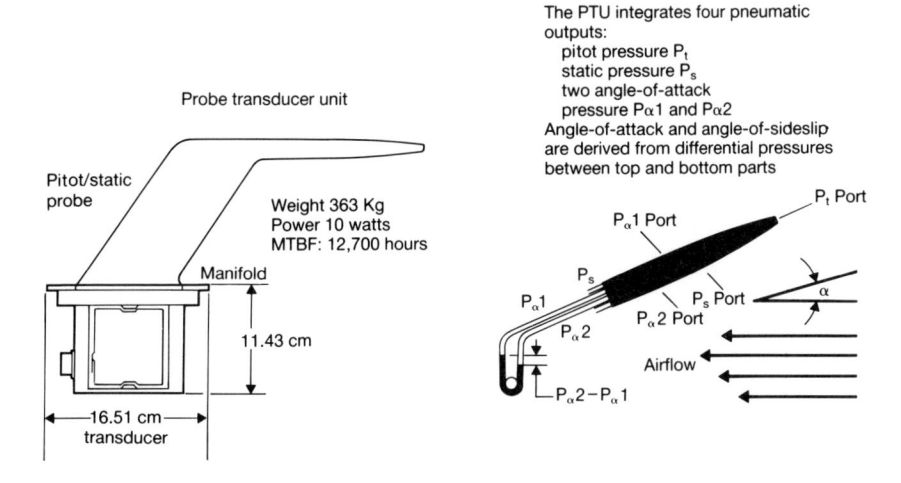

Principles of flight control

Fig. 5.7 Pitot static multi-function air data probe'. *Rosemount*

The PTU integrates four pneumatic outputs:
  pitot pressure P_t
  static pressure P_s
  two angle-of-attack pressure $P\alpha 1$ and $P\alpha 2$
Angle-of-attack and angle-of-sideslip are derived from differential pressures between top and bottom parts

the on-coming airflow. Slots in the outer tube permit the static pressure to be sensed and in the example shown differential sensing across the probe gives a measure of airstream detection. It is common to use pitot minus static (Pt − Ps) to schedule aircraft control laws. Static pressure scheduling (Ps) may be used where the performance of the control loop requires improvement as the aircraft altitude increases to compensate for shortcomings in the aerodynamic performance of the aircraft at high altitude.

Certain systems require measurement of the direction of the airstream relative to the aircraft axes. The quantities measured are angle of attack (α) and angle of sideslip (β). Refer to Fig. 5.8. The angle of attack is that between Ox, the

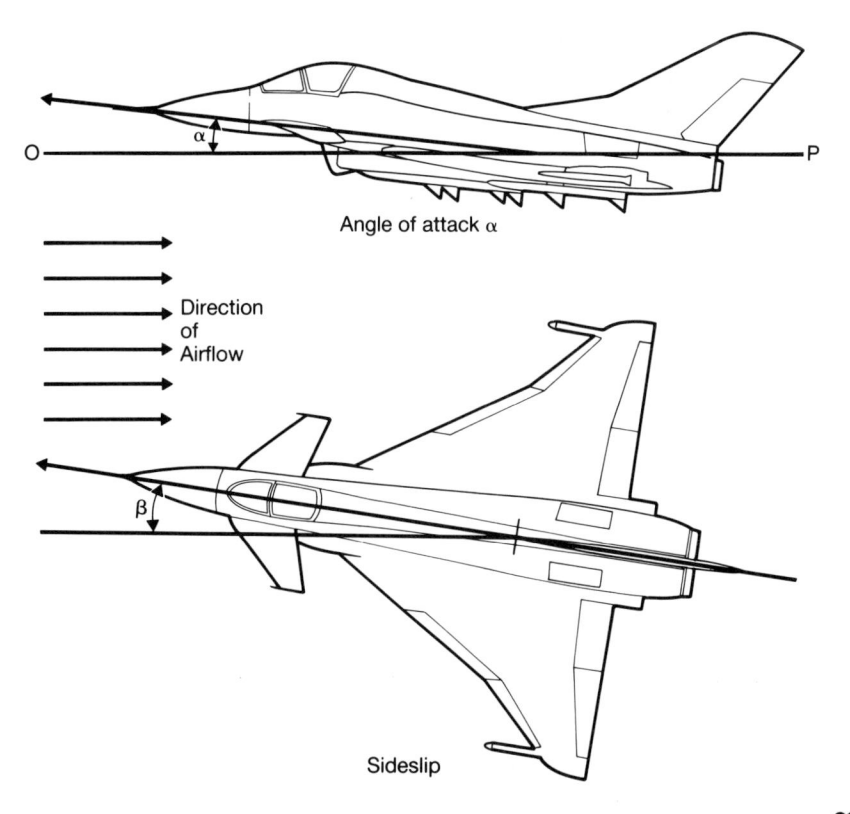

Fig. 5.8 Airstream detection – angle of attack and sideslip

Flight Control Systems

longitudinal axis and the line OP. An aircraft will always fly at some positive angle of attack relative to the airflow in order to generate lift. Angles as high as 25 to 30 degrees may be reached by a high performance fighter aircraft.

Usually the angles of sideslip will be small or zero. However, for highly agile fighter aircraft manoeuvring concurrently at high rates of pitch and roll, sideslip may be significant. The standard method of measuring incidence and sideslip is by means of small vanes which follow the direction of the airflow and therefore allow the airflow relative to aircraft axis to be determined.

The information derived from air data sensors for the purposes of flight control gives the following parameters:

- indicated airspeed;
- true airspeed;
- air density;
- air pressure resulting in barometric height;
- outside air temperature;
- angle of attack;
- angle of sideslip;
- Mach number;
- rate of change of height (vertical speed indication).

It is often the case that these parameters are calculated or derived in one or more air data computers. However the increased levels of integration being experienced in flight control systems, as well as a tendency towards 'smart' probes and sensors (e.g. the Rosemount probe shown in Fig. 5.7), mean that some tasks historically associated with the air data computers may be undertaken elsewhere in the system.

Inertial sensors measure quantities relating to the motion or dynamics of the vehicle. Inertial measurement is independent of the medium in which the aircraft or vehicle is operating. Such sensors will therefore give information relating to the vehicle motion irrespective of whether it is operating in water, air or a vacuum. If the reference set chosen for the inertial sensors are the same axes as are chosen for the aircraft in Fig. 5.1 then the following inertially derived information may be useful in determining the dynamics of the aircraft and assisting in control. Refer to Fig. 5.9.

Fig. 5.9 Inertial sensors axis set

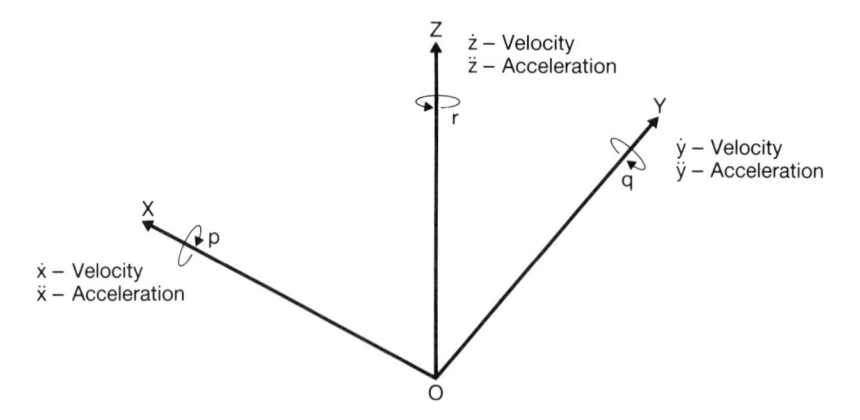

$\dot{z}$ – Velocity
$\ddot{z}$ – Acceleration

$\dot{y}$ – Velocity
$\ddot{y}$ – Acceleration

$\dot{x}$ – Velocity
$\ddot{x}$ – Acceleration

The roll rate p, pitch rate q, and yaw rate r may be measured using roll rate gyros. Similarly roll, pitch and yaw attitude may be determined by attitude sensing gyros. Accelerometers suitably mounted will sense accelerations in the Ox axis (longitudinal acceleration or $\ddot{x}$), Oy axis (lateral acceleration or $\ddot{y}$), and Oz axis (vertical acceleration or $\ddot{z}$). Integrating these accelerations will yield the longitudinal, lateral and vertical velocities ($\dot{x}$, $\dot{y}$ and $\dot{z}$). Therefore by sensing aircraft attitude and rotation rates with gyros and acceleration using accelerometers, integrating acceleration to derive velocities, a range of dynamic sensor information may be made available to control the aircraft.

Attitude information is commonly used in autopilot modes to control pitch, roll and yaw attitude (heading). Of these modes heading hold or heading acquire modes are perhaps the most common though some fighter aircraft have pitch and roll attitude hold modes; the Tornado autopilot uses such modes. Body rate and acceleration information is usually used for auto-stabilization and more sophisticated control loops of the type most often found on board the latest generation of fighter aircraft.

In the examples which follow later in this chapter several flight control system architectures will be described and the use of both air data and inertial sensors will be identified.

ESSENTIAL ELEMENTS OF CONTROL

The typical closed loop control system described briefly in Fig. 5.6 has a number of yardsticks by which performance is specified and measured. The most common criteria specified relate to accuracy, transient performance and stability.

Accuracy may typically be defined by a requirement which specifies that a certain control parameter must be maintained within specified limits. For example, a heading hold mode of an autopilot may be specified as requiring ±0.5 degree accuracy, or an altitude hold mode may demand an accuracy of ±200 feet. Therefore the autopilot will be expected to control the heading to within 0.5 degrees and the altitude to within 200 feet of the respective datums set by the pilot. The nature of the problem posed for the designer in respect of transient performance may be illustrated by a simple example depicted in Fig. 5.10. This diagram shows the transient response to a step input of three notional systems: A, B and C. System A rises rapidly to satisfy the input, however it overshoots and undershoots the datum several times before settling within prescribed limits. This eventually occurs at point A on the diagram. System B is slower to respond than system A. It settles within the limits at point B. System C is still slower to respond and it settles within limits at point C. System B clearly has the best transient response of the three systems; it should also be noted that system C although slowest to respond of all, actually settles within limits before the faster rising but over-oscillatory system A.

Fig. 5.10 Comparison of transient performance

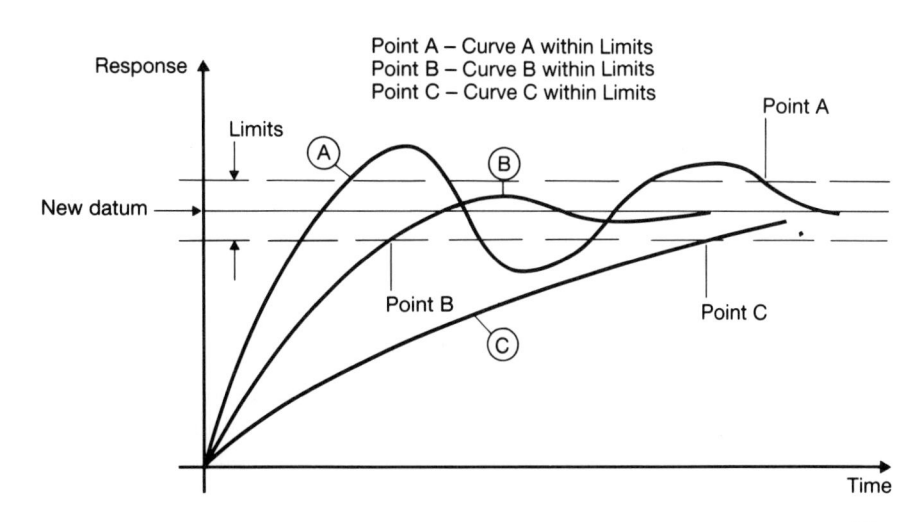

Flight Control Systems

This simple example is indicative of the problems that the control systems engineer has to face. Furthermore, the response indicated by system B may be perfectly acceptable at one point in the flight envelope; at other points where conditions differ markedly this may not be the case and variations in the control laws will be required for acceptable performance to be attained.

Fig. 5.11 Comparison of stable and unstable systems

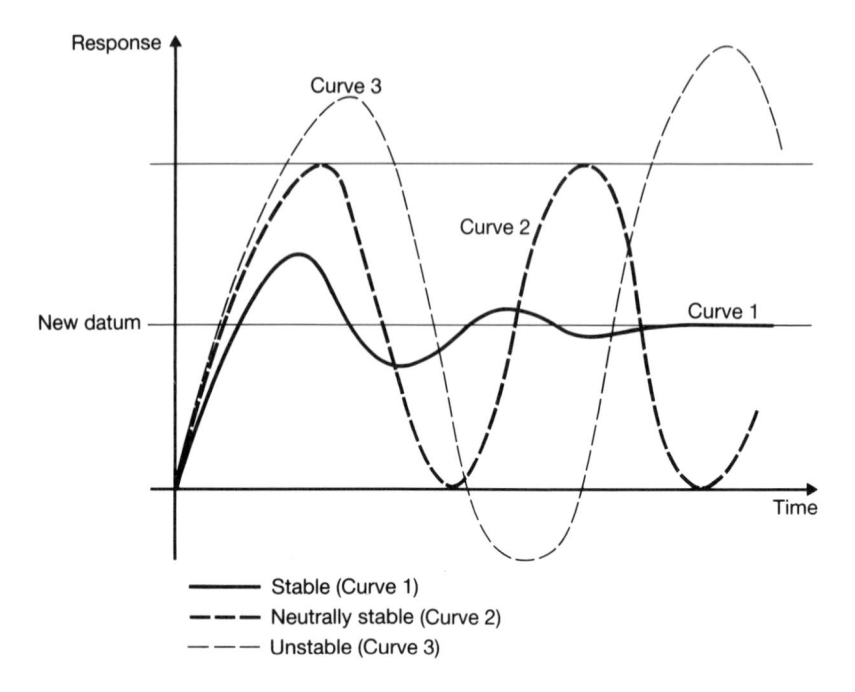

——— Stable (Curve 1)
– – – Neutrally stable (Curve 2)
– — – Unstable (Curve 3)

The differences between stable and unstable systems may also be simply shown. Fig. 5.11 shows three different responses to a control input or system perturbation and how each attains the new system datum. Curve 1 illustrates a response of the type shown by system A in the earlier example. The response is oscillatory but is damped and will eventually settle at the desired datum. Such a system is said to be stable. Curve 2 is more oscillatory than Curve 1, however the amplitude of the oscillation remains constant. This response is neutrally stable. Curve 3 is highly oscillatory, furthermore the amplitude of the oscillation increases with each swing of the system. Left to its own devices, this oscillation will diverge to infinity or some practical limit. Curve 3 represents an unstable system and unless additional control methods were employed could lead to spectacular results if used in a flight control system.

The relationship between where the lift and weight forces act on an aircraft gives an indication of whether it is stable or unstable in pitch and to what extent. Fig. 5.12 is a simple diagram showing the relationship of lift and weight forces. The aircraft weight is a downwards force acting through the aircraft centre of gravity. The aircraft lift force is an upwards force acting through a point called the aerodynamic centre of pressure. In the example shown a downwards trim force is required of the tailplane to keep the aircraft in trim, that is, preventing it from pitching nose down. This aircraft configuration is stable and represents the overwhelming majority of aircraft flying today. The prime disadvantage is that in order to manoeuvre in a turn, i.e. pitch-up, the taileron down force must increase and this downwards force acts against the total lift force. Therefore although the

Fig. 5.12 Relationship of lift and weight forces – stable configuration

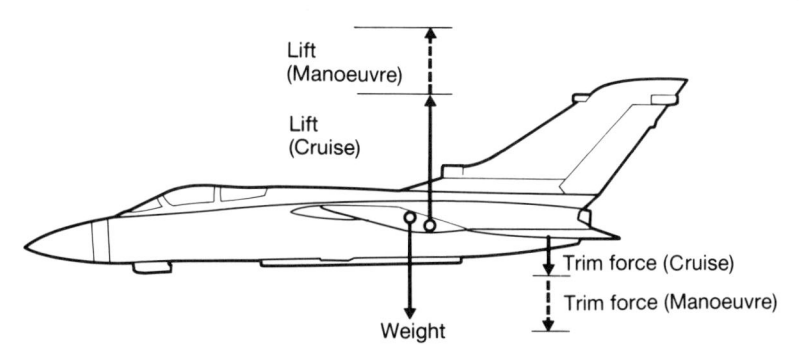

Fig. 5.13 Relationship of lift and weight forces – unstable configuration

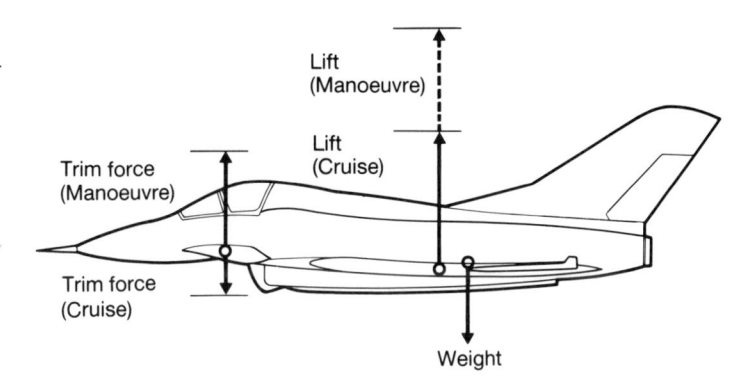

configuration is stable it does not offer optimum manoeuvrability.

The more unorthodox canard configuration adopted by four recent fighters is shown in Fig. 5.13. In this case the centre of pressure is located forward of the centre of gravity. The foreplanes are required to supply a downwards trim force to prevent the aircraft from pitching nose up. In this case when the aircraft is required to manoeuvre by pitching up the foreplane downwards trim force is reversed, now becoming an upwards force acting in the same direction as the lift force. This is an unstable configuration and cannot fly safely without computer control. It has the advantage of improved manoeuvrability and this may be at a premium for a fighter aircraft.

Many of the control characteristics described have been known and understood since the early days of flying. However the means by which the necessary control techniques could be implemented were not generally available until the advent of modern reliable micro-electronics. Electronic components and techniques which satisfy the integrity requirements of the necessary control systems are now available.

CIVIL SYSTEMS

Of the large number of civil flight control systems in evidence today, two will be briefly described, the British Aerospace 146, Fig. 5.14 and the Airbus Industries A320, Fig. 5.15.

The autopilot for the BAe 146 is produced by Smiths Industries and is known as the SEP 10 (Smiths Electric Pilot 10) and is a logical development of the SEP 1, itself derived from the RAE Mark IX. The system schematic is shown in simplified form in Fig. 5.16. The main system components are:

Flight Control Systems

Fig. 5.14 British Aerospace BAe 146 *British Aerospace, Hatfield*

Pilot controls:
- autopilot controller
- mode selector
- altitude controller

Sensors:
- lateral accelerometer
- yaw rate sensor
- altitude sensor
- indicated airspeed sensor
- normal accelerometer

Computers:
- yaw computer
- autopilot computer
- monitor computer (for Category 2 operation only)

Actuators:
- yaw, roll and pitch servomotors
- pitch trim servomotor

The basic modes provided by the autopilot include:

- pitch attitude hold with datum adjust
- roll attitude hold with datum adjust
- altitude hold
- altitude preset
- airspeed hold
- mach hold
- vertical speed hold
- glide slope capture and tracking
- directed go-around
- heading preselect
- VOR/ILS localiser any angle capture and tracking
- back beam coupling.

Perhaps the best way of outlining what these modes offer the pilot is to describe a typical flight leg when a number of these features will be used.

Following take-off the pilot will engage attitude hold while accelerating towards the desired climbing speed. At this point he may also have a VOR beacon selected

Civil systems

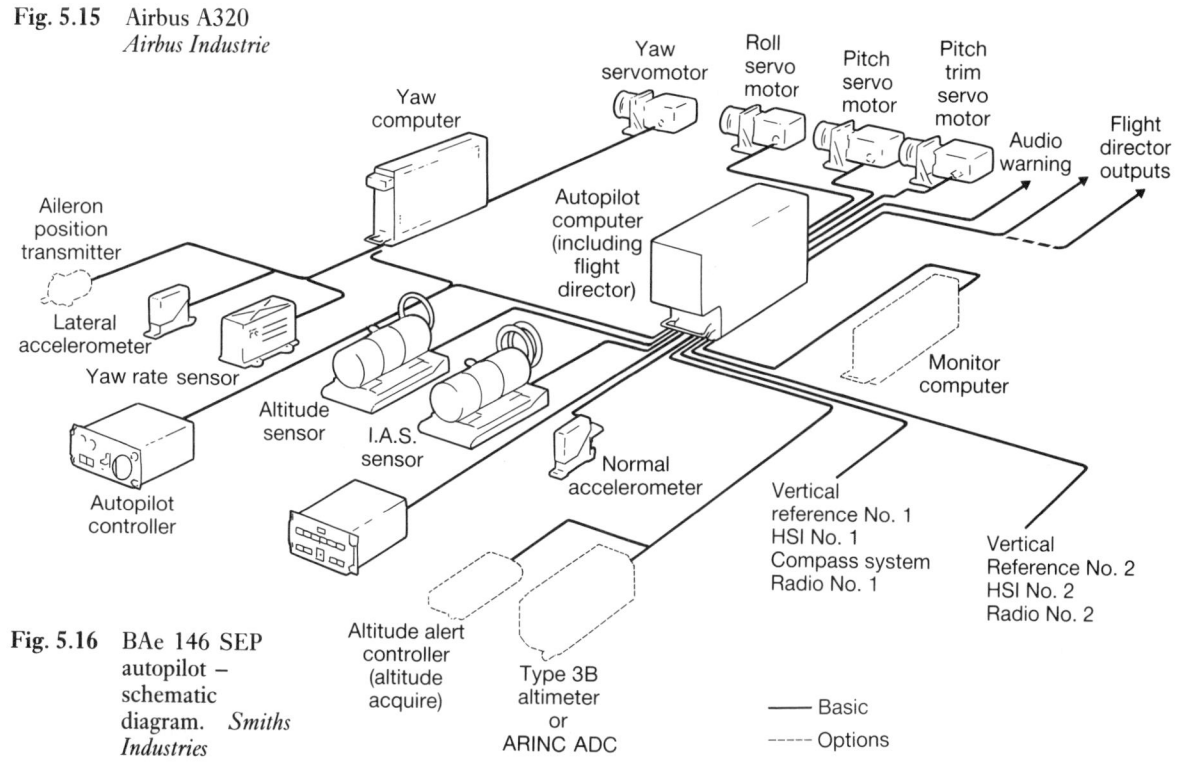

Fig. 5.15 Airbus A320
Airbus Industrie

Fig. 5.16 BAe 146 SEP autopilot – schematic diagram. *Smiths Industries*

Labels in diagram:
- Aileron position transmitter
- Lateral accelerometer
- Yaw rate sensor
- Autopilot controller
- Altitude sensor
- I.A.S. sensor
- Yaw computer
- Yaw servomotor
- Roll servo motor
- Pitch servo motor
- Pitch trim servo motor
- Autopilot computer (including flight director)
- Normal accelerometer
- Audio warning
- Flight director outputs
- Monitor computer
- Altitude alert controller (altitude acquire)
- Type 3B altimeter or ARINC ADC
- Vertical reference No. 1 / HSI No. 1 / Compass system / Radio No. 1
- Vertical Reference No. 2 / HSI No. 2 / Radio No. 2

—— Basic
----- Options

95

to aid navigation during the departures procedure phase. Alternatively, he could be using the back beam mode coupled to the ILS localiser. During the climb out to cruising altitude, the aircraft may be required to hold at a specific altitude to avoid conflict with other air traffic. At this stage the pilot could select altitude hold mode together with VOR coupled mode or heading hold to provide guidance in azimuth. Given clearance to continue the climb to cruise altitude would involve re-selection of the airspeed hold mode together with the selection of further VOR beacons as necessary. Upon attaining the assigned cruise altitude the altitude hold mode would be engaged together with airspeed or mach hold and the necessary navigation mode selections. At the conclusion of the cruise phase of the flight the descent phase would commence, probably using the airspeed hold mode to control airspeed during the descent. At various stages during the descent the pilot may be requested by Air Traffic Control to hold altitude when the altitude hold mode would be re-engaged. During busy traffic periods he may be instructed to enter a 'stack' at various altitudes. During this holding pattern he may be called upon to fly a race-track pattern relative to a VOR beacon. This could be accomplished by using the altitude hold mode in conjunction with the VOR coupled mode. When instructed by Air Traffic Control the captain would exit the holding pattern and commence his approach to the airfield. Initially this could involve the selection of altitude hold together with the ILS localiser capture, later the ILS glideslope capture could also be selected to initiate a full ILS coupled approach. At this stage speed would also be controlled by the autopilot. Depending upon the weather and aircraft fit this could be continued to a full Category 2 approach and landing.

This example gives a simple indication of how the pilot may use the various autopilot modes to assist him throughout typical phases of the flight. It can be appreciated that the availability of these autopilot features is of great benefit as the crew execute all the procedures and checklists associated with flying the complex aircraft through the busy airspace which surrounds airfields in the 1980s.

One of the most progressive flight control systems to be built for a civil airliner is the one developed for the Airbus Industrie A320. This system is the first fully fly-by-wire control system designed for an airliner; the requirements for correct performance and in-built handling protection features include design for better than 1×10^{-9} failures per hour. The aircraft is also novel in that it is the first airliner to be designed with sidestick controllers in lieu of the conventional control column.

The simplified block schematic of the A320 flight control system is shown in Fig. 5.17. The aircraft has three independent hydraulic systems called blue (B), green (G) and yellow (Y); the segregation of these hydraulic systems by various flight control actuators is shown. The flight control system comprises a total of seven computers of three different types. Two elevator aileron computers (ELACs) supply inputs to the four elevator and aileron sections. Three spoiler elevator computers (SECs) supply the four elevator and ten spoiler sections. Two flight augmentation computers (FACs) effectively provide yaw damper and yaw guidance demands for the three rudder sections. During normal operation the control surfaces are controlled by the appropriate numbered computer as shown. From this diagram the combination of failures of hydraulic systems and computers may be deduced. The trimmable horizontal surface actuator (THS actuator) is a triplex trim motor. The moding of the spoiler and aileron sections is as follows:

- Ground spoilers (lift dump) are exercised by all ten spoiler sections.
- The speed brake function is exercised by the three inboard spoiler sections on each wing.

Fig. 5.17 Airbus A320 flight control system
Interavia

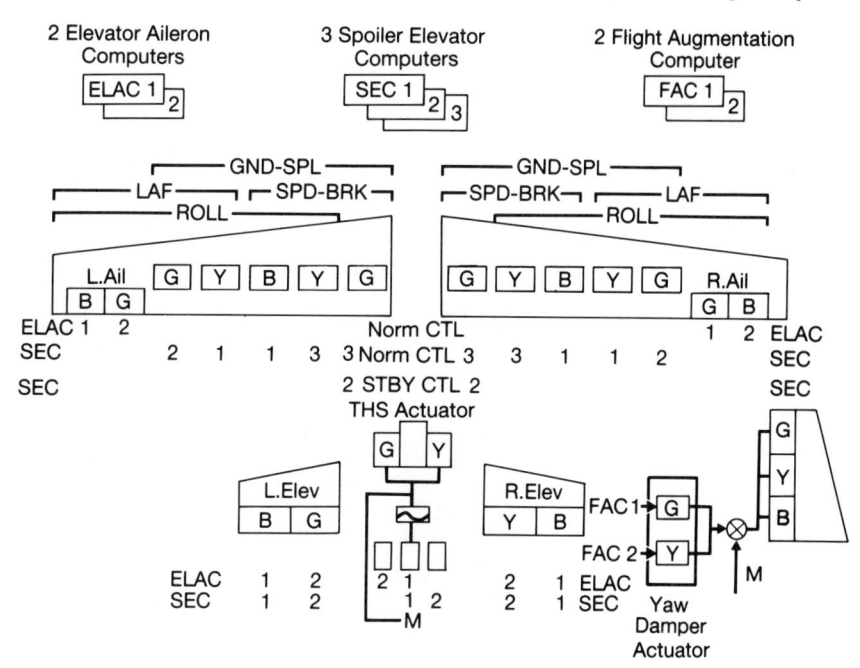

- Roll control is exercised by all four aileron sections and by the outer four spoiler sections on the inside or lower wing.
- Load allevation is achieved by using all four aileron sections and the two outboard spoiler sections on each wing.

In the unlikely event of a total loss of electrical power and therefore electronic control, the input shown M represents a mechanical link with control over the variable incidence tailplane and the rudder. Flight tests on a range of Airbus models have demonstrated that it is possible to land the aircraft using just these controls. However this mechanical reversion mode is intended as a back-up control while the crew execute emergency procedures to restore the primary electronic flight control system. The manufacturers involved in the development of the three types of flight control computer are SFENA and Thomson-CSF. Both were constrained to use different processors and programming languages; the complete segregation of design teams was also decreed. The system has therefore been designed in accordance with the dissimilar processor/dissimilar software principle to avoid subtle though possibly unsafe generic or single mode system failures. (Fig. 5.18 and Ref.1)

The concept of sidestick control was demonstrated by fitting sidesticks to a modified A300 which was used to prove the concept in flight to some twenty airlines, six certification authorities and the aviation press. Initial pilot reaction to the implementation of the sidestick controllers was very favourable and there is an acceptance that it is the logical way to fly a digital flight control system. For an account of flying the A 300 sidestick demonstrator aircraft refer to the 'Flight International' article,[2] a later one[3] describes the flight characteristics of the A320.

UK MILITARY DEVELOPMENTS

The improved performance of military aircraft following the Second World War, particularly those powered by jet propulsion, led to the need for dutch roll

Fig. 5.18 The world's first 'intelligent' avionic system developed by GEC Avionics to control the powerful slats and flaps of the A300 series Airbus. It incorporates two different types of microprocessor to ensure that it will 'fail-safe'. The slats and flaps control system computer is shown with, on the left, the Intel 8085 and right, the Fairchild 6800 microprocessors. Other 8-bit microprocessors can be used, Motorola in place of Fairchild and AMD in place of the Intel *GEC Avionics*

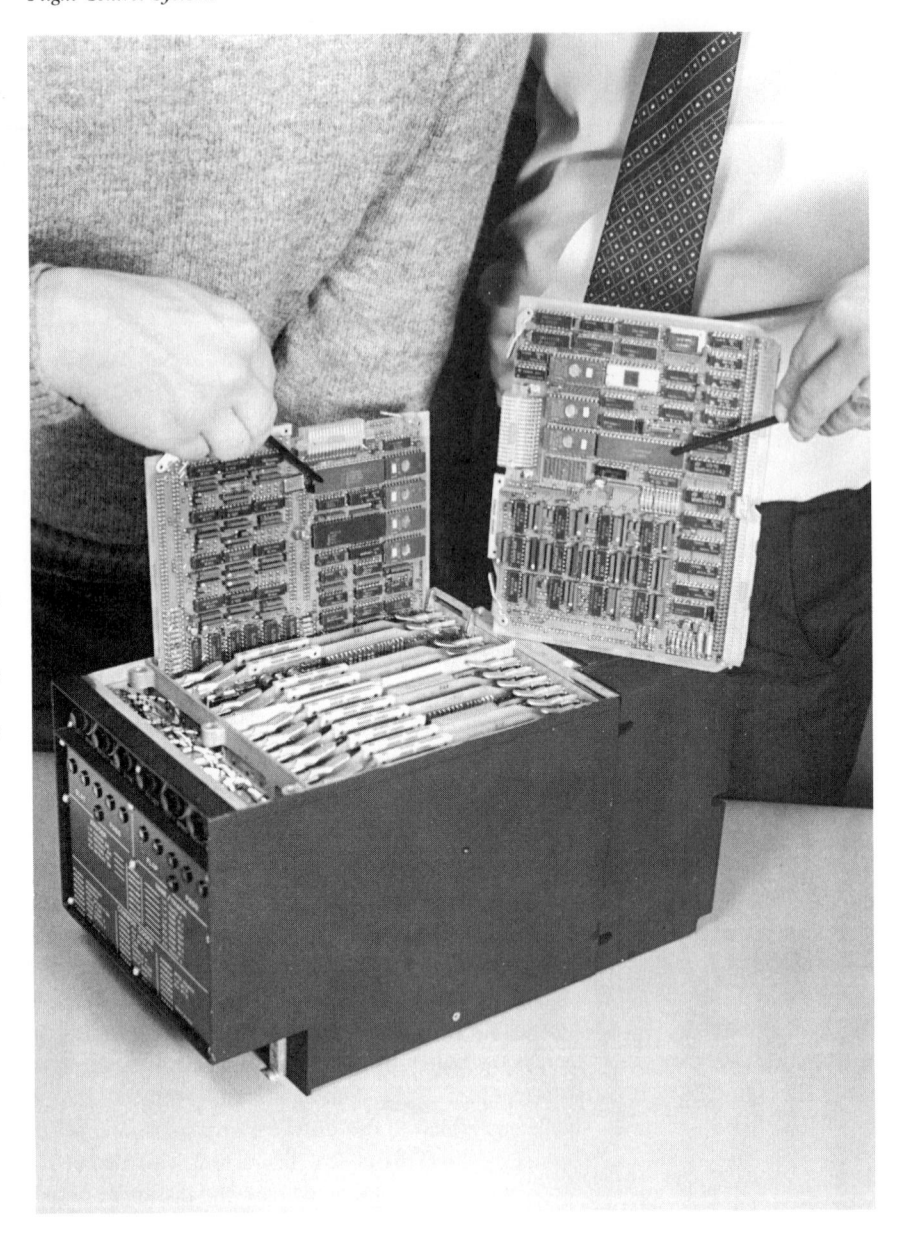

damping. Dutch roll is a periodic oscillation in pitch and yaw in which the aircraft axes trace an elliptical path with the magnitude of the yaw oscillation around two or three times that in pitch. The solution to the problem was to incorporate yaw damping; this was the forerunner of the fly-by-wire flight control systems used by many of today's high performance fighter aircraft.

The introduction of high pressure hydraulic systems led to the development of electrically signalled hydraulic actuators to satisfy the requirements of high performance aircraft. As fly-by-wire flight control systems have evolved these actuators have developed into complex units with multiple channel redundancy matching that of the flight control computation.

The availability of cheap, high speed, high reliability micro-electronics has played a significant role. The replication of different lanes or channels of analogue

UK military developments

Fig. 5.19 Panavia Tornado IDS P12 with wings in fully swept position *Arthur Gibson*

or digital computing has become more reliable and cost-effective, thereby allowing triplex and quadruplex computing implementations of flight control laws to become reality. The high speed logic capabilities of modern micro-electronics also permit monitoring, cross-monitoring and moding logic to be incorporated without significant hardware overheads or penalties.

The main UK military developments in the fly-by-wire flight control systems area have encompasssd the last 13 years. Developments have progressed from the fly-by-wire (FBW) system on the Tornado (Fig. 5.19) in the early 1970s, through the Jaguar FBW demonstrator to the EAP. The Tornado first flew in 1974, the Jaguar FBW (Fig. 5.21) in 1982 and the EAP (Figs 5.23, 5.24) in 1986. In a little over a decade the fly-by-wire system has advanced from a first-generation triplex analogue computing system with a mechanical reversion to a fully digital quadruplex system with total authority. The main capabilities and features of these systems are outlined below.

The Tornado Control and Stability Augmentation System (CSAS) fly-by-wire system was the first of its type to be utilised on an operational combat aircraft in Europe. The Tornado is not an inherently unstable aircraft as has been described earlier, where the aircraft would be totally uncontrollable without the provision of electronic control. However, due to the variable sweep configuration there would be enormous variations in aerodynamic characteristics, and therefore handling qualities throughout the flight envelope, were electronic assistance not offered to the pilot. The aircraft could be flown without augmentation, but at the penalty of an unacceptable workload on the pilot of an operational aircraft. It was therefore the first British military aircraft to employ a fly-by-wire system to improve stability, provide the pilot with sensibly predictable control forces (constant stick force per 'G') and enhance the weapon-aiming capabilities for the greater part of the flight envelope. To achieve these objectives the system has the following features:

- Triplex autostabilisation in roll, pitch and yaw utilising feedback signals from triplex roll, pitch and yaw rate gyro packages.
- Gain scheduling in the forward computing paths to normalize the aircraft handling qualities. This gain scheduling will typically neutralise or minimise the effect of changes in dynamic pressure ($P_t - P_s$), wing sweep and trim changes (for example, due to airbrake deployment), so that the pilot may effectively ignore them during his handling of the aircraft.

A simplified block diagram of the Tornado CSAS is depicted in Fig. 5.20. The pilot's inputs to the system are signalled via triplex potentiometers connected to the control column (roll and pitch) and rudder pedals (yaw). The pilot may also feed trim inputs into the system. Triplex inputs to the computers are: air data – dynamic pressure; inertial data – roll, pitch and yaw rate; and aircraft configuration data – wing sweep and airbrake position. Pitch computations are carried out in the triplex pitch computers which compute pitch rate demands using analogue computing techniques. The triplex lateral computers calculate roll and yaw rate demands in a similar fashion and there are cross-coupling links between both sets of computers. Quadruplex demands in roll and pitch are signalled; the taileron actuators acting differentially for roll demands and in unison for pitch demands. A lesser level of redundancy is employed when demands are signalled to the rudder and spoiler sections. Extensive cross-monitoring and voter-monitoring is employed at various computational stages to disable or vote out signals outside the normal

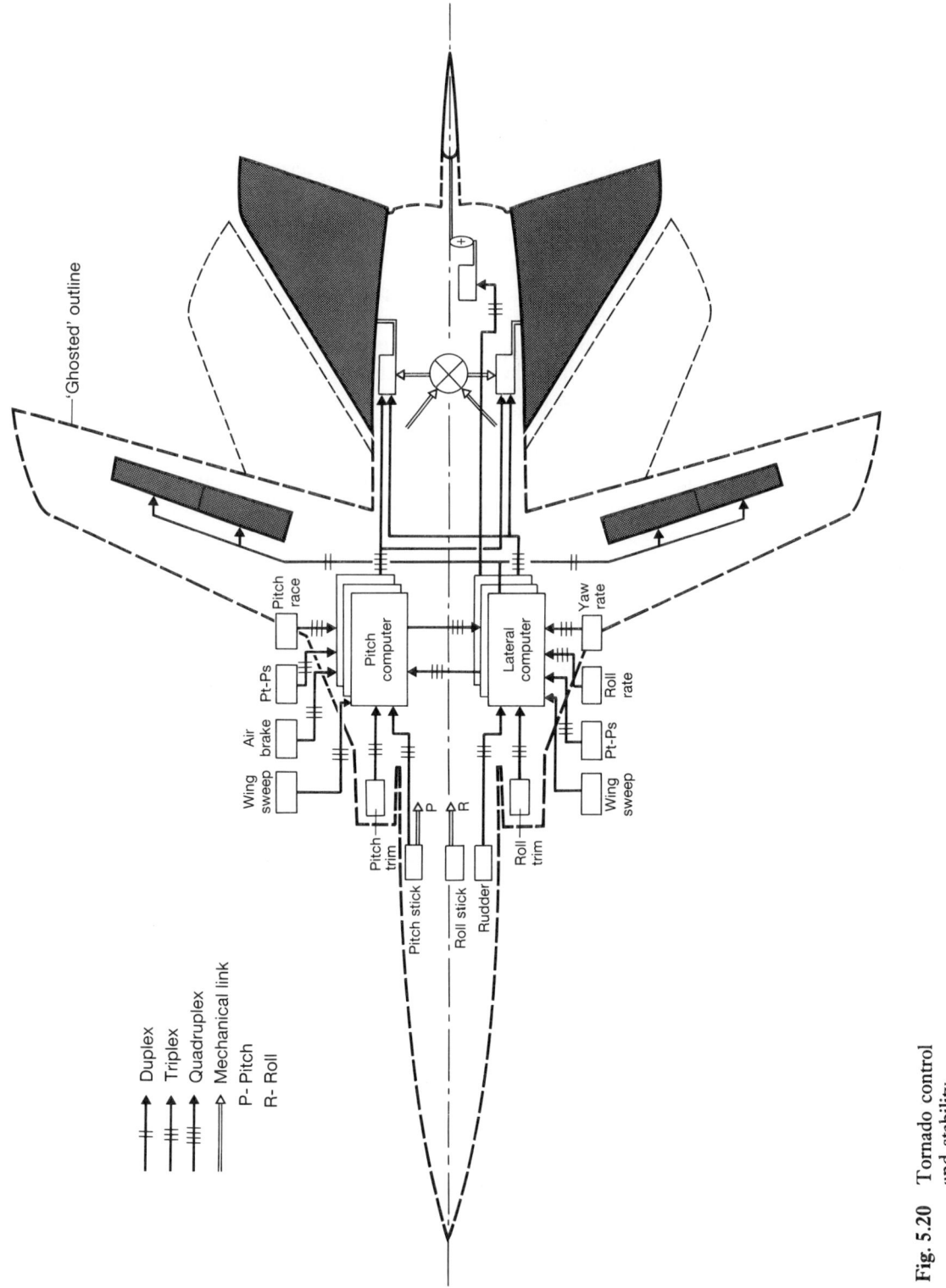

Fig. 5.20 Tornado control and stability augmentation system (CSAS) – simplified block diagram

range of operation. A number of reversionary modes may be used in the event of system failures. The primary modes of operation of the CSAS are as follows:

- Full CSAS operation.
- Reversion to direct electrical link in pitch and roll following second lane failure in the autostabilization loop. In the event of such failures in the yaw channel the rudder is fixed in the neutral position.
- In the event of a second lane failure in the forward computing (or gain scheduling) path the system may be flown with a direct mechanical link in pitch and roll.

In addition to his primary inputs into the system in pitch and roll (stick) and yaw (rudder), the pilot has a CSAS control panel to assist in controlling and monitoring the system. The control panel allows to initiate system built-in test (BIT) to check for correct operation prior to flight. The control panel also serves to indicate to the pilot first or second lane failures in each axis. In certain circumstances it also offers the facility of selecting failure states in order to be able to demonstrate these for training purposes. For the Air Defence Variant (ADV) or Tornado F2 and F3, additional facilities are available by the addition of an Automatic Configuration Control Processer (ACCP). This will automatically control the wing sweep, flap and slat settings at various stages in flight and offers the pilot the automatic selection of optimum configuration of high lift devices whilst engaged in close air-to-air combat.

In 1977 British Aerospace was awarded a contract to develop and demonstrate the benefits of Active Control Technology (ACT). The prime aims of the programme were to design, develop and demonstrate the safe and practical use of a primary flight control system for a modern combat aircraft. Among the prime assumptions was that the production models have the airworthiness capability to prove safe software in flight. The additional control procedures to design, test and qualify the high integrity system were also required. Therefore the decision was taken to dispense with a mechanical reversion and rely upon a quadruplex digital system with common high integrity software, and it was designed from the outset without the inclusion of either mechanical or direct electrical link reversions which were features of the Tornado CSAS. The Jaguar was chosen as the ACT demonstrator and modified appropriately (Fig. 5.21).

In view of these ambitious aims certain precautions were taken to increase the redundancy of the system to sextuplex, or more correctly, duo-triplex by the means of further consolidation within the system. Quadruplex inputs to the four flight control computers were extended to duo-triplex by the use of supplementary Actuator Drive and Monitor Computers (ADMCs) as shown in Fig. 5.22. Therefore each Power Flying Control Unit (PFCU) contained six servovalves which accepted the duo-triplex demands (FCC1 + FCC2 + ADMC1) and (FCC3 + FCC4 + ADMC2). Mechanical consolidation of these duo-triplex demands within the PFCUs effectively assured that any two-lane failures could be absorbed. Therefore the term 'failure absorption' was coined to describe this ingenious implementation. The precise engineering in this area was undertaken by Dowty Boulton Paul[4,5]. Control in roll, pitch and yaw depends upon the comparison of pilot inputs with aircraft body rates in a similar fashion to the Tornado CSAS. The forward gain of the control loop is scheduled in accordance with certain variables: airspeed, altitude, incidence and undercarriage position. The aims of the control laws were to minimise the variation in stick force per G as was the case for the Tornado. The aircraft was made aerodynamically unstable

Fig. 5.21 ADV Tornado in formation with ACT Jaguar demonstrator. Note position of Tornado (a very stable aircraft) control surfaces. Maximum negative lift on the tailerons while the Jaguar, a totally unstable aircraft, shows slight positive lift *Arthur Gibson*

in the later phases by the fitting of large strakes on each side of the Jaguar intakes. These had the effect of moving the aerodynamic centre of lift much further forward than on the conventional Jaguar: it will be recalled that movement of the centre of lift forward of the centre of gravity makes an aircraft unstable. After the strakes had been fitted the stability of the aircraft could still be varied by fitting large quantities of ballast in the rear of the fuselage to vary the centre of gravity. The combination of those and other modifications allowed the centre of gravity to be moved back to -10% $\bar{c}$ during the flight test programme. ($\bar{c}$ is the mean aerodynamic chord. This is an aerodynamically defined parameter and depends upon the wing planform. For most conventional flighter wing planforms the mean aerodynamic chord approximates in value to the geometric mean chord.)

Flight testing of the FBW Jaguar was split into five distinct phases, commencing in October 1981 and culminating just over three years later.[4] The programme provided significant information upon the performance of an unstable aircraft. It also yielded valuable experience in the management and proving of a programme utilizing high-integrity software driven flight control laws.

In early 1981, work at British Aerospace Warton was focusing upon a British Industry funded project called P 110. In 1982 this British initiative came to an end but was replaced by collaborative work undertaken by the Tornado partners – British Aerospace, Aeritalia and Messerschmitt-Bölkow-Blohm (MBB). This project was known as the Agile Combat Aircraft (ACA) and was undertaken on the basis of a 40/40/20 percent workshare between the UK, German and Italian airframe companies respectively. A Joint Flight Control System development team

Flight Control Systems

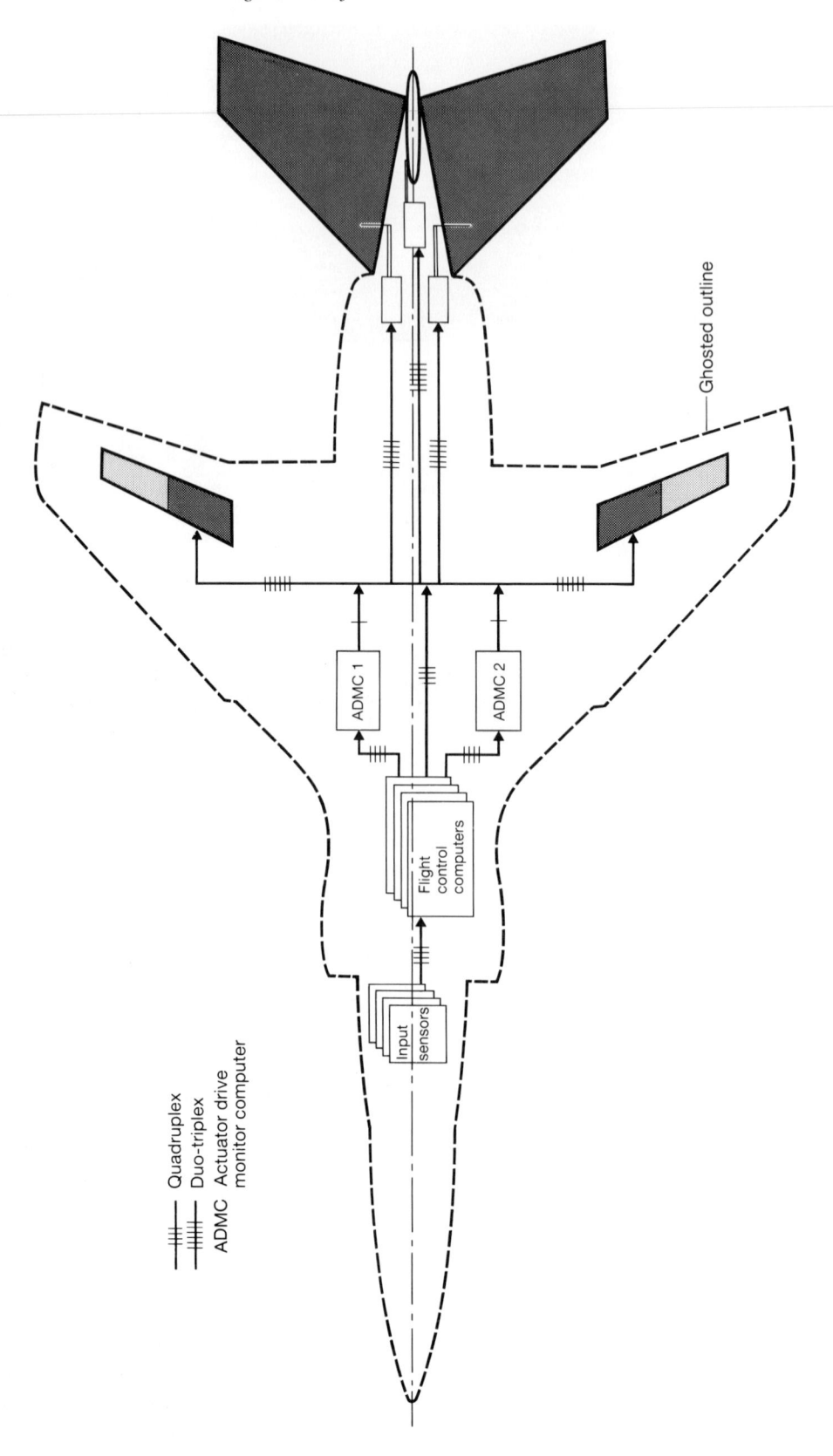

Fig. 5.22 Jaguar ACT (fly-by-wire) – simplified system block diagram

was established at the MBB plant in München which continued flight control system design based upon the experience gained by British Aerospace with the FBW Jaguar and the MBB CCV F-104. Unfortunately, withdrawal of support by the German Ministry resulted in the ACA being dropped in 1983. Instead, a joint UK MoD and British industry development, partly assisted by some German and Italian firms, continued with a single aircraft demonstrator programme: the Experimental Aircraft Programme or EAP. Much of the flight control design undertaken by the ACA Joint Team in München was incorporated into the EAP flight control system.[7]

The aerodynamic configuration of the EAP (Figs 5.23, 5.24) differs fundamentally from the more conventional configuration already described for the Tornado and Jaguar. The aircraft is of canard configuration with a cranked delta mainplane and two foreplanes or canards located ahead of the mainplanes. Control of the aircraft is exercised by:

– foreplanes for pitch and roll control;
– trailing edge flaperons for pitch control and trim;
– rudder for directional control;
– leading edge flaps, primarily to optimise control in pitch.

Fig. 5.23 EAP demonstrator *British Aerospace, Warton*

Flight Control Systems

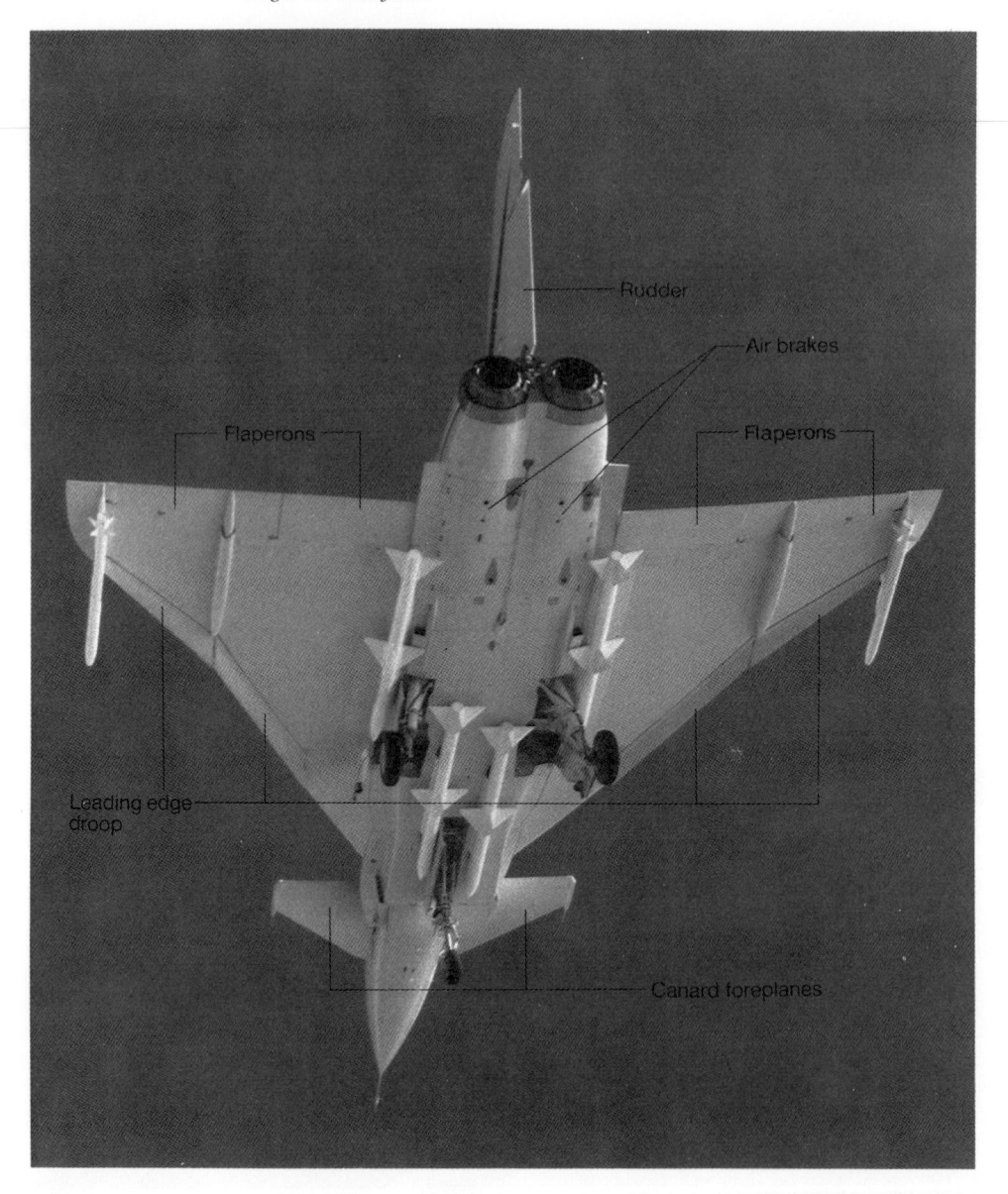

Fig. 5.24 EAP demonstrator showing canard foreplanes, leading edge flaps and flaperons *British Aerospace, Warton*

The ability to control the leading edge flaps and flaperons effectively allows the wing contour to be altered for various flight conditions, either improving the mainplane lift characteristics or minimising adverse effects such as pitch-up. Therefore in a simplistic way there are some similarities between these features and the Mission Adaptive Wing (MAW) concept being tested in the US and described later in this chapter.

In addition to these primary flight control surfaces, control is also exercised over the left and right intake ramps and the nosewheel steering system. The prime avionic contractor is GEC Avionics who were also the avionics supplier for the FBW Jaguar.

A simplified schematic of the EAP flight control system is shown in Fig. 5.25. It will be noted from this diagram that there are no discrete rate gyro or accelerometer packages. Instead, aircraft motion is sensed by means of four Aircraft Motion Sensing Units (AMSUs) which are effectively strap-down inertial units

UK military developments

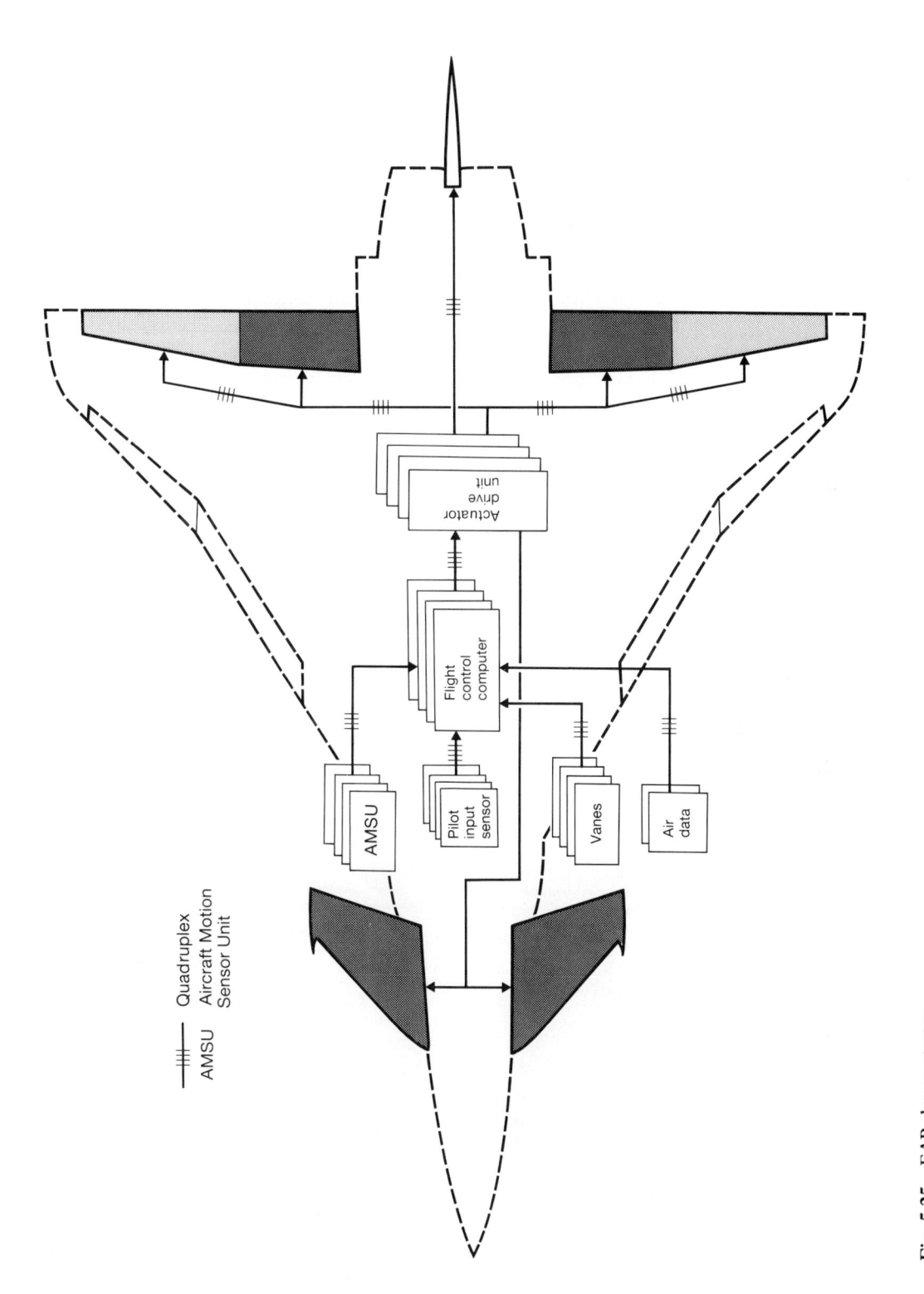

Fig. 5.25 EAP demonstrator – simplified system block diagram

Flight Control Systems

which are used to provide three axis body rates and accelerations as well as three axis attitude and inertial information.

Air data from the pitot/static system is provided by two dual channel air data computers. Four Airstream Direction Detectors (ADDs) feed incidence and sideslip information into each of the four Flight Control Computers (FCCs). The pilot's inputs to the system are independently sensed by pitch, roll and yaw sensors and fed in quadruplex to the four FCCs. Computed demands from the FCCs are fed to four actuator drive units which drive the left and right flaperon sections and the rudder; signalling is quadruplex. The foreplane actuators are driven directly in quadruplex from the FCCs. The links between AMSUs, ADCs, FCCs and ADUs is by means of digital transmission. All computing is digital with the control laws being executed by high integrity software. The system is purely electrically signalled, there being no mechanical reversion. The aircraft operates in a totally unstable configuration with a manoeuvre margin of around $-12\frac{1}{2}\%\ \bar{c}$.

The EAP first flew as a high technology demonstrator for key systems in the European Fighter Aircraft project (Fig. 5.26) on 8 August 1986 and successfully concluded its initial flight clearance programme before the SBAC Show at Farnborough in September 1986. Further flight testing with some modifications for high speed and high angle of attack test sorties were made before the Paris Show in

Fig. 5.26 European Fighter Aircraft project – Artist's impression *Arthur Gibson*

the following year. Since that date the programme has continued with few snags and is providing considerable data for the EFA which is a collaborative project involving British Aerospace, MBB in Germany, Aeritalia in Italy and CASA in Spain (see also Chapter Six).

ADVANCED DEVELOPMENTS

The preceding section described the evolution of UK and European fighter fly-by-wire systems over the past one and a half decades. It is interesting to review some of the corresponding developments by the American aerospace industry in recent years. The F-15, F-16 and F-18 all utilize flight control systems based upon multiple channel fly-by-wire systems. The F-15A/C versions utilise a three axis stability augmentation system supplied by General Electric. The more recent F-15E Dual Role Fighter variant has a digital flight control system supplied by Lear Siegler Inc. The F-16A initially employed a quadruplex analogue system, however later F-16Cs will have a quadruplex digital system developed by Bendix. The F/A-18 is fitted with a quadruplex digital flight control system supplied by General Electric.

More recently a number of programmes have been initiated by US Governmental Agencies to examine a range of advanced techniques by means of technology demonstrator programmes. These programmes include:

The Defense Advanced Research Projects Agency (DARPA)/Grumman X-29 (Fig. 5.27).
Advanced Fighter Technology Integration (AFTI) F-111 Mission Adaptive Wing (MAW) (Fig. 5.28).
F-15 STOL Manoeuvre Technology Demonstrator (Fig. 5.29).
DARPA/ROCKWELL/MBB X-31 (Fig. 5.30)

The forward swept wing of the Grumman X-29 is surely one of the most startling aircraft of recent times, though aerodynamically it is not novel in concept: aerodynamicists were always aware of this possibility. See Fig. 5.27. The benefits

Fig. 5.27 Grumman X-29 configuration

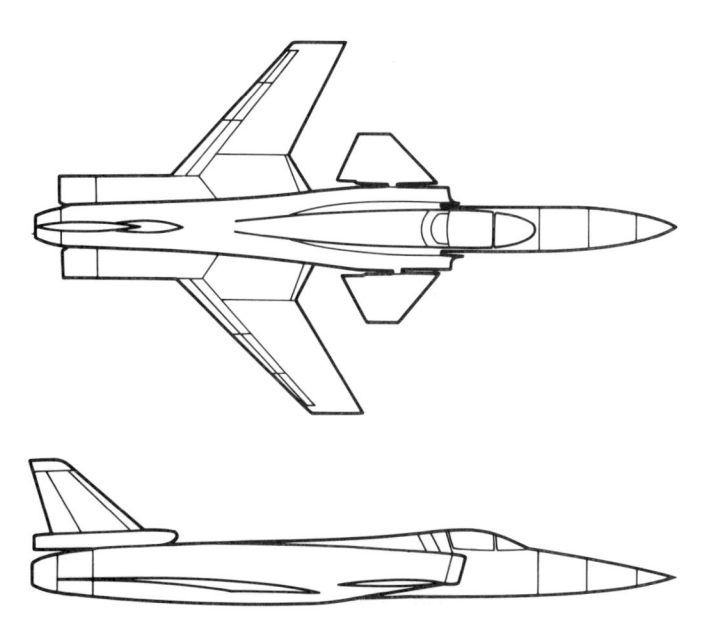

Flight Control Systems

Fig. 5.28 F-111 mission adaptive wing – primary modes

Variable camber envelope

Manoeuvre

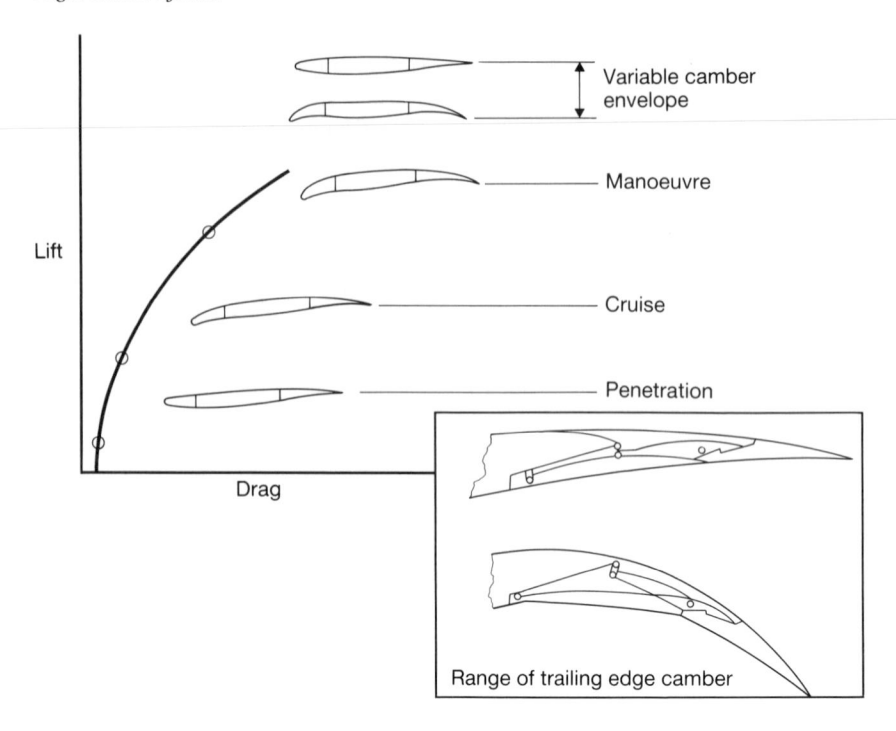

Lift

Cruise

Penetration

Drag

Range of trailing edge camber

Fig. 5.29 F-15 STOL manoeuvre technology demonstrator configuration

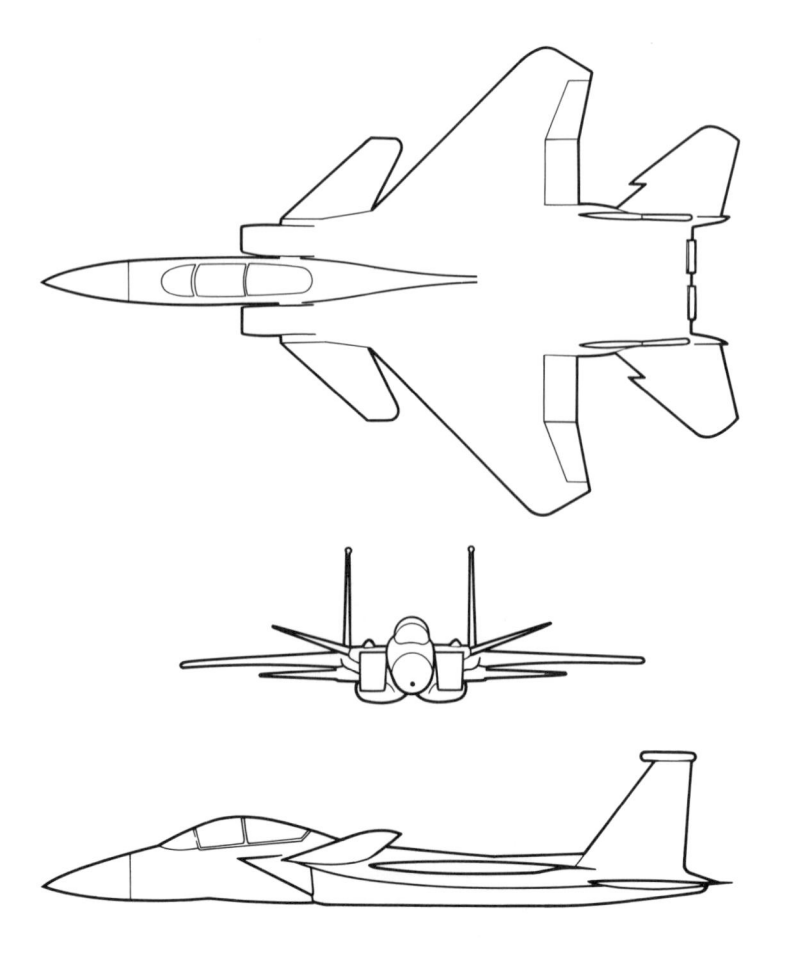

Advanced developments

Fig. 5.30 DARPA/Rockwell/MBB X-31 configuration

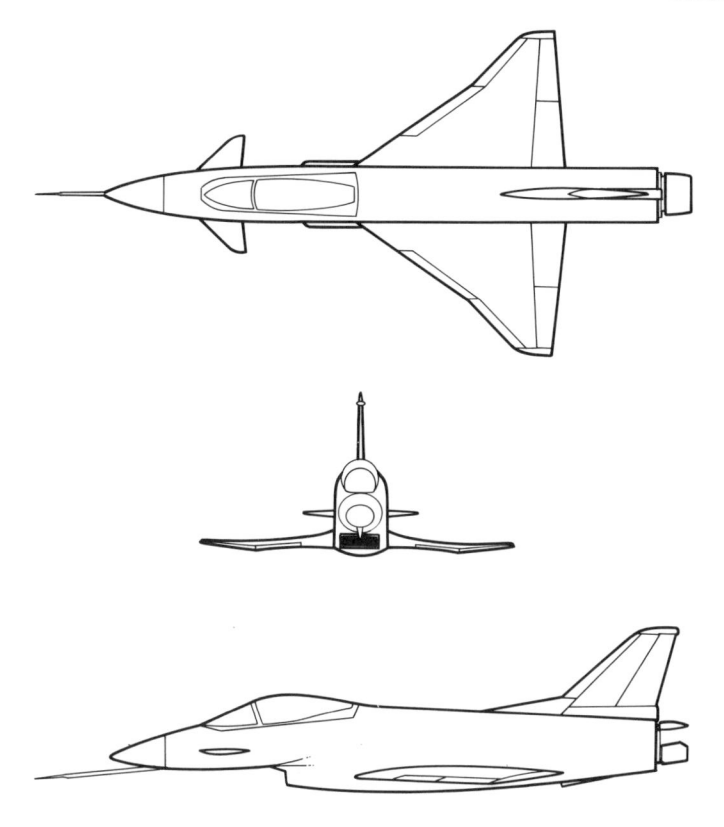

of such a forward swept wing are difficult to quantify without a detailed aerodynamic dissertation. For the purposes of outlining the aims of the X-29 program the advantages may be assumed to be:

- greater lift slope, that is a better life : drag ratio;
- lower induced drag;
- improved low-speed handling; in other words, less pronounced pitch-up when operating near the stall;
- improved wing area, or smaller structural span for a given wing loading with consequent saving in aircraft weight.

The technical capabilities needed to successfully operate a forward swept wing are due in part to advanced structural and composite techniques to make the wing sufficiently 'stiff'; the other prime contributor is advanced micro-electronics. Some of the structural and construction features of the X-29 augment the fundamental aerodynamic advantages. The first is the use of a supercritical wing which delays shock-wave formations at high Mach numbers. The second is the use of carbon composite material in the wings which increases stiffness and reduces wing aeroelastic effects. This therefore reduces structural weight and to a lesser degree, reduces drag. It has been estimated that for a given mission drag may be reduced by between 10 and 20 per cent.

Aircraft control is effected by the forward canards, flaperons on the rear of the swept wing, strake flaps adjacent to the rear fuselage section and the rudder. The aircraft is probably one of the most unstable ever flown with a manoeuvre margin of −35% mean aerodynamic chord (compared with −10% for the FBW Jaguar

Flight Control Systems

and $-12\frac{1}{2}\%$ for EAP). The flight control system is based upon a Honeywell system developed for the SR-71 strategic reconnaissance aircraft. Triplex redundant digital flight control computers are used with an integral analogue system to satisfy integrity requirements.

The DARPA/US Air Force funded programme began in 1981 and the first flight took place on 14 December 1984. Two experimental aircraft are being used in the programme to expand the flight envelope to establish the true benefits of this unique configuration.

The possibility of altering wing camber was mentioned during the description of the EAP system. An ambitious programme of development and demonstration involving NASA Dryden and the US Air Force has been under way since 1979 to apply such a technique in a more advanced form to a heavily modified F-111. This has been termed mission adaptive wing and the aims are to combine a smoothly variable camber wing with a system which already employs automatic flight control. (The F-111 employs a triplex analogue computing system to augment the primary mechanical flight control linkage). The main phases of operation being examined are:

- Manoeuvre Camber Control: This mode is intended to set the wing camber to achieve maximum lift : drag ratio according to a pre-programmed schedule of mach number, altitude and 'G' loading.
- Cruise Camber Control: In this mode the system will inch the flaps in small increments during flight to optimize speed for a given power setting.
- Manoeuvre Load Setting: The movement of the outboard trailing edge sections are modified to minimize the wing root bending moment to pre-set limits.
- Manoeuvre Enhancement/Gust Control: This feature uses vertical (normal) acceleration to excursions in the vertical plane.

For a ready comparison of these modes refer to Fig. 5.28.

The X-29 concentrates development upon a new, and to some extent, limited airframe. The F-111 AFTI/MAW development relates specifically to lift and load control on a specific airframe – a very heavily modified F-111. The F-15 STOL Manoeuvre Technology Demonstrator is in contrast in the sense that it demonstrates technology in an active front line US Air Force aircraft, albeit with significant modification. It is more relevant to operational Air Force requirements in that, as well as exploring advanced control techniques the vehicle is required to demonstrate the ability to land and take-off in 1500 feet in low visibility conditions. This requirement is essential to operating off cratered runways in poor weather conditions and could well represent a realistic wartime scenario. It is also possible that some of the technologies demonstrated could be embodied in future versions of the F-15 and variants of the Advanced Tactical Fighter (ATF). The ATF is presently under way in a Demonstration-Validation (Dem/Val) phase between competing teams led by Lockheed with the YF-22A and Northrop with the YF-23A.

The F-15 SMTD program involves significant modification to a standard F-15B two seat aircraft. The aircraft planform is modified as shown in Fig. 5.29. In relation to the flight control system these modifications are:

- the addition of canard control surfaces adjacent to the intakes;
- the inclusion of 2-dimensional (2D) thrust vectoring and reversing nozzles to add an additional means of manoeuvring in pitch by means of thrust vectoring.

- As well as these flight control system alterations, the aircraft will demonstrate lightweight structural materials on the upper wing surfaces and be fitted with an electro-optic sensor to prove the concept in low visibilities.

The demonstration of the 2D thrust vectoring/thrust reversing nozzle is an important feature. These nozzles were developed by Pratt and Whitney to be used in conjunction with the F-100 engines. The flight control system has been evolved from the General Electric fly-by-wire system used for the F/A-18. Each of the quadruplex processors utilizes a MIL-STD-1750A instruction set with software written in the US standard J-73 (JOVIAL) software language. Certain elements of the programme – probably the newer and most innovative modules – are programmed in ADA.

The F-15 SMTD is certainly one of the most exciting flight control developments presently under way in the US. This is particularly so because of the choice of demonstrator vehicle and the relevance of the programme aims to future fighter aircraft requirements.

The DARPA/ROCKWELL/MBB X-31 programme was initiated in 1987 by DARPA in conjunction with Rockwell and MBB. The prime aim of the two aircraft demonstrator programme is to provide a technology feasibility fly-off whereby unique high manoeuvre techniques may be examined. The programme is jointly funded by the US and Germany, with formal go-ahead given in February 1987. Test flying is intended to commence in 1989. The aircraft layout is shown in Fig. 5.30. There is a great deal of emphasis placed upon thrust vectoring in this program, as for the F-15 SMTD. In this case the thrust vectoring is in the yaw plane and is implemented by means of vanes which extend into the engine jet efflux. Movement of the vane from side-to-side causes the jet efflux to be deflected in the yaw plane hence causing the aircraft to yaw. This feature is expected to be of benefit at low speeds and very high angles of attack (*circa* 80 degrees) in allowing the pilot to out-manoeuvre an opponent in close-in dogfight situations. The principle of using vanes to vector aircraft thrust in this way is presently being proved on a modified F-14A under a programme funded by the US Navy. Initial test results have indicated that the technique is yielding better performance than was predicted.

The use of light transmission by means of fibre-optic techniques has long offered attractions to the flight control system designer. The particular advantages of fibre-optic signalling include higher data rates and a considerable reduction in the effects of electro-magnetic interference or lightning strikes upon the sensitive signalling lanes between computers and actuators. These effects are potentially more bothersome in modern aircraft where large sections of structure may comprise composite rather than metallic material. The use of these composite materials in the structure denies the avionic systems a degree of natural screening normally given by conventional metallic materials.

Despite all of the foregoing descriptions of high technology and high integrity systems being applied to current front line fighters and technology demonstrators, it may come as something of a surprise to realise that the only fly-by-light system presently in service is on an airship. Airships pose particular problems in that they possess non-rigid structures which continually flex during flight. Therefore a traditional mechanical cable and pulley signalling system is unsuitable. Similarly, the presence of long screened signalling wires within a largely non-metallic structure has the disadvantage of potentially acting as a lightning conductor, positively encouraging lightning strikes.

Consequently, a fly-by-light system was adopted for the Airship Industries Skyship 600. This airship uses a duplex system developed by GEC Avionics and was first demonstrated in 1982. The Skyship 600 utilises a duplex system since the pitch response of an airship is much less dramatic than that of a conventional aircraft – hence a duplex system offers sufficiently high integrity. Nevertheless, the fly-by-light techniques must be considered to be a viable option for future military and civil aircraft when some of the existing component problems have been overcome.

REFERENCES

1. **Lambert, M.** 'A 320 Dual Sidestick System Takes Off', *Interavia* 12/1986.
2. **Hopkins, H.** '*Flight* flies fly-by-wire Airbus', *Flight International* 13 September 1986.
3. **Hopkins, H.** '*Flight* tests the A320 Airbus' *Flight International* 12 December 1987.
4. **Hilton, R. G., Steed, D. J.** 'Fly-By-Wire Actuation for Combat Aircraft', *Aerospace*, March 1984.
5. **Hilton, R. G., Steed, D. J.** 'Fly-By-Wire Actuation for Combat Aircraft', *Aircraft Engineering*, February 1985.
6. **Yeo, G. R.** 'Fly-By-Wire Jaguar' *Aerospace*, March 1984.
7. **Kaul, H.-J., Sella, F., Walker, M. J.** 'The Flight Control System for the Experimental Aircraft Programme (EAP) Demonstrator Aircraft', 65th Flight Mechanics Panel Symposium on 'Active Control Systems', Toronto, 15–18 October 1984.

CHAPTER 6 AIRCRAFT MANAGEMENT SYSTEMS

IAN MOIR BSC, CEng, MIEE, MRAeS

Senior Multiplexing Systems Engineer, Smiths Industries. Mr Moir undertook his graduate training as an Engineering Officer in the Royal Air Force being awarded a BSc (Eng) 2nd Class Honours Degree in Aeronautical Engineering. Service in the Far East was followed by responsibility, at RAF Brize Norton, for first line servicing of VC 10 and Belfast aircraft.

In 1970, in the rank of Flight Lieutenant, he took a post-graduate Course at RAF College, Cranwell, having already been awarded an Air Training Corps Flying Scholarship and the de Havilland Flying Trophy for the best pilot in RAF Officer Cadet Entry (1965).

In 1972 he joined the RAF Project Team with responsibility for the digital avionics system of the MRCA (Tornado) at British Aircraft Corporation (later BAe) at Warton.

In the rank of Squadron Leader Mr Moir served at the Government Communication Headquarters Cheltenham, was Officer Commanding, Electronic Engineering Squadron, RAF Leuchars and the Ministry of Defence (Air) Directorate of Tornado Engineering and Supply.

In 1980 he joined Smiths Industries, becoming involved in systems integration using MIL-STD-1533B data bus hardware on the BAe Experimental Aircraft Programme. From 1984 he has been Group Leader of the Utilities Systems Management Group.

Mr Moir has presented many Technical Papers at home and abroad and written a number of Technical Articles. In 1979 he was awarded the RAF Strike Command Smallwood Trophy for a submission on Electronic Warfare. He is a Member of the Royal Aeronautical Society, a Chartered Engineer and a Member of the Institution of Electrical Engineers.

The flight control systems described in Chapter Five are associated with aircraft primary flight control, that is the control of the aircraft flight path. However there are other control systems – sometimes termed aircraft management systems – which are necessary for the successful completion of a route leg or mission. These systems fall broadly into the following categories:

- Engine or Propulsion Control – those systems associated with the control of the engine or reheat control and involving the metering of fuel, control of air bleed, reheat nozzle position and so on.
- Flight/Performance Management – those systems which manage the flight profile of the aircraft, selection of flight routes and navigation beacons, procedural direction, and advise upon the performance of the aircraft in certain phases of flight.
- Utilities Management Systems – those systems which control aircraft systems other than flight control or propulsion are called utilities systems. These systems

Aircraft Management Systems

include fuel, hydraulics, environmental, secondary power, electrical and some smaller systems.
- Health and Usage Management – this embraces the monitoring of primary aircraft systems to check performance, identify incipient failures before they result in catastrophic damage and to generally assist in maintaining the good health of the aircraft.
- Weapon/Stores Management – those systems associated with the safe carriage and accurate release of a wide range of weapons or stores.

ENGINE AND PROPULSION CONTROL

To illustrate the application of electronics to engine control, two examples of the typical engine control task will be outlined. Both the examples chosen relate to turbofan engines – one military and one civil. However the techniques described may apply equally to turboprop or helicopter engines.

For many years the metering of fuel to the engine was accomplished by engine mounted hydro-mechanical systems. Primary throttle demands to the engine were signalled by means of control rods and cranks, in some cases flexible cables called Teleflex were used. During the 1950s, Ultra Electric Controls – now part of Dowty Electronics – developed an electrical throttle system which was used on the Proteus engine used to power the Bristol Britannia and the Tyne used to power the Vickers Vanguard and Short Belfast. These signalling links were, in fact, speed and temperature limiters which did not break into the engine inner control loop.

The desire for better, more economical fuel control, and the evolution of more demanding control loops meant that some improvement over the hydro- mechanical methods was needed. The implementation of greater control was leading towards authority engine control and it was natural that as electronic components became more reliable, electronics would be employed. Initially these developments used transistors in full authority analogue systems, first for the Olympus 320 designed for the TSR 2 (Fig. 3.1) cancelled in 1965. Later, the Olympus 593 used to power the Anglo-French Concorde (Fig. 6.1) utilized a similar system. Later modifications introduced a micro processor based Built-In-Test (BIT) system to this design.

The evolution of miniaturized, high reliability solid-state electronic circuits in the form of Large Scale Integration (LSI) integrated circuits gave the technology required to implement digital rather than analogue control. This technology became available during the 1970s, furthermore the steep increase in crude oil and hence fuel prices during 1974 gave a strong impetus to save fuel. This pressure to seek more fuel-efficient solutions was felt by military and civil operators alike. Further engine hardware developments also required engine control to be more precise and more reliable than the existing analogue systems. The availability of digitally controlled, software driven systems also yielded significant benefits of flexibility during development and for maintenance once in service.

A typical engine control layout for a high performance military engine is as shown in Fig. 6.2. The main elements of control are:

- Intake control – control of the intake is required to match a wide range of air intake conditions to provide a smooth flow of subsonic air into the engine.
- These systems are usually supplied by electronic manufacturers not associated with the engine manufacturer; systems designed in recent years are almost invariably digitally controlled.

Fig. 6.1 The Anglo-French Concorde

- Main engine control – the main engine control function is concerned with metering of fuel to ensure smooth running and acceleration characteristics throughout the flight envelope. The correct relationship between the low pressure (NL) and high pressure (NH) compressor and turbine shafts must be maintained in twin-spool engines. During certain phases of operation it may be necessary to trim or 'top temperature limit' the engine to avoid overheating and damage to the turbine blades.
- Reheat/nozzle control – reheat control is the metering of fuel to the reheat or afterburner section of the engine to increase engine thrust for take-off or combat; it is used almost exclusively for military engines with the exception of Concorde which uses reheat for take-off. Nozzle control assures the optimum nozzle area to maximise propulsive thrust for given flight conditions.

Now that the 'typical' military engine has been described it has to be stated that two of the foremost front line fighter aircraft in service with the Royal Air Force differ from the engine described. These are the RB 199 which powers the Tornado GR1, F-2 and F-3 aircraft and the Pegasus which powers the Harrier GR1 and GR5.

The Tornado RB 199 differs in having a triple-spool engine. Engine control is exercised by a Main Engine Control Unit (MECU), a dual channel analogue control system produced by Lucas Aerospace. The MECU is fitted to the GR1

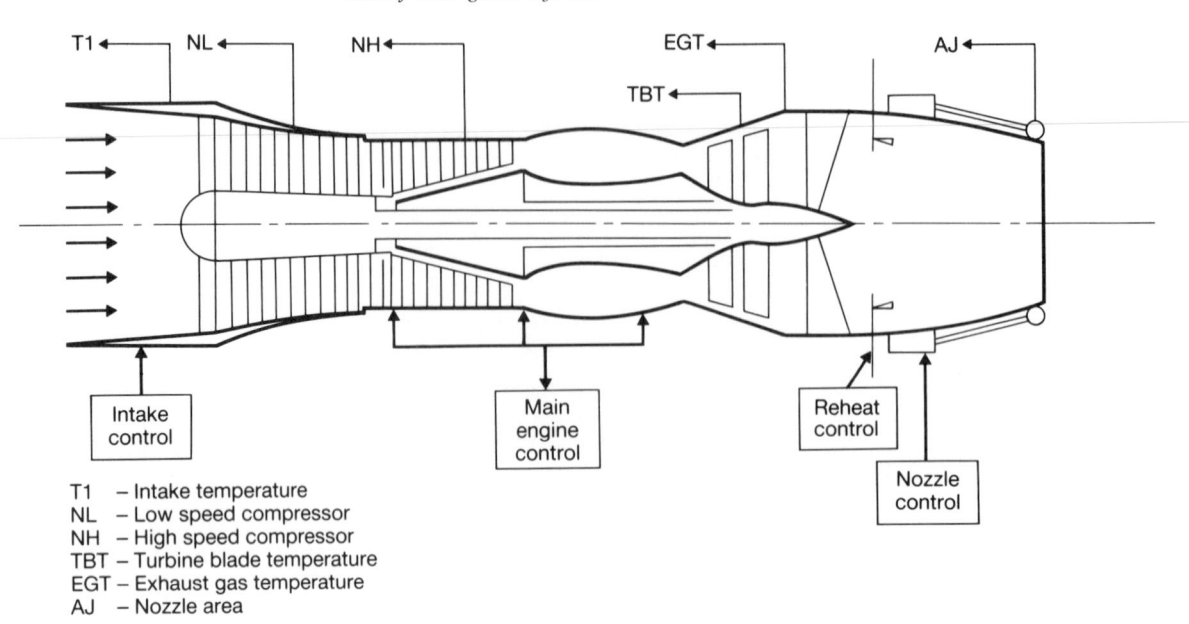

Fig. 6.2 Typical engine control for a military aircraft engine

T1 — Intake temperature
NL — Low speed compressor
NH — High speed compressor
TBT — Turbine blade temperature
EGT — Exhaust gas temperature
AJ — Nozzle area

and F-2 Tornado aircraft and performs well after initial problems. The MECU is not engine-mounted, but is located in a conditioned avionic bay forward of the undercarriage. The improved environment of the conditioned bay as opposed to being mounted on the engine improves the reliability of the unit because of the lower ambient temperatures and a more sympathetic vibration environment.

During the late 1970s and early 1980s, the UK developed a digital control unit which is a form, fit and function replacement for the MECU. This unit is called a Digital Engine Control Unit (DECU) and it is fitted to the later Marks of RB 199 installed in the Tornado F-3 and the Experimental Aircraft Programme (EAP) demonstrator aircraft. This system employs a high integrity dual channel digital controller with a third simplex channel controlling reheat.[1]

The Pegasus is a twin-spool engine with the spools rotating in opposite directions to balance out torque effects in hover mode. It does not employ reheat, so far, owing to the difficulty in establishing stable re-heat combustion conditions with the four vectored nozzles.

The latest versions entering service on the GR 5 (AV-8B) employ a Dowty & Smiths Industries Controls (DSIC) Digital Engine Control System (DECS). This unit consists of a dual lane system in which each lane utilises dual 16 bit microprocessor control – one channel being in control while the other acts as a monitor. This is known as a dual-dual configuration (dual lane/dual channel) and it clearly serves to improve integrity. A failure in the active control channel will be detected by the monitor channel and control will be switched to the other lane. Full control may therefore be exercised without any degradation in performance. Finally, in the unlikely event that a second failure renders the second lane unserviceable, a simple mechanical reversion exists to act as a 'get-you-home' system. A further difference between the MECU/DECU fitted on Tornado and the DECS is that the latter is engine-mounted rather than being located in an avionic bay. The high temperatures experienced in the engine bay necessitate the use of fuel cooling to maintain the electronic components at a respectable temperature.

The apparent differences in the integrity and reversion philosophy between the RB 199 and the Pegasus are explained by the fact that the Tornado is powered by two RB 199s while the Harrier is powered by only a single Pegasus. Both the RB 199 DECU and the Pegasus DECS are full-authority digital engine control (FADEC) implementations.

The future trend for military aircraft engine control units is to embed a MIL-STD-1553B data bus remote terminal which allows communication with the rest of the aircraft avionic and aircraft systems. The European Fighter Aircraft (EFA) in Europe and the Advanced Tactical Fighter (ATF) in the US are almost certain to adopt this solution by way of increased systems integration. Other advanced developments include the demonstration of integrated flight/propulsion control on the F-15 STOL Manoeuvre Technology Demonstrator described in Chapter Five.

The high by-pass turbofan engines presently used by most of the modern airliners represent a different control problem to that posed by the high performance military fighter. The main differences are that the civil engine will generally have a higher by-pass ratio and will have no need for intake and reheat control. The emphasis for the civil operator is primarily one of fuel economy. A typical high

Fig. 6.3 Typical engine control for a civil aircraft

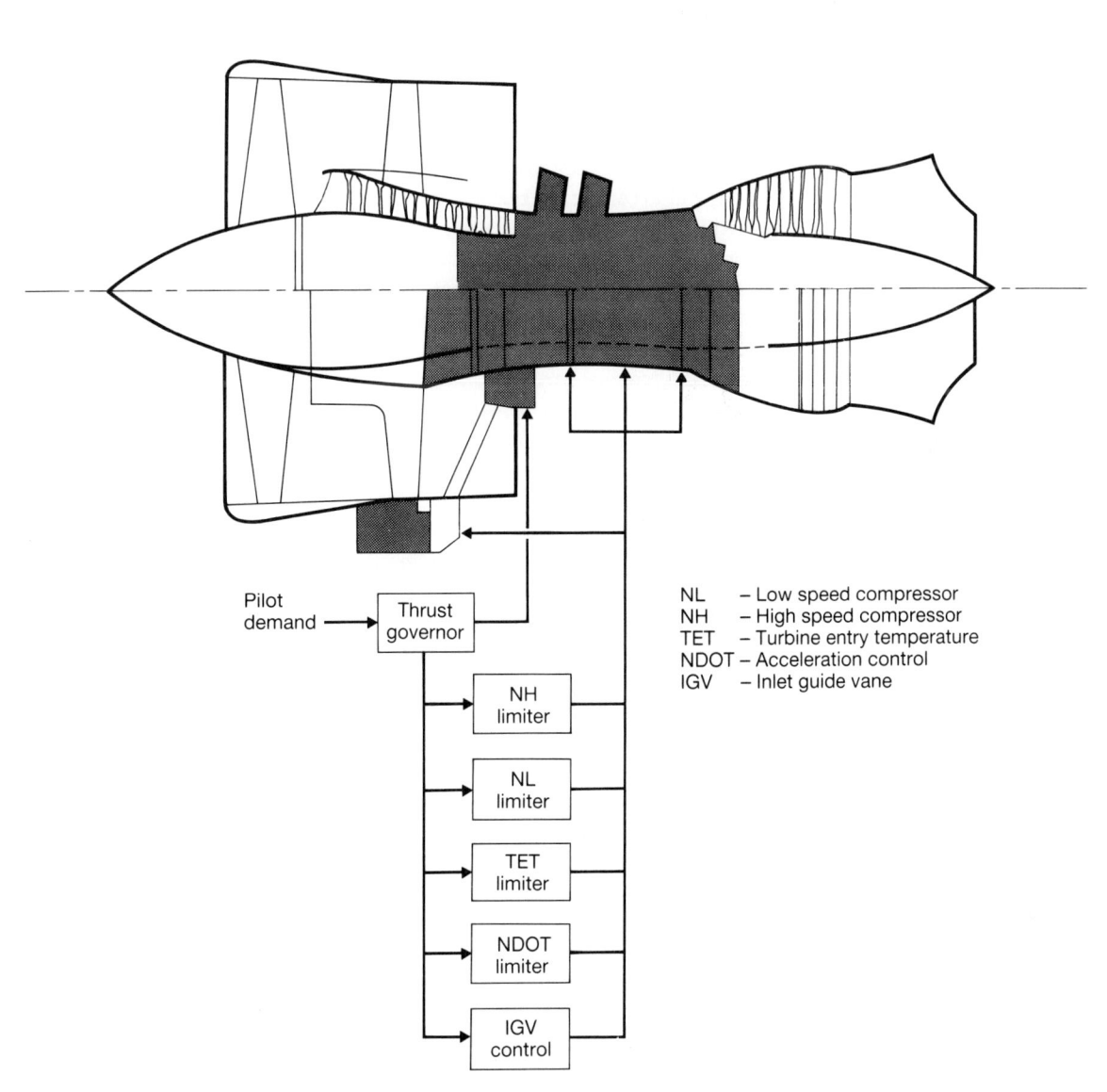

NL   – Low speed compressor
NH   – High speed compressor
TET – Turbine entry temperature
NDOT – Acceleration control
IGV – Inlet guide vane

by-pass engine together with the associated control parameters is shown in Fig. 6.3.

The pilot input power lever is used to apply a power demand into the engine control system. This demand is effectively aiming to control LP shaft speed. This demand may be affected or modified by a number of limiters which alter it to ensure that certain engine parameters do not exceed specific limits. An HP compressor (NH limiter) will prevent NH from falling below set limits. Conversely, LP compressor (NL) and HP turbine entry temperature (TET) limiters will avoid excess temperature and prevent damage to the LP compressor and HP turbine respectively. The purpose of the modified power demand is to effectively exercise NDOT ($\dot{N}$) or acceleration control. This demand will be mixed with the engine fuel schedules; it may also be subject to NH and intermediate shaft (HI) limiters. Certain engines may also exercise control over inlet guide vanes (IGVs) though this control is different in nature from the air intake control used on fighter aircraft.

The move towards FADEC for civil engines has been more cautious than for military engines. However it has been recognized that FADEC systems are lighter that their electro-mechanical predecessors. Also by employing a FADEC system it is far easier to modify the engine performance by means of software changes. During the development of the Pratt and Whitney PW2037 engine used on some Boeing 757s it was common to make minor changes in software from one day to the next, and to completely rewrite (re-compile) the control software within a week. Furthermore, the electronic control system has access to far more data than was previously possible thereby making development testing a quicker learning experience than before. This additional data may also be used for engine health monitoring and to ease the maintenance burden once the aircraft has entered service.

The Hamilton Standard unit fitted to the PW2037 is said to have a mean time between failure (MTBF) of over 30,000 hours, equivalent to about ten years' operation in a typical transport aircraft. Also the in-built redundancy permits the unit to continue to operate once a single failure has occurred. In the case of the PW2037 the US Federal Aviation Administration (FAA) has decreed that the control unit may be operated for a further 150 hours after a single failure, before the unit must be replaced. This allows the airline the flexibility of flying the aircraft to a main base, locating a spare or carrying forward the defect until another servicing activity before the item is replaced. The use of BIT aids diagnosis and it is claimed that the Pratt and Whitney FADEC may be replaced in around 15 minutes as opposed to three hours for the hydro-mechanical equivalent.[2]

New propulsion devices such as the Un-Ducted Fan (UDF) and the V2500 Super Fan will place further demands upon the use of electronic means for the solution of the engine control problem. The search for better controlled, more fuel efficient engines will be a powerful stimulus towards achieving more cost-effective, profit bearing operations.

FLIGHT/PERFORMANCE MANAGEMENT

The development of digital technology introduced the possibility of using computers on the flight deck to improve navigation, aid fuel efficiency and reduce crew workload. It became possible to use computers to guide the aircraft along complex routes using lateral guidance. It also became feasible to include vertical guidance to calculate optimum cruise altitudes and to determine the best combi-

Flight performance management

Fig. 6.4 Typical airline route section

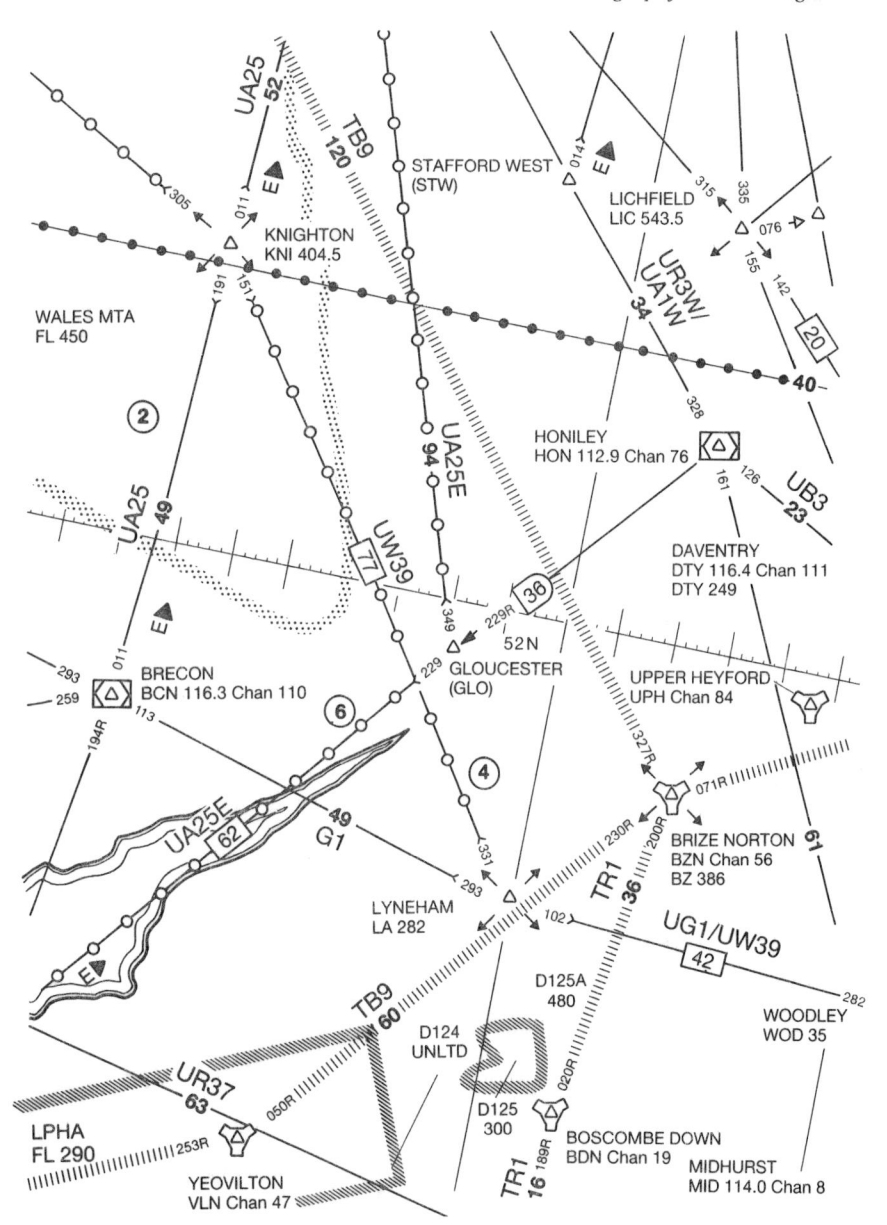

nation of autothrottle setting and speed during climb and descent. An example of a typical route section is shown in Fig. 6.4.

The first system to attempt flight management was that designed by Lockheed in association with Arma for the Lockheed L-1011 TRISTAR. This system used a combination of radio and inertial sensors for lateral guidance but possessed only limited vertical guidance.

The recognition of the need for a flight management computer system (FMCS) led to the definition of this type or system as ARINC 702. The basic architecture of a typical system is shown in Fig. 6.5. The computing tasks are undertaken by a Flight Management Computer Unit which interfaces via ARINC 429 data links to a range of aircraft sensors which will usually include the following:

Aircraft Management Systems

Fig. 6.5 Flight management control system (FMCS) – block schematic (*Note:* H = High rate L = Low rate ARINC 429) *Royal Aeronautical Society*

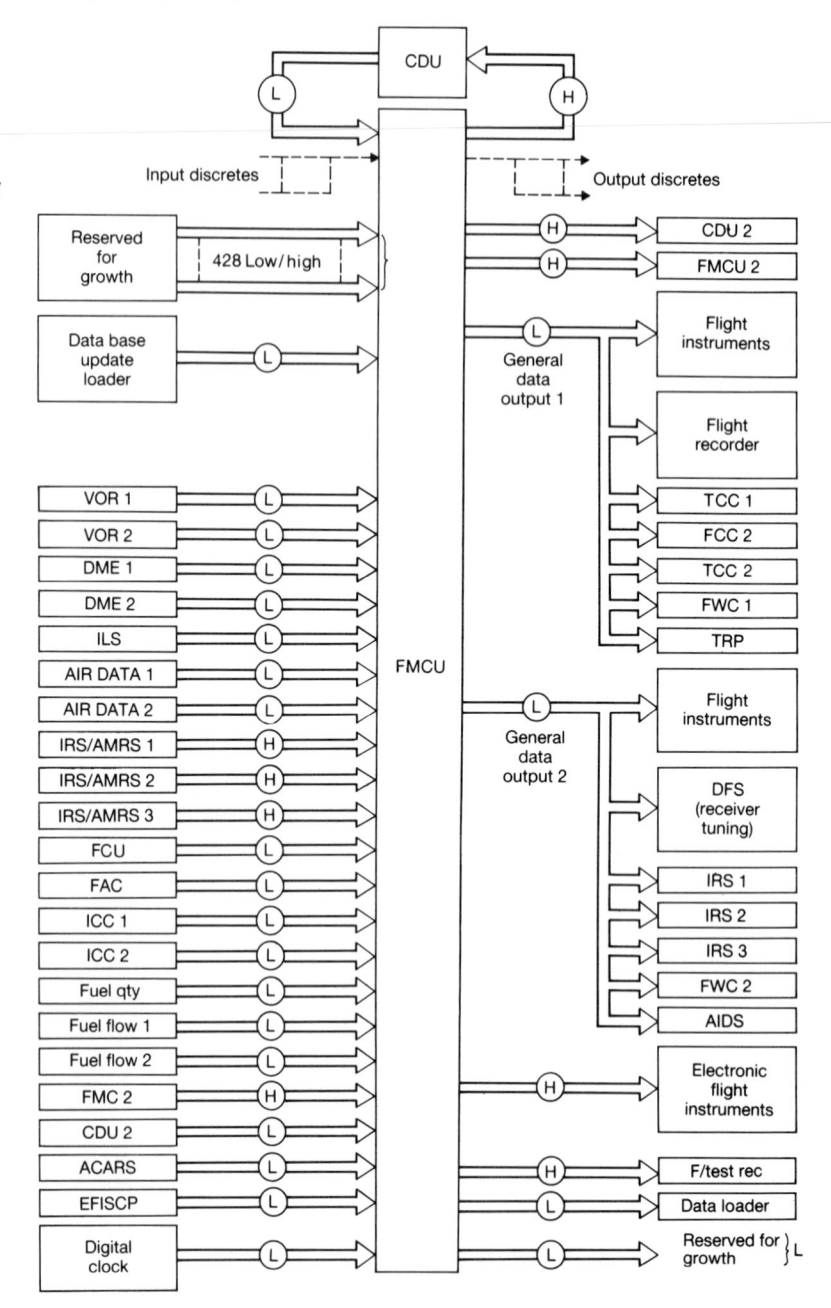

Inputs:
VOR 1 and 2
DME 1 and 2
ILS
Air Data 1 and 2
Inertial reference set/attitude & heading reference set 1, 2 and 3
Fuel flow/fuel quantity
Data loading (data base renewal or modification)

Outputs:
Flight instruments
Flight recorder
Inertial reference set
Flight control computers
Flight warning computers
Electronic flight instruments
Other EMCS (if a dual installation is fitted)

Flight performance management

Inputs:
Other FMCS (if a dual installation is fitted)
Chronometer

The system conveys information to and receives information from the crew by means of a control and display unit (CDU). A typical CDU – that used on the Smiths Industries FMCS is shown in Fig. 6.6. The display comprises a CRT

Fig. 6.6 Control and display unit (CDU) of the Smiths Industries flight management system which provides primary control and display interface between the pilots and the flight control and navigation systems of the Airbus. *Smiths Industries.*

display, alpha-numeric keyboard, dedicated function keys, and alongside the CRT, line keys to give access to lateral (left keys) and vertical (right keys) guidance information.

The primary functions of the FMCS are defined as follows[3]:

- preparation of flight plans including diversion to alternative airfields;
- synthesis of navigation information from all available sources;
- automatic frequency selection and tuning of VOR/DME receivers;
- lateral guidance by coupling the FMCS to the automatic flight control system;
- optimization of the flight path in the vertical plane to achieve minimum cost or shortest flight time;
- guidance in the vertical plane by means of coupling the FMCS to the auto-throttle and automatic flight control system;
- prediction of flight parameters along the aircraft route;
- display management by coupling the FMCS to the electronic flight instrument system (EFIS).

The Smiths Industries FMCS uses a distributed architecture of the type shown in Fig. 6.7[4]. Performance and navigation functions are each undertaken using a dual INTEL 8086 configuration used in conjunction with an INTEL 8087 maths co-processor plus the associated program memory (UV PROM) and working memory (CMOS RAM). The magnetic bubble-based data bank memory is managed by an INTEL 8088 and a further INTEL 8088 is used as an ARINC 429 input/output controller.

Fig. 6.7 FMCS – computing architecture *Royal Aeronautical Society*

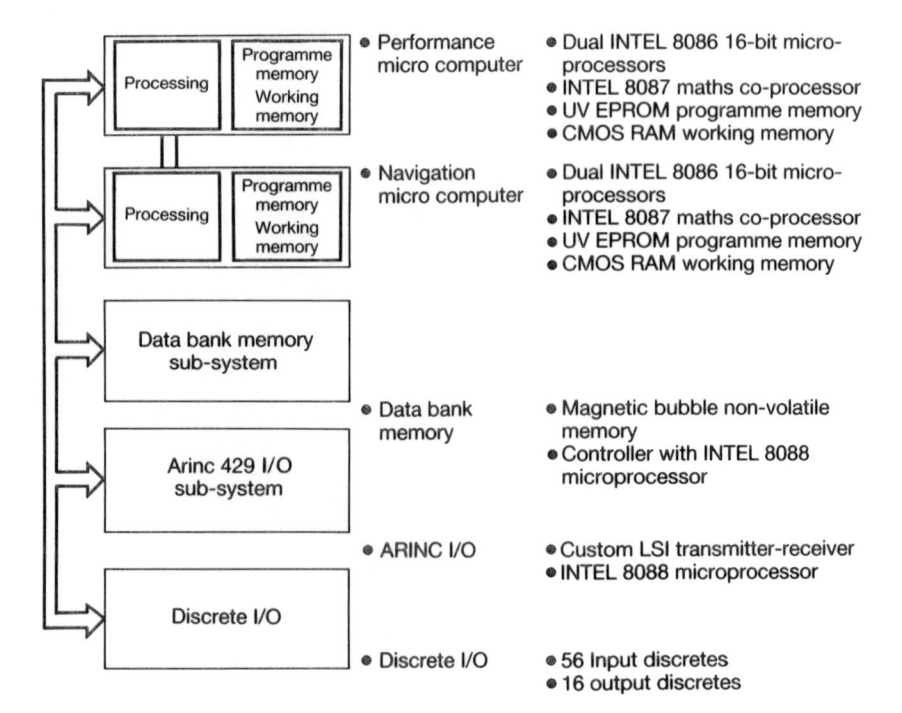

The Electrically Programmable Read Only Memory (EPROM) hosts the operational program for the system and cannot be changed without removing the memory devices from the unit. The Random Access Memory (RAM) provides a scratchpad facility for dynamic data and calculations. The bubble memory is a memory type which has the capability of retaining the aircraft, engine and route data in a non-volatile form (that is the memory contents are not lost when power to the unit is removed). The bubble or data bank memory is the only one which may be altered on the aircraft. This is necessitated by the fact that the data bank has to be updated every 28 days in order to keep navigation data up to date.

As the system is powered up, the crew will be given confirmation of the aircraft and engine type, program data base details (identification) and an indication that the data base is still valid. Operation of the start key will enable the flight plan to be constructed with the insertion of route particulars: alternative airfield, wind velocity, temperature, cruise altitude and others. Insertion of the start and destination airfields will enable the crew to select a stored route or enter a new one by means of airways designators and waypoints. In most cases it is likely that a stored (airline preferred) route will be used with the crew having the option of selecting alternative airfields based upon weather forecasts in the destination region. At this stage insertion of precise aircraft position at the departure gate will enable the latitude and longitude to be inserted into the inertial reference system to commence platform alignment. Further checks by the crew will ensure that aircraft fuel load, centre of gravity position, fuel allowances, reserve fuel, take-off and landing weight are all correctly entered and within prescribed limits. Once this is complete the crew may review the entire plan for correctness and accept it as valid. If required the crew may also present information relating to the departure airport such as active runway identification, runway bearing and length, primary approach aid frequency and so on.

Once lined-up on the departure runway ready for take-off the crew will then be able to call upon a range of displays for different situations some of which are summarised as follows:

– Take-off performance – data may be displayed which relates to take-off including V1, VR, V2; flap and slat retraction speeds; climbing speed, noise abatement height, acceleration height and normal and engine-out departures.
– Departure/climb routeing – during climb-out the estimated (and actual) times of arriving at waypoints, distance to a waypoint, expected flight level at a waypoint, etc., may be displayed.
– Performance – performance may be determined according to a number of criteria, perhaps the two most obvious being maximum performance and economy performance. In the cruise, options may be given together with predictions of destination arrival time and fuel state. This may indicate the desirability of climbing to a new flight level after a certain distance or time flown. Similarly during the descent, a range of options may be presented showing the effects of an early or late descent and an indication of any slight alterations in power settings necessary to maintain optimum performance.
– Fuel management – the aircraft fuel state and the effect upon routeing, holding in a pattern or on diversion options can be displayed to the crew. Fuel consumption may be compared with that which was predicted and extra (or lesser) holding time calculated and displayed.
– Engine-out performance – should an aircraft engine need to be shut down for any reason, the system has the facility of presenting vital information to the crew regarding performance (engine-out ceiling), local terrain information relative to that ceiling and any guidance information which may be necessary to avoid high ground. This enables the crew to proceed with confidence that the aircraft performance is adequate for the terrain being overflown.

- Descent/approach phase – similar information will be available to the crew during the descent mode to that existing during climb and cruise. The aircraft may be controlled down to touch-down in accordance with the standard approach procedures for the airport. This mode will also assist the aircraft in engaging an approach aid such as the ILS localiser. The approach phase will also display to the crew target values of airspeed, power setting and rate of descent to achieve the necessary approach speed when established on the glideslope. Wind velocities along and across the runway may be calculated and displayed.
- Go- around performance – in the event of a missed approach the pilot may be presented with information relating to the overshoot; this is likely to be similar to that displayed during take-off. Lateral guidance information will provide the crew with the necessary data to initiate a second approach or to commence a diversion to an alternative airport.

The development of the FMCS was a new experience for the manufacturers producing this type of system. It is fair to say that most of them grossly underestimated and then struggled with the size of the software development task. Not the least of the problems is the enormous data base which is required, bearing in mind that a particular aircraft type in an airline may be required to fly a wide range of routes, each with particular navigation and airport information. This problem is compounded by the fact that the data base needs to be updated every 28 days to maintain its currency. This in turn generates an on-going software update and maintenance task of significant proportions. However against these difficulties, the continuous pressure to extract additional performance coupled with economy and the need to reduce crew workload mean that the FMCS will be more widely used and enhanced in forthcoming years. For the experiences of a supplier at the forefront of FMCS development see Ref[4].

UTILITIES MANAGEMENT SYSTEMS

The term utility system or utilities systems is somewhat loosely applied to those airframe systems not associated with avionics or displays, flight control or weapons release. In practice these systems perform many of the house-keeping functions associated with flying the aircraft. Typical systems are:

- Fuel quantity indication
- Fuel management
- Hydraulic system control and indication
- Environmental control
- Secondary power system control
- Landing gear control and indication
- Wheel brakes and anti-skid
- Electrical power generation control and monitoring
- Engine and associated services
- Aircraft lighting
- Probe heating

Some of these systems are vital for the safe operation of the aircraft. Incorrect operation of the fuel system could cause a total loss of engine power; loss of hydraulic pressure could cause the loss of the aircraft no matter how sophisticated the flight control system. Total loss of electrical power would be disastrous for many modern aircraft. The importance of these systems is attested by the redundancy features inherent in system design. An aircraft may possess two, three or even four hydraulic systems depending upon the integrity required and the degree of dependence of aircraft airworthiness upon the correct functioning of these systems. Other systems may be duplicated to avoid inconvenience following system failure. Still others may only be simplex in implementation as their failure may induce only minor difficulties. In practice, for a passenger-carrying aircraft these systems are likely to be at least dual or standby redundant, and possibly triplex or quadruplex redundant.

Control of these systems has previously been exercised in discrete or independent sub-systems. In other words, each system has had its own dedicated control and indication panels, switches, wiring control units, etc. While this has been acceptable in the past the moves towards modern avionic systems are leading to greater integration and more efficient use of electronic resources. The developments of the modern electronic or 'glass' cockpits are causing major rationalisation in cockpit design with the former cluttered cockpits increasingly inclined to use electronic flight instruments systems (EFIS) and flight management computer system (FMCS) displays. The disappearance of the flight engineer from the crew complement of many modern airliners has removed the need for the conventional engineer's panel with banks of instruments. This has placed a greater reliance on cockpit automation to keep pilot workload within reasonable limits. Similarly the imposition of greater workloads in the modern combat aircraft is dictating that tasks and decisions regarding aircraft systems must be increasingly removed from the crew. These tasks and functions require centres of intelligence which allow many tasks to be undertaken on behalf of the aircrew without any knowledge or intervention on their part.

Many of these conventional control systems require immense quantities of wiring to interconnect the control units, displays and components. Wiring is heavy, requires a large number of connectors and is expensive to build and install in the aircraft. Therefore aircraft manufacturers are increasingly turning towards multiplexed data buses as a way of reducing the weight and quantity of wiring and connectors. Typical data buses in common usage are ARINC 429 in the civil field and MIL-STD-1553B in the military; these data bus standards are described in more detail in Chapter Two. It has been reported by the Boeing Company that a weight saving of around 2000 pounds might be possible if multiplexing techniques were widely used on a Boeing 747. Experienced aircraft engineers are well aware of the disproportionate number of problems caused by poorly mating aircraft connectors. Recent work by the US Air Force Wright Aeronautical Laboratories has suggested that around 40 per cent of the defects incurred by a modern aircraft avionic system may be attributed to connector problems. Whatever the savings, clearly they will differ from one aircraft to another for a variety of reasons; there is little doubt that these advantages are seen as increasingly attractive.

The use of redundant data buses (MIL-STD-1553B) and multiple processing elements assists in system redundancy and availability. For the civil operator this means increased flexibility in operating the aircraft. Modern electronics are less likely to fail in the first place; the better Built-In-Test (BIT) that digital systems offer also result in better diagnosis when a failure does occur. Better redundancy allows the operator greater flexibility in returning the aircraft to a main operating base for repair when a fault occurs.

For the military operator, improved reliability increases the chances of the aircraft being available to conduct a specific mission. In military parlance this is called a 'force multiplier effect'. Furthermore, the availability in combat of redundancy increases the battle damage tolerance of the aircraft, allowing it to return to base with damage which might previously have caused its loss.

The use of multiplexed data buses and distributed processing elements also lends the system naturally to the Health and Usage Monitoring System (HUMS) described later in this chapter. There is a ready access to a wide range of system parameters and better monitoring of these signals can lead to an improvement in flight safety and airworthiness.

These pressures and advantages are therefore causing airframe manufacturers increasingly to consider the integration of the aircraft utility systems into an integrated Utilities Management System (UMS). Whereas this type of system has initially been considered for combat aircraft, serious consideration is being given to applications on civil aircraft. The first system of this type is flying on the UK Experimental Aircraft Programme. The system has flown many hours during a most rigorous test programme and has been acclaimed by development pilots and groundcrews alike for the reduction in workload and ease of diagnosis and maintenance.

As this system is relatively new, and as the development is so typical of the fusion of many disciplines required on modern development programmes, a joint account of the development seen from the eyes of the senior British Aerospace and Smiths Industries engineers involved is adduced[5].

The utility systems of a combat aircraft are those which are responsible for the continuing and safe operation of the aircraft rather than mission critical avionic systems or those concerned with flight control and engine control. Over a period of time these predominantly mechanical systems have incorporated electronics to achieve increasing accuracy of control, better performance, improved monitoring and failure detection.

In the late 1970s when successors to Tornado were being sought it appeared from a study of customer requirements that a small high performance, highly manoeuvrable aircraft was required. Taking account of the small airframe associated with such an aircraft, and the emergence of cheap and readily available digital computers, it was felt that smaller, lighter electronic equipment would be required and must be available within the timescale of a new project.

The use of serial multiplex data transmission for avionic systems and the new techniques of cockpit display and control presentation at the time was leading to major discrepancies between the control philosophy of utility systems and the techniques used in avionics. This is illustrated in Fig. 6.8 where the discrepancy between the many single-function control units and the data bus connected displays may be seen.

In addition the customer was seeking major improvements in performance, reliability and maintainability to reduce the cost of ownership.

The utility systems of the aircraft pose special control problems because of their inherent 'single' nature. They do not fall into the category of information processing systems such as avionics nor of the high integrity servo control, multiple redundant category of flight control. Falling somewhere in between, they are a collection of systems with varying integrity requirements, some simplex in nature, others dual redundant or duplex. A feature of the systems is the large number of hardware derived inputs and the need to switch power to many output devices.

British Aerospace embarked on a pilot study to examine the feasibility of centralising the control and management of utility systems. As a result of this initial work a programme was initiated with MoD funding to examine utility systems as a total concept to see if a universal solution could be achieved. This study started by considering the control of utilities on an in-service aircraft to determine what shortcomings existed.

Fig. 6.8 Problems inherent in interfacing with the modern cockpit
Royal Aeronautical Society

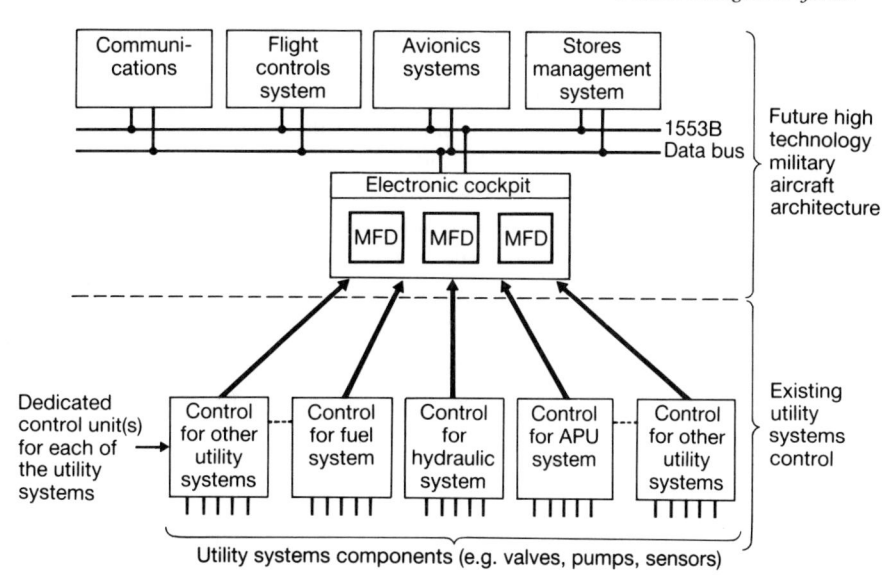

The study found that an individual approach involved numerous design engineers sometimes from different departments, which resulted in a lack of consistency and in interface problems where systems had to inter-communicate. Control units were separately specified. A large number were required. Utilization was low and the cost of installing the equipment was relatively high.

It was noted that some 25 items of equipment were used to control the utility systems on one particular aircraft. All of these systems were single-function. For example the equipment used to start the engines was used for only two minutes per flight. Units were installed in the airframe wherever a space could be found and connected to their sensors, other equipments, and the cockpit with bundles of discrete wiring which was heavy, space consuming, costly and vulnerable to external radiated fields.

The study proceeded by visualising the control of the set of utility systems as a single control problem, as shown in Fig. 6.9, to which a single solution would be pursued to meet the following requirements:

Fig. 6.9 Solution to interfacing with the modern cockpit
Royal Aeronautical Society

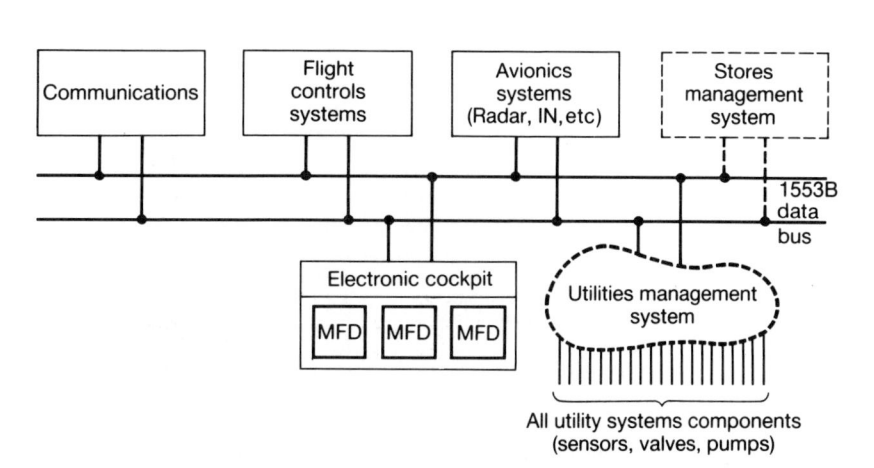

Aircraft Management Systems

- To provide a means of controlling and monitoring for the set of utility systems
- To provide sufficient automation to reduce pilot workload whilst maintaining the integrity of the individual systems
- To reduce the aircraft wiring and to provide a suitable interface with the data bus
- To reduce the number of items of equipment required and to minimize their mass, volume and dissipation
- To capitalise on technology to provide improved monitoring of the system's performance by use of continuous built-in test.

In the work which followed alternative system architectures were examined and the resulting solution is shown in Fig. 6.10. This shows an interface with an

Fig. 6.10 Simplified EAP architecture *Royal Aeronautical Society*

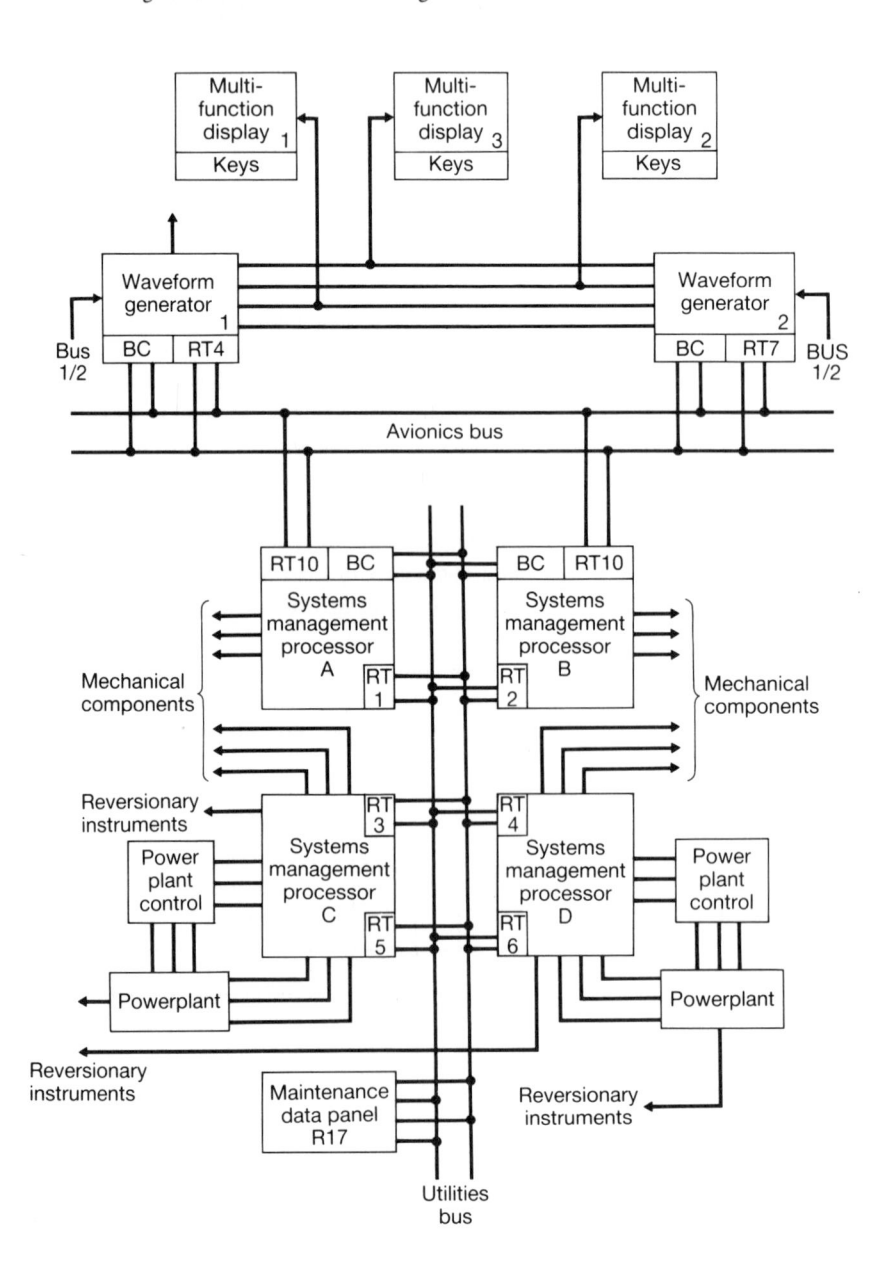

Fig. 6.11 Comparison of conventional and integrated utilities control *Royal Aeronautical Society*

Conventional control system

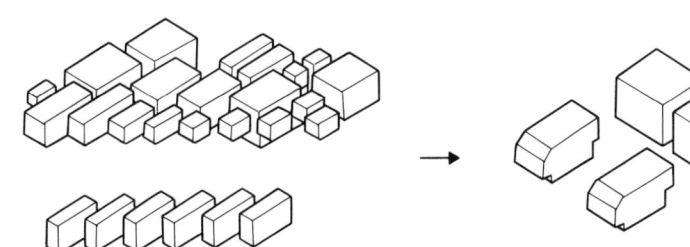

Features
– 20 to 25 Dedicated LRU's
– 6 Power switching relay units
– Extensive overhead of wiring, connectors, mounting trays
– Dedicated instruments/display panels

Utilities system management

Features
– Four dedicated LRU's
– Dedicated 1553B data bus
– Distributed intelligence
– Integral power switching devices
– Interfaces with modern digital cockpit and multi-function displays

avionics data bus and the displays suite. The intention was that the solution should comprise a number of equipments inter-connected on a MIL-STD-1553B data bus. This system would be located in the airframe so that inter-connecting wiring would be reduced to a minimum. The equipments were to be designed to provide suitable interfaces for input and output devices for the data bus and to provide a means of control using a microprocessor. Figure 6.11 illustrates the key differences between conventional control of utility systems and an integrated systems management approach.

EAP SYSTEMS

British Aerospace at Warton adopted a Utilities Management System as one of the prime areas of high technology demonstration for introduction into the EAP (see Chapter Five).

The programme objectives of the EAP were to demonstrate the following key areas of technology:

– Carbon fibre composite (CFC) structures
– Active control technology (ACT)
– Advanced avionics including multi-function colour displays, data buses and UMS
– Advanced aerodynamic design.

The major utility systems of EAP are listed on Fig. 6.13 which also shows the location of the four Line Replaceable Units (LRUs) which control them. These LRUs are termed Systems Management Processors (SMPs).

The system block diagram is shown in Fig. 6.12. In it the four LRUs are shown connected to a dual redundant data bus. Bus control is located in SMP A and SMP B.

Also connected to the data bus is a maintenance data panel (MDP) and two computers of the flight control system (FCS). The bus is terminated at each end and a spare stub is brought out to the panel of the MDP to allow connection of test equipment to monitor bus traffic.

Aircraft Management Systems

Fig. 6.12 EAP – major utility systems *Royal Aeronautical Society*

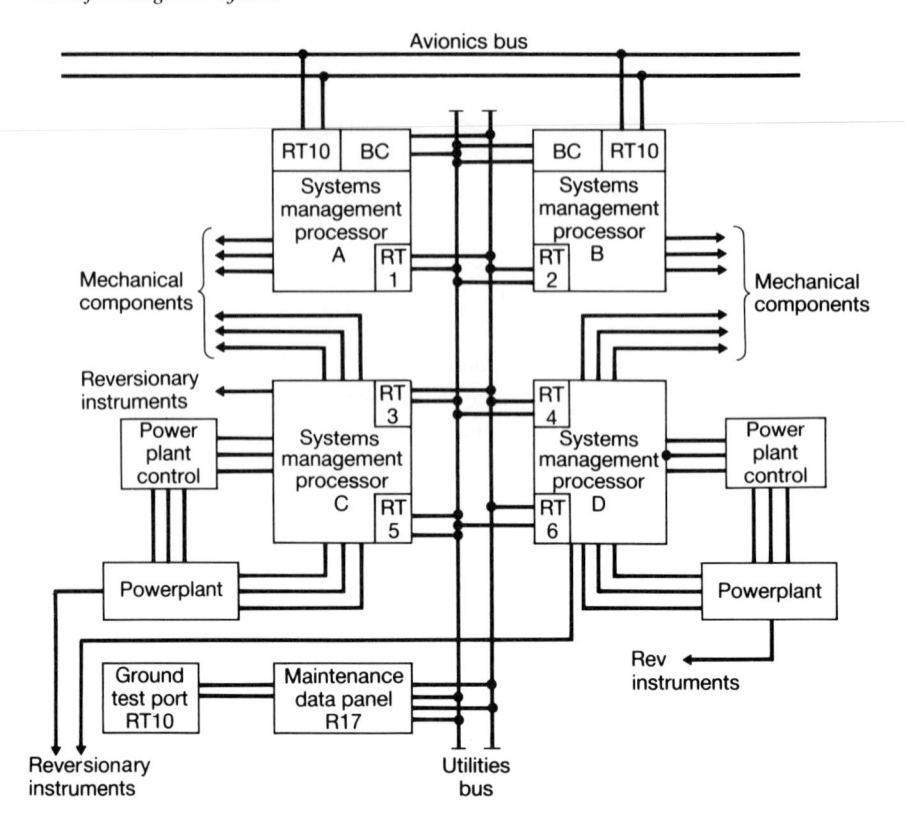

Fig. 6.13 EAP – utilities architecture *Royal Aeronautical Society*

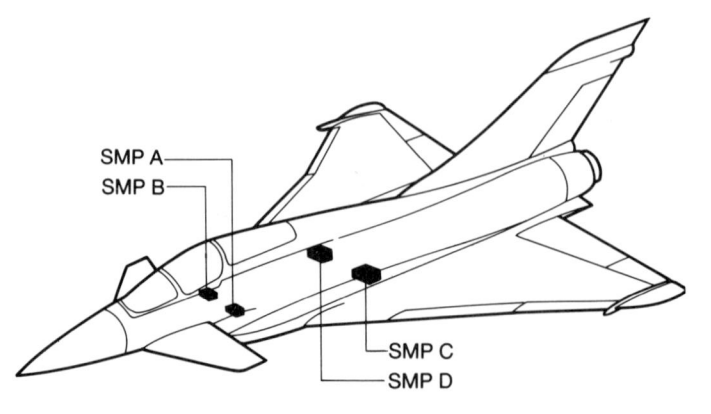

Main functional groups

- Engine control and indication
- Fuel management and fuel gauging
- Hydraulic system control and indication, undercarriage indication and monitoring. Wheel brakes
- Environmental control system, cabin temp. control
- Secondary power system
- Lox contents, electrical generation and battery monitoring, probe heating, emergency power unit

The lines shown directed to mechanical systems are the multiple connections to the components of the systems. An indication of the numbers of signals and power drives is shown in Fig. 6.14.

The input/output characteristics of the utility systems are typified by a large

Fig. 6.14 EAP – summary of parameters *Royal Aeronautical Society*

- Engine and associated systems — 141 Signals/33 Power drives
- Fuel management and gauging — 206 Signals/36 Power drives
- Hydraulic systems — 146 Signals/13 Power drives
- ECS/cabin temperature — 59 Signals/14 Power drives
- Secondary power systems — 46 Signals/ 7 Power drives
- Miscellaneous systems — 34 Signals/ 7 Power drives

Total: 632 Signals
110 Power drives

number of discrete states: analogue inputs relating to rotational speed, linear and angular position, temperature, fluid level and pressure. In cases where a control loop is closed, analogue drives may also be required. The use of distributed microprocessors permits an almost unlimited capability of polling the large number of interface signals. Access to the aircraft systems data is therefore far greater than ever previously possible, allowing greater scope for trend and system health monitoring. The provision of maintenance data logging or recording systems interfaced with the utility systems data bus further increases this capability.

Typical input/output signals are:

Inputs:	Outputs:
Discretes	DC discretes
Potentiometers	DC power drives
Pulse probes	Low voltage analogues
Platinum resistance bulbs	High current servodrives
Thermistors	Low current servodrives
Pressure transducers	
AC synchro/resolvers	
Frequency modulation	

To illustrate how each SMP handles the data and performs its task a generalized SMP architecture is shown in Fig. 6.15.

Inputs from system sensors are acquired on a number of interface cards segregated according to system function to maintain inter-system separation. Signals are acquired from position sensors, pressure and temperature sensors, etc., and are used by the CPU to perform its control algorithms.

Fig. 6.15 Generalised systems management processors (SMP) architecture *Royal Aeronautical Society*

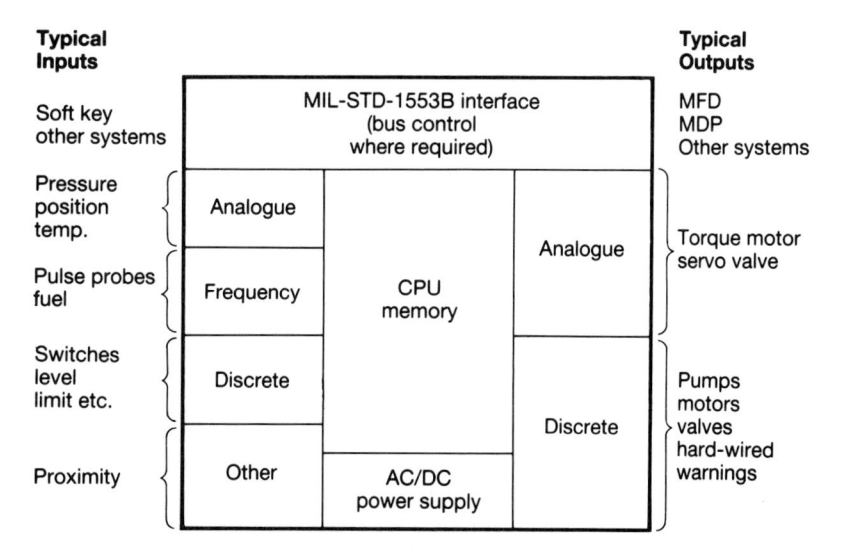

Aircraft Management Systems

Outputs to pumps, valves, actuators, etc., are switched by relays. Transistor drives are used to drive servomotors.

Inputs and outputs to data bus connected systems are routed through a remote terminal associated with each processor Segregation of the input/output circuitry associated with different utility systems was a prime aim. A situation where one interface card failure could affect more than one system, possibly with disastrous results, was not permissible.

Throughout the design, care has been taken to maximise fault tolerance. As well as the segregation of the systems input/output circuitry already mentioned it was also recognised that different systems required different levels of redundancy. Accordingly, Smiths Industries and British Aerospace worked closely together in the definition stage to ensure that these levels were met within the six distributed computing and data bus elements. The simpler solution of replicating all the computing independently for each system would have been more expensive and would have significantly increased system weight. The system availability and resistance to battle damage offer significant advantages over conventional systems.

A typical example of interfacing with the hardware of a major system is shown in Fig. 6.16. Here an exploded 3-D diagram of the fuel system is depicted, together with the main system functions. Also shown is the UMS hardware associated with the interfacing, control and indication of the fuel system. Figure 6.17 shows the EAP cockpit with the starboard display panel indicating how the fuel system information is portrayed to the pilot using data transferred from the utilities data bus to the avionics data bus and thence to the display suite. A further example is shown for undercarriage and brake systems in Fig. 6.18.

Fig. 6.16 EAP – Fuel system example *Royal Aeronautical Society*

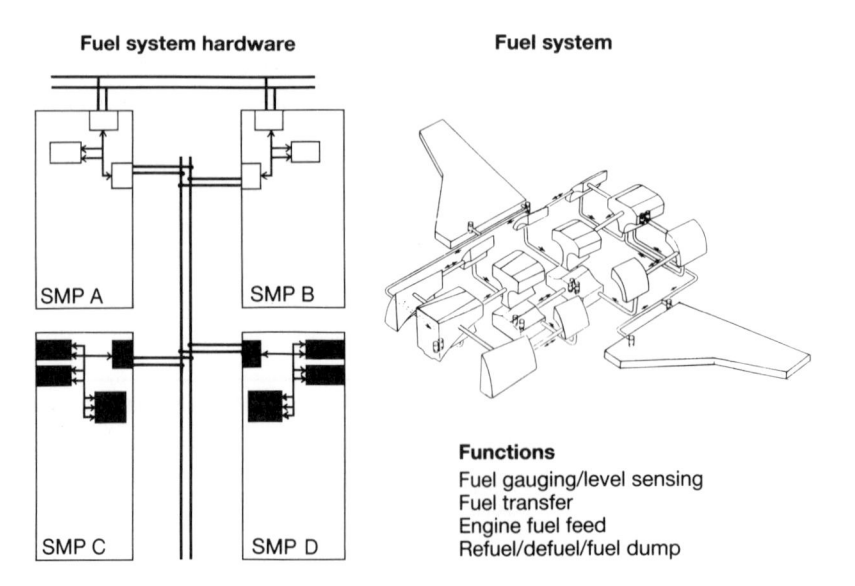

The steps in the design process and the relationships between the documents produced were controlled using a structured software design methodology. It was important at all times not only to achieve a very precise specification of hardware and software requirements, but also to ensure that both hardware and software were integrated precisely and, further, that interfaces with other systems were rigidly defined.

Software for system control was controlled rigidly from the beginning by the

EAP systems

Fig. 6.17 EAP cockpit layout showing engine display (port), attitude display, (centre) and fuel system display (starboard and inset). *Royal Aeronautical Society*

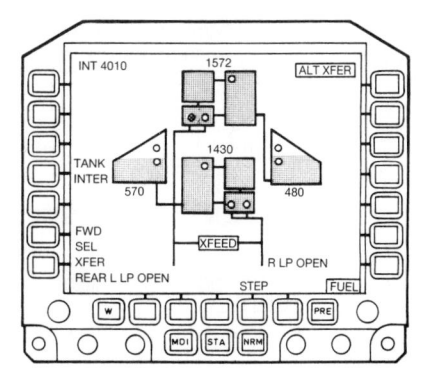

application of a formal method of obtaining requirements. This technique was known as Semi-Automated Functional Requirements Analysis (SAFRA). A key feature to this method was Controlled Requirement Expression (CORE) which specified system information paths.

The software for the systems was controlled entirely by British Aerospace in terms of specification of a tool set, integration and configuration control. After selection of suppliers for various systems' components and for some software from specialist suppliers the actual allocation of responsibility for production resulted in the workshare shown in Fig. 6.19.

In all some 400 K words of software are installed in the UMS which required the generation of approximately 190 K words of unique code. This code was produced fully documented from the specifications with a productivity of 4500 words per man-year. The number of changes produced at each stage of the software life cycle shows that one of the objectives was achieved – detecting the greatest majority of problems at an early stage of design and production to minimise cost and programme disruption.

135

Aircraft Management Systems

Fig. 6.18 EAP – undercarriage and brake system example *Royal Aeronautical Society*

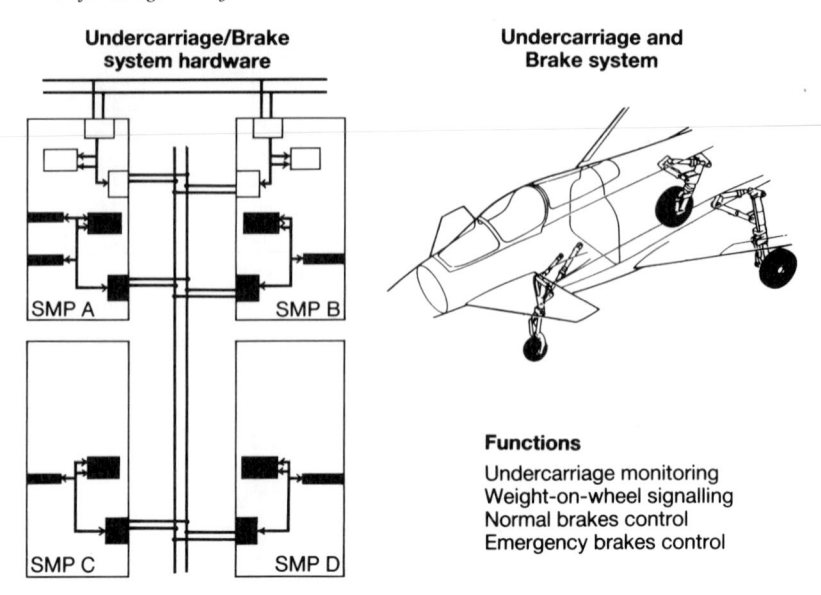

Fig. 6.19 EAP – software development workshare *Royal Aeronautical Society*

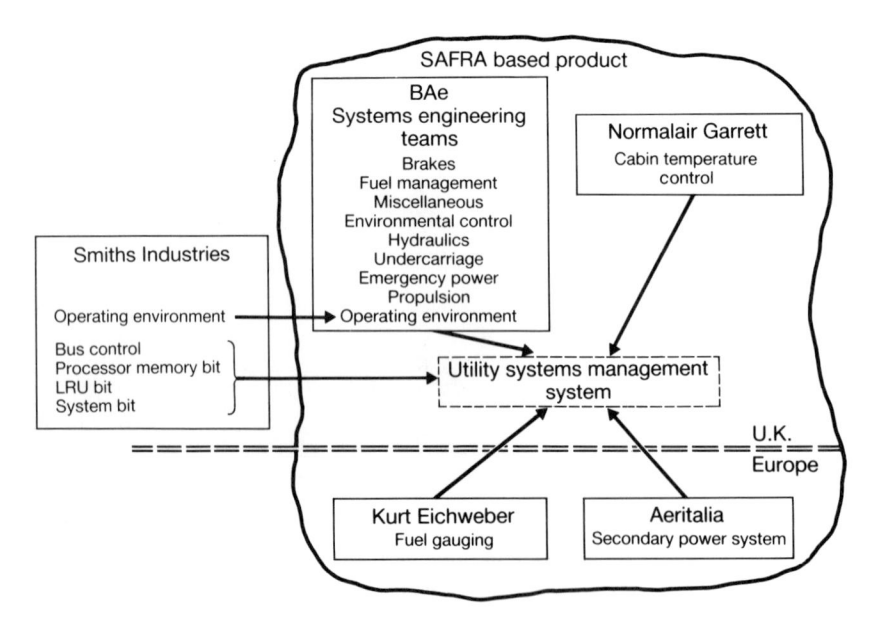

Testing was conducted at a number of stages in the life cycle. Initially testing was conducted in the host machine, then in the target in a complete set of brassboard hardware, in individual mechanical system rigs for some systems, and finally in integration test using aircraft hardware. This final test allowed all the integrated systems to be constructed in rig form and the entire systems operated on the ground to satisfy quality assurance requirements.

The hardware and software was then released to the aircraft and each utility system functionally tested prior to flight.

To enable this comprehensive test programme to be achieved, the following equipment was provided by Smiths Industries:

– Complete brassboard set of electrical model equipment
– Part-populated brassboard electrical model systems as part of sub-system test

rigs for hydraulics, environmental control systems and secondary power systems.
– Four complete flight capable sets of equipment.

The system hardware design was based upon a range of standard electronic cards and modules which were used in as many places as possible through the system. In this respect the approach was identical in conception to the PAVE PILLAR initiatives being pursued by the USAF. Out of a total of 72 cards utilised throughout the system, the requirements are satisfied by 15 different designs.

The use of common card designs in the various locations considerably reduced the design task and overall programme risk. For example, the Z8001 CPU card is used eight times within the system while the discrete I/O card is utilised in eleven different locations.

The basic building block for electronic packaging is the standard 3/4 ATR module shown in Fig. 6.20. This module comprises two 3/4 ATR electronic cards clamped together on to a mechanical wedgelock assembly which also provides the means of attaching the modules within the LRUs. Copper heat sink overlay on the surface of the card, together with the wedgelock assemblies, ensures a low thermal resistance path to outside walls or cold walls. Relays were utilised in preference to solid state power switching devices for discrete power switching purposes.

Fig. 6.20 EAP – standard electronic module *Royal Aeronautical Society*

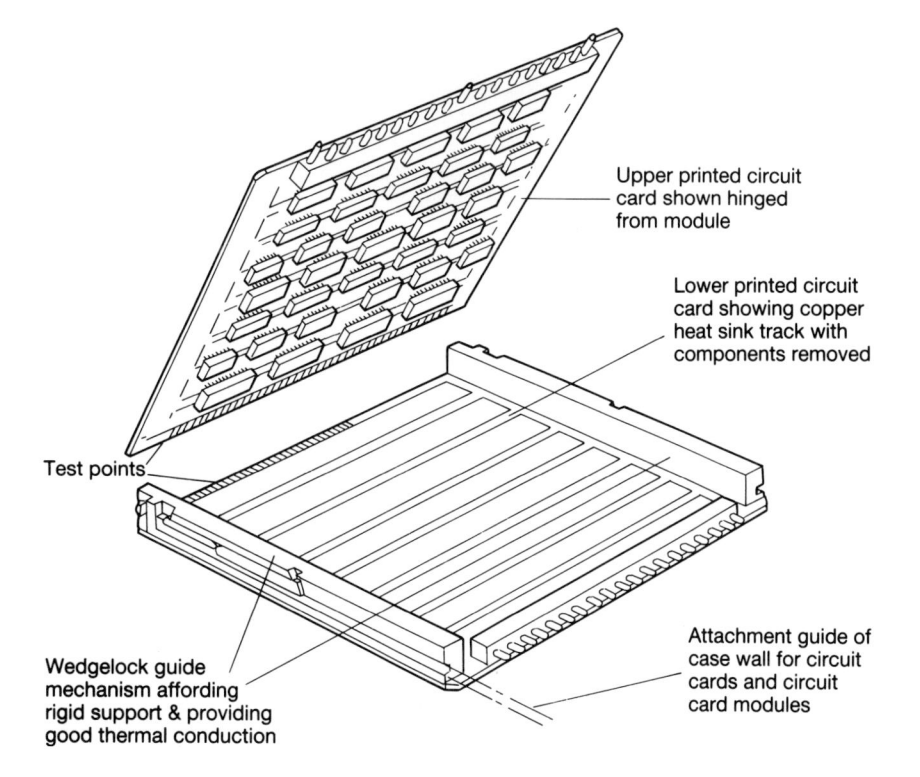

While the choice of relays for a high technology demonstrator may be surprising, several practical considerations dictated their selection.

Figure 6.21 is a photograph of one of the front LRUs (SMP A or B). Access to 10 electronic modules is from the top of the LRU. The AC power supply unit is located at the front of the LRU, the DC power supply unit at the rear. The aircraft interfaces connect via a quadruple DPX connector at the rear of the LRU.

Aircraft Management Systems

Fig. 6.21 EAP – line replaceable unit SMP A/B *Smiths Industries*

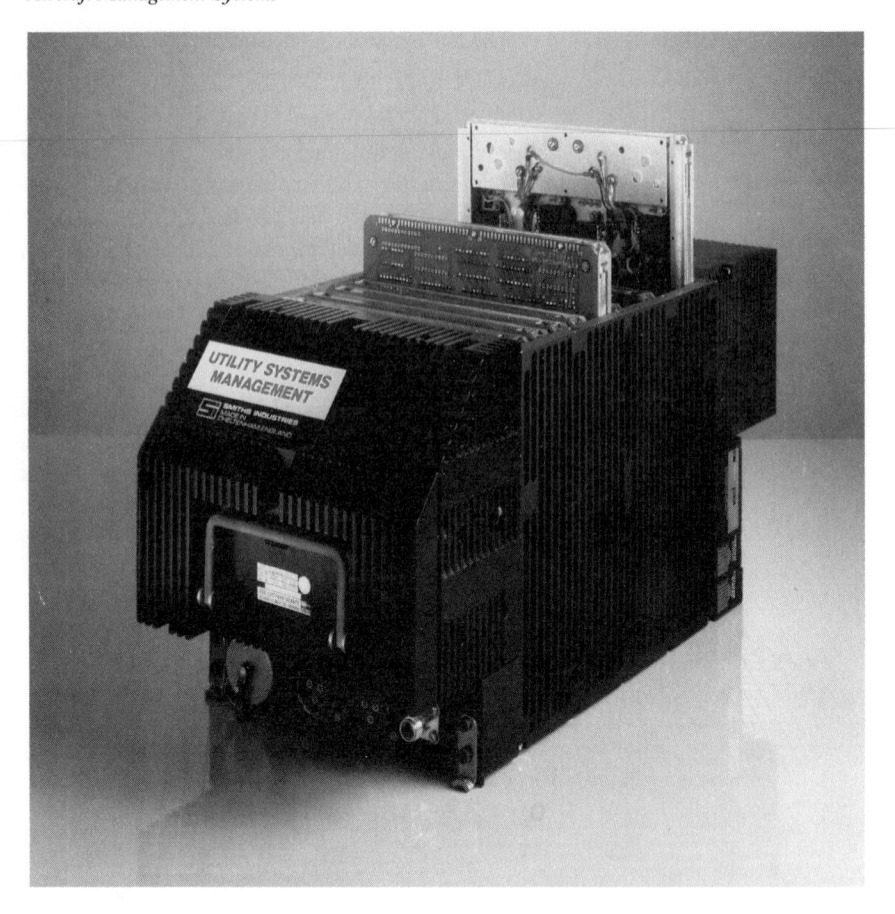

The data bus connectors, test connector, bond connector and hold-downs are positioned on the front face. The unit is cooled by means of an external air-wash over the external finned surface. The LRU weight is 12 kg with a primary power dissipation of 140 watts.

Figure 6.22 shows one of the rear LRUs (SMP C or D). Access to the 12 electronic modules is from the base of the unit. The power control enclosure forms the top of the unit, and AC and DC power supply units are located on the right hand and left hand sides respectively. As with the front LRUs, the aircraft systems connectors are located at the rear, while the data bus, test, bond connectors and hold-downs are located on the front face. The LRU is cooled indirectly by forced air entering through two cooling inlets at the base of the unit and exhausting from two outlets on the front face. These cooling air ducts form cold walls for the electronic modules, and for each of the inner faces of the two power supply units. The LRU weight is 20 kg with a primary power dissipation of 230 watts.

The reduction of the number of LRUs associated with the control of utility systems offers obvious weight savings for the total system. The utilisation of modern micro-circuit technology including data buses, microprocessors and very high density memory technology gives significant reduction in electronic packaging with an improvement in reliability. The adoption of an integrated systems approach offers marked improvements in fault tolerance and in access to the aircraft systems data. System trend and health monitoring capabilities are greatly improved.

The main benefits of adopting the UMS for a typical high performance fighter have been assessed as more than 50 per cent reduction in installed weight and

Fig. 6.22 EAP – line replaceable unit SMP C/D *Smiths Industries*

operating costs, and availability which is eight times as good.

The first lesson learned was that the job could be done and done efficiently. The task was unique – a distributed system employing digital computing to control a large number of systems and with a very large estimated software content, to be performed on an experimental aircraft with a very tight timescale. It is a credit to all who worked on the UMS project teams that the task was completed on time. In addition to this, software and hardware from companies throughout the UK, Italy and Germany was integrated successfully at a single site and qualified for flight.

The key to the production of good software with high productivity was undoubtedly the insistence on adherence to a rigid method of software control and management. CORE especially proved to be a powerful tool, not only for design, but also for fault finding. It was found to be very versatile in assisting the engineering teams to rapidly locate a problem area in software and rectify it.

The drive to achieve commonality has obvious benefits for any potential customer – the reduction in third line spares, etc. However the modular approach yielded benefits that were unforeseen at the beginning of the project. That was the flexibility of using the different sets of equipment as a supply of spares by exchanging from rigs currently in use on lower priority tasks. This enabled faults to be isolated to particular cards, to confirm the existence of faults in software or hardware and generally helped to keep the job going when rig downtime seemed unavoidable.

The use of computer-aided tools was a significant aid to document production and modification. The tools associated with SAFRA were used as well as mainframe text processors, minicomputer word processors, standard forms, and data base tools. For future projects an engineering data base with mass storage, standard form documents and configuration control will be used.

HEALTH AND USAGE MONITORING (HUM)

Health monitoring and usage monitoring have evolved over the years to improve the methods of monitoring critical aircraft components. The principle was first applied to the monitoring of aircraft engines. This was for two prime reasons: the most obvious was the need to prevent engine damage and possible hazard to the aircraft following a catastrophic failure; the second was that such failures as well as creating a hazard, also are very expensive to rectify. The detection of failures incipient, before any real damage was incurred therefore became very important. As the principle has evolved, monitoring has been extended to gearboxes and transmission systems – particularly on helicopters – and for the monitoring of aircraft structural life consumed.

The first British application of engine health monitoring was on British European Airways Trident fleet. Initially simple discrete inputs giving engine speed, temperature, pressure and vibration levels were monitored and recorded on a suitable on-board tape recorder. This data could then be removed from the aircraft for the necessary analysis using ground based equipment.

The availability of cheap microprocessors in the early 1970s enabled a further development to be enbodied: that of including intelligence within the data-gathering or aquisition unit. This permitted more exotic measurements and calculations to be made of which low cycle fatigue is a typical example. Variations in engine speed induce variations in stresses in the rotating assemblies of the engine. Variations in stress in those parts of the engine most highly stressed can lead to fatigue failure. Ideally, the best way to minimise variations in stress in an engine is to run the engine as near to constant speed as possible at a throttle setting which causes the least stressful regime. However, though this may be practical for an engine powering a ship or an electrical generation turbine, it is not practical for an aircraft engine. The aircraft engine stresses will vary from start-up from cold, through taxying, full power take-off, climb, cruise, descent, holding pattern, approach, landing and shut-down. It follows that, due to the manner in which an aircraft is operated, these variations cannot be eliminated though they may be minimised by sympathetic handling of the engine. The next best thing is to monitor and measure stress cycles so that the effects of stress are better understood and catastrophic failures may be forestalled.

The Low Cycle Fatigue Counter (LCFC) accepts inputs from the engine for such parameters as engine speed (NL and NH) of compressors and turbines and processes this information to calculate engine damage cycles. These damage cycles are not related to actual damage but rather are a measure of the component life being consumed by these critical items. Calculation of fatigue cycles based upon this type of equipment fitted to the Royal Air Force Red Arrows aerobatic team Hawk aircraft have shown wide variations in the consumption of engine life. The team leader who has to perform gentle manoeuvres so that the remainder of the team may readily formate on him uses during a display around a tenth of the engine life of one of the outside wingmen who have to continuously make throttle corrections to maintain station in the formation.

Typical of the parameters monitored by a modern health and usage monitoring system are:

Engine speeds	Accelerations
Engine temperatures	Vibration levels
Engine pressures	Stress
Engine torque	

The extension of HUMS is extended to the monitoring of gearboxes and transmission trains on helicopters where the continued operation of the power train is essential to airworthiness. This will enable the following parameters to be recorded and analysed:

Engines:	Transmission:
Power performance index	Torque usage
Low cycle fatigue	Diagnostic vibration
Thermal creep	Qualitative wear debris monitor
Exceeding of limits (speeds and temperatures)	

A typical HUMS may be as shown in Fig. 6.23. The engine and other health parameters are conditioned and converted into a suitable digital format for use by the microprocessor. After the necessary calculations and algorithms have been executed the data is stored in memory until the conclusion of the flight or until a suitable route point is reached. Then the data may be extracted by means of a data transfer unit (DTU), usually by means of an ARINC 429 or MIL-STD-

Fig. 6.23 Typical health and usage monitoring system (HUMS)

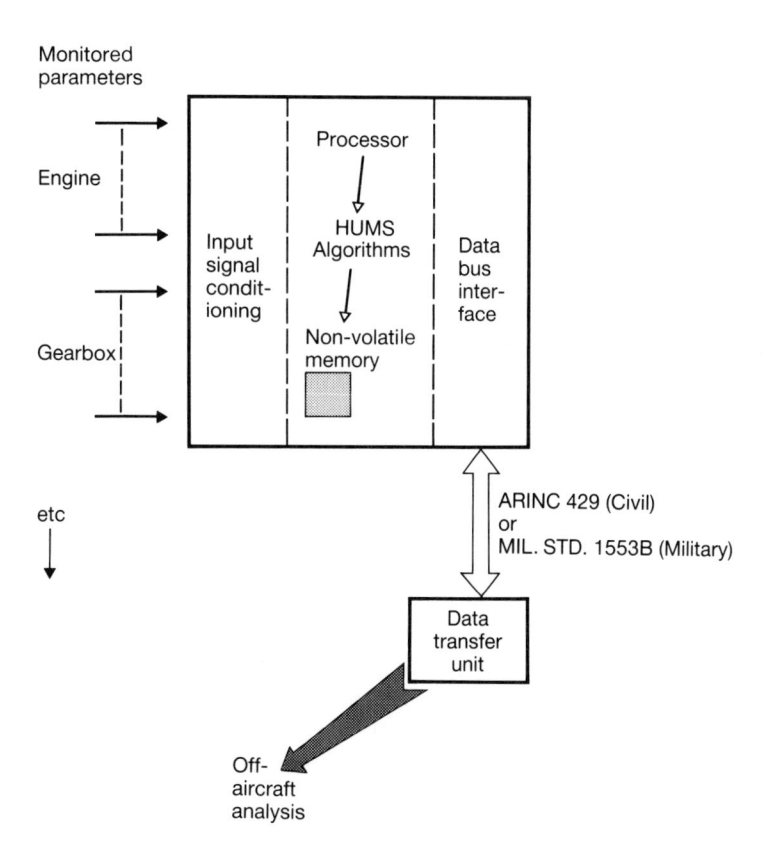

1553B data bus; refer to Chapter Three. The DTU is then used to transfer the health/usage data to a ground station for off-line analysis. More sophisticated aircraft may have the capability of transferring data on request via an automatic communications and recording system (ACARS).

For a number of years aircraft have used similar techniques to measure the consumption of fatigue for the airframe itself. The unit which undertook this function was called a fatigue meter and was located in the aircraft close to the centre of gravity. The fatigue meter measures the number of normal accelerations sensed within certain 'G' thresholds and displays the aggregate number of counts registered in each band. These counts were read from a series of counters on the front face of the unit after each flight and registered in the aircraft servicing documents. These counts were then analysed to assess the degree of fatigue being consumed by the aircraft.

The problem with the fatigue meter was that although a very useful measure of fatigue, it only measured the accelerations being experienced at the point in the airframe where it was mounted. More likely, fatigue would be occurring at critical locations in the aircraft structure such as the wing root attachment points or the main spar or at hard points on the wing and fuselage where fuel tanks or external stores may be carried. The fatigue meter therefore was only able to present part of the picture.

There is now an increased tendency towards systems which utilise strain gauges at various points in the airframe so that monitoring may be applied in a more selective fashion and include areas where structural stress may reasonably be expected to be high. Such systems are sometimes called Structural Usage Monitoring Systems (SUMS).

A development akin to the HUMS principle which has been in use for a number of years is the accident data recorder which is associated with the rather gruesome task of aircraft investigation. Accident data recorders or crash recorders are rugged units which record vital aircraft parameters during flight and which can be recovered from a crash site and used to construct the behaviour of the aircraft in the minutes before the accident occurred. These 'black box recorders', which are in fact coloured orange to assist in identification at the crash site, are often keys in establishing the true cause of an accident. The Royal Aerospace Establishment at Farnborough has specialised facilities capable of reading the recorder traces and displaying them on a composite display which allows the aircraft flight path to be quickly understood. It is also usual for aircraft to be fitted with a Cockpit Voice Recorder (CVR) which records the conversations in the cockpit and which may also assist in accident investigations.

STORES MANAGEMENT SYSTEMS

During the First World War, when the protagonists began to realise the fighting potential of the flying machine, attempts at arming aircraft were fairly crude. Initially bombs were dropped by physically releasing them over the side of the cockpit. Before the introduction of the first interrupter gear, guns were fired straight through the propeller arc with crude deflector plates fitted to deter the odd errant round which coincided with the propeller blade. Later, during the Second World War, release systems became more sophisticated, though many bombs were dropped by mechanically activated bomb hooks.

The increasing use of electronics in the fire control systems of the second generation jet fighter aircraft was to lead to increased integration of sensors and weapon release systems. The US North American F-86D (Sabre Dog) was the first aircraft to boast the capability of automatically engaging a target blind without the pilot needing to see the target. In the UK, the development of the Firestreak and then Red Top air-to-air missile demanded increasingly sophisticated fire control systems. The transition to the modern multi-role fighter aircraft with multiple sensors and a wide range of weapon fits has demanded that the Stores Management System (SMS) employ state-of-the-art-technology. This is needed both because of increasing performance and the need to upgrade the overall weapon system from time to time with improved or totally new sensors. These sensors may either require or make possible the carriage of different weapons options. If the SMS cannot handle the new weapons then the weapons system as a whole is impotent.

A prime requirement of a weapon release system is that of accuracy. The timing of the release pulse is clearly critical: at an aircraft speed of 400 knots a delay of 100 milliseconds (1/10 second) will equate to an along-track error of almost 70 feet. The system is also commonly required to account for the different ballistic characteristics of various types of weapon.

As well as accuracy, safety is a prime requirement. First, there is usually a high integrity requirement placed upon the system which constrains the designer to make safeguards against an accidental or inadvertent release during peace-time training. However if the pilot needs to jettison his stores and fuel tanks in an emergency – because of a loss in engine power, for example – then it is essential that this feature should operate when required. Therefore the system will also have important safeguards embodied to ensure that the Emergency Jettison of E/J facility operates, otherwise the aircraft could be lost. Finally the system will need to inform the pilot of the fusing (or priming) and arming state of the weapons being carried so that incorrect or invalid selections are avoided.

A notional SMS which depicts the key elements is shown in Fig. 6.24. This system presumes a single-seat fighter ground attack aircraft with a head-up display (HUD) for primary weapon aiming. No radar or electro-optic sensors are assumed available to aid weapon release; the system is, by present standards, a fairly basic one. It has also been assumed that the host aircraft has four wing pylons and a centreline carriage position: five weapon stations in all. The system is also assumed to have provision for a gun and comprises the following units:

> Pilot's weapon aiming mode selector (WAMS) panel;
> Control stick and throttle mounted controls;
> Master armament selector switch;
> Emergency jettison switch (guarded);
> Dual-channel SMS computer;
> Safety break;
> Weapon station encoder/decoders:
> – Left wing inboard and outboard station
> – Centreline station
> – Right wing inboard and outboard stations
> Gun electronics unit.

The pilot's WAMS panel presents all the information necessary regarding the various weapon stations and the gun: weapons being carried, type and number, fusing characteristics, etc. This panel enables the pilot to make the selections required for a particular engagement; type of weapon or bomb, single weapon or salvo, single bomb or stick and so on. The WAMS panel will also confirm to the

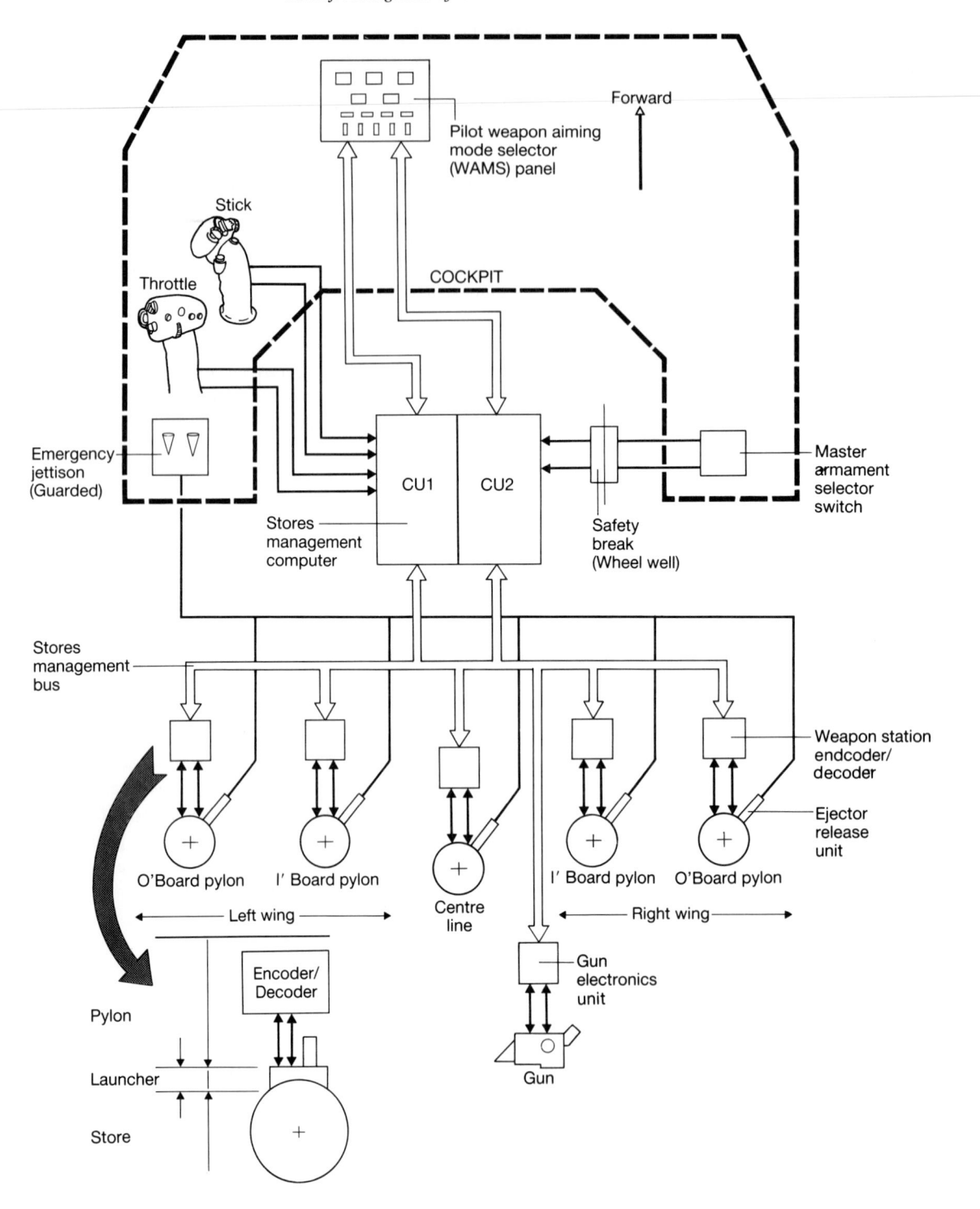

Fig. 6.24 Notional stores management system architecture (SMS)

Stores management systems

pilot that the selections made are valid for the weapons fit pertaining and that the correct fusing or arming actions have been made. At the top of the control stick and throttle the pilot will have a number of switches which enable key selections to be made without recourse to the WAMS panel. These controls permit the pilot to fly with hands-on-throttle-and-stick and are referred to as HOTAS controls. The principle behind HOTAS is the simple maxim that in combat the pilot cannot afford to keep scanning instruments and feeling for switches inside the cockpit, nor can he afford to take his hands off either throttle or stick. Figure 6.25 shows the HOTAS controls for the Hawk 200 aircraft[5] which are typical though not specifically related to this notional system. The purpose of the HOTAS controls shown in the figure is in each case self-explanatory (see also Fig. 4.14).

Fig. 6.25 British Aerospace Hawk HOTAS controls *Flight International*

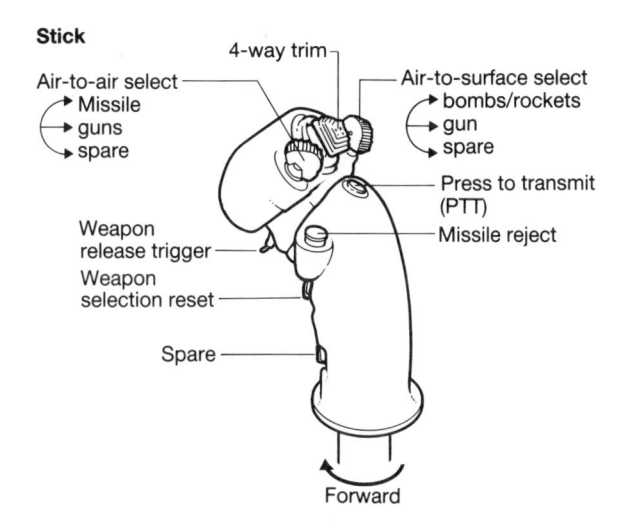

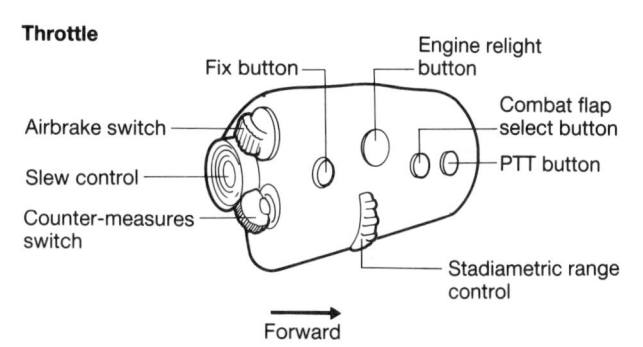

Stick
- 4-way trim
- Air-to-air select
  - Missile
  - guns
  - spare
- Air-to-surface select
  - bombs/rockets
  - gun
  - spare
- Press to transmit (PTT)
- Weapon release trigger
- Missile reject
- Weapon selection reset
- Spare
- Forward

Throttle
- Fix button
- Engine relight button
- Airbrake switch
- Combat flap select button
- Slew control
- PTT button
- Counter-measures switch
- Stadiametric range control
- Forward

Also in the cockpit the pilot will have a Master Armament Selector Swith (MASS) which is effectively a master on/off switch which is used to power-up and arm the system. Without the selection of the MASS the system is effectively inert and for safety reasons the MASS is only selected immediately prior to combat or when entering a firing range. In series with the MASS is a safety break – often located in an undercarriage bay – which is the last safety device to be removed on the ground. This is usually done by the groundcrew immediately prior to the pilot taxying for take-off. The safety break is a further safety feature intended to give additional protection on the ground while the aircraft is being armed. Finally, the pilot has an Emergency Jettison (E/J) switch which is guarded to prevent inadvertent selection. This switch when operated will apply power to the Ejector

Release Units (ERUs), small explosive devices which ignite and throw the weapon launchers, stores and fuel tanks clear of the aircraft. This feature is intended to 'clear wings' of external stores if the aircraft encounters an emergency. The E/J line circumvents all weapon release mechanisms. Inadvertent operation of E/J on the ground is prevented by the fitting of safety pins which prevent the ERUs from being fired. The safety pins are removed immediately prior to a sortie. For safety during servicing operations the ERUs are electrically disconnected.

The system shown is assumed to have a dual channel digital SMS. The dual computers interface to the WAMS panel, weapon station encoder/decoders and gun electronic unit by means of a dedicated data bus. At the pylon or weapon station, information between the computers and stores is converted from digital to discrete or analogue signals depending upon the requirements of the store. Therefore the pilot selections made on the WAMS panel are validated by the computer(s) and signalled to the pilot as valid or otherwise. Weapon station and weapon selections are sent to the encoder/decoders or GEU as appropriate. Weapon status information is fed back to the computer(s) and therefore to the pilot. A typical example of the weapons/stores options available on the Hawk 200 is shown in Fig. 6.26.[6]

Fig. 6.26 BAe Hawk weapon options *Flight International*

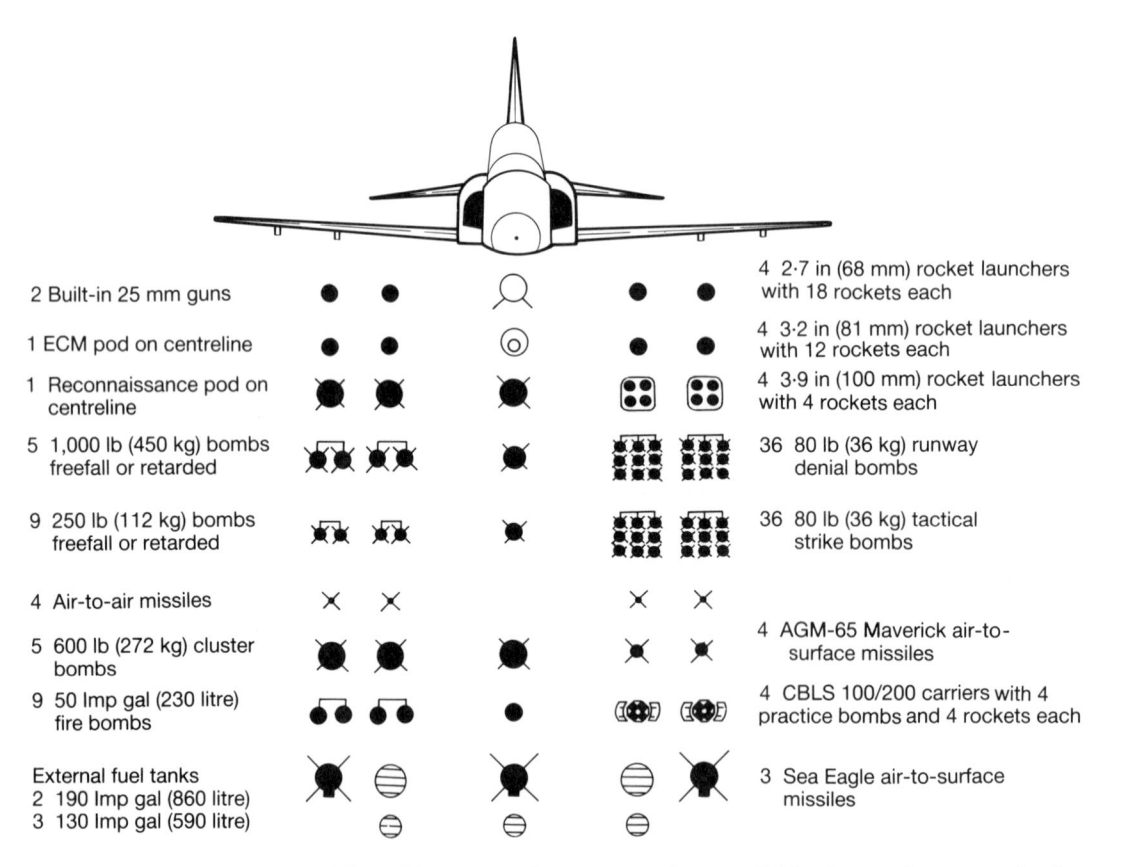

- 2 Built-in 25 mm guns
- 1 ECM pod on centreline
- 1 Reconnaissance pod on centreline
- 5 1,000 lb (450 kg) bombs freefall or retarded
- 9 250 lb (112 kg) bombs freefall or retarded
- 4 Air-to-air missiles
- 5 600 lb (272 kg) cluster bombs
- 9 50 Imp gal (230 litre) fire bombs
- External fuel tanks
  2 190 Imp gal (860 litre)
  3 130 Imp gal (590 litre)

- 4 2·7 in (68 mm) rocket launchers with 18 rockets each
- 4 3·2 in (81 mm) rocket launchers with 12 rockets each
- 4 3·9 in (100 mm) rocket launchers with 4 rockets each
- 36 80 lb (36 kg) runway denial bombs
- 36 80 lb (36 kg) tactical strike bombs
- 4 AGM-65 Maverick air-to-surface missiles
- 4 CBLS 100/200 carriers with 4 practice bombs and 4 rockets each
- 3 Sea Eagle air-to-surface missiles

The wide range of weapon options available for carriage can clearly cause logistic and inter-operability problems, especially across a wide range of aircraft and weapon options as typified by the NATO Air Forces. Attempts are being made to improve inter-operability by means of standardising the weapon/pylon interface. The standard concerned, MIL-STD-1750, attempts to standardise signal types and formats for this interface, thereby enhancing US Tri-service and NATO Air Force inter-operability. The standard also makes provision for future 'smart' weapons which may require standard data bus and video bus interfaces with the aircraft.

REFERENCES

1. **Gilson, C.** FADEC for fighter engines, *Interavia*. 11/1986.
2. **Condom P.** Using electronics to save fuel, The development of FADEC, *Interavia*. 8/1985.
3. **Condom P.** Flight management systems: wanted but only when they work, *Interavia*. 5/1984.
4. **Blackman A. L.** Flight management computer systems, *Aerospace*. February 1984.
5. **Moir I., Seabridge A. G.** Vol 13 No 7. Management of Utility Systems in the Experimental Aircraft Programme, *Aerospace*.
6. **Gaines M.**, Combat-ready Hawk, *Flight International*. 20 June 1987.

CHAPTER 7 NAVIGATION SYSTEMS

SIMON J. K. WALKER M B A.

Mr Walker is an aviation consultant in a specialist company with offices in Washington, London and Hong Kong.

Qualified as a commercial pilot he has worked for corporate flight departments and airlines in UK and was a Regional Technical Sales Manager for British Aerospace Civil Division. He followed this post with one as Marketing Manager for the Sperry Flight Systems Group of Honeywell.

Mr Walker has an honours degree in Economics and a Masters Degree in Business Administration.

Until recently navigation and voice communication were entirely separate functions for the crew. Navigation was carried out by a combination of pre-flight planning, often with incomplete weather information, with the location of the aircraft during its flight being predicted by 'dead-reckoning', a technique which is based upon calculations of speed, course, time, wind drift effect and the previous known position. Trans-ocean and transcontinental flights would rely upon more sophisticated stellar navigation using sextants.

The rate of growth of airline operations, both in terms of the number of aircraft, their much higher speed and the complexity of their flight planning has introduced the vital element of coordination with a large group of ground-based experts, such as air traffic controllers who will assist the crew as well as taking vital decisions which affect the safety of the aeroplane.

RADIO WAVE PROPAGATION

This subject is dealt with in some detail in Chapter Eight so it is sufficient to say here that, in general, navigational systems are based upon classic radio theory with the curve depicting the movement of transmitted energy having two basic characteristics – amplitude, the range of fluctuation from the mid-point to the extreme point of the wave, and wavelength which is the measure of the linear distance of a wave cycle. The frequency is, of course, measured by counting the number of cycles completed in an appropriate timespan, usually one second. The characteristics of radio wave propagation vary according to the degree of resistance to the energy. The earth can be considered to be the greatest resistor, with the strength of the radiated energy being attenuated in proportion to the distance travelled.

As radio waves per unit of time involve extremely high numbers, radio frequencies are usually referred to in multiples of kiloHertz (kHz), which equates to a thousand cycles per second, or megaHertz (MHz), which equals 1000 kHz.

Radio waves are produced by sending a high frequency alternating current through a conductor – the aerial – which radiates a wave with the same frequency as the current that generated it.

Radio frequencies are classified in groups of 'bands' because of the different characteristics which particular frequency levels display. The wave bands and their corresponding frequencies are listed below:

Band	Frequency range
Very low (VLF)	below 30 kHz
Low (LF)	30–300 kHz
Medium (MF)	300–3000 kHz
High (HF)	3000 kHz–30 MHz
Very high (VHF)	30–300 MHz
Ultra high (UHF)	300 MHz–3000 MHz (3 GHz)

These characteristics have a profound effect on the design and use of navigation equipment, imposing limitations on specific uses.

Low frequency waves are radiated in all directions, and comprise both a ground wave element that dissipates outwards from the transmitter, and a sky wave that is reflected back in the ionosphere. These latter signals can be received at varying distances from the source of the transmission depending on physical effects as well as environmental conditions.

High frequency radio waves reflect back and forth between the ground and the ionosphere several times. This effect means that they can be used over considerable distances, albeit with the characteristic of a continual background noise. HF radio is traditionally used for voice communication in parts of the world where ground stations are not feasible, such as over oceans and deserts. In addition, they are a crucial element in the present level of ACARS technology (see Chapter Eight).

The VHF and UHF bands do not normally display any reflection from the ionosphere, and hence are only useful in what are termed 'line-of-sight' conditions. Under these conditions, the height of the transmitter and the receiver play a crucial role in determining useful distance. Various formulae exist to estimate the effective range of a transmitter based on the sum of the aircraft and transmitter station elevations above ground. Frequencies in these two bands are used for both communication and navigation.

NAVIGATION EQUIPMENT

To transmit and receive electronic information that allows for accurate navigation, the following essential equipment is necessary:

(a) a method of determining the characteristics of the generated signal;
(b) a transmitter with associated aerials to generate and transmit the radio wave;
(c) aircraft receiver units and aerials to intercept, select and interpret the signals received;
(d) suitable visual displays for the pilot to make appropriate assessments of the signals received.

The most common method of short–medium range navigation in use today relies on VHF signals being received on the aircraft, and displayed on a suitable medium. In addition, there are a number of long-range techniques employed, using inertial and hyperbolic systems. These are covered in the appropriate sections of this chapter.

AIR TRAFFIC CONTROL

The complexity of modern aviation, caused by the volume of users in a finite amount of air space, has created the requirement for an ever-expanding and sophisticated Air Traffic Control service. Navigation systems can thus be described as having two complementary functions: accurate navigation of aircraft from departure to arrival airport; and secondly the means to define and follow specific tracks in space in order to comply safely with the requirements of the control system.

Air traffic control can be split up into a number of segments. These include: local airport Tower control; Approach control; and *en-route* Airways control. Each has a predefined function to ensure adequate separation, terrain clearance guidance and smooth flow of aircraft within its jurisdiction.

Controlled airspace is divided into two main categories: Control Zones and Control Areas. The first extends from ground level at an airport to a particular height, typically 3000 feet. The control area covers the remainder of the airspace under positive control, including the airways network. A transition level separates the two.

The safe and expeditious use of controlled airspace relies on the accurate display and use of navigational data. The most vivid example of this is in Terminal Control Areas, where complex arrival and departure paths are created (see Fig. 7.1). These standard instrument departures (SIDs) and standard terminal area arrival routeings (STARs) are defined for each runway and conceivable arrival/departure route at major airfields. They will normally involve a sequence of radials and distances to fly, utilising a range of navigation aids that have been preselected prior to take-off or arrival.

The intricate pattern of a modern air traffic control system is displayed on radio navigation charts produced and updated, on a weekly basis, by a number of firms. The two that meet the world-wide needs of the professional pilot are Jeppesen and Aerad.

NAVIGATION DISPLAYS

A more detailed description of the form of navigation displays has been given in an earlier chapter. The purpose of this section is to set out those characteristics appropriate to an understanding of methods of navigation.

The four displays that will be briefly described include one relatively old display, plus three modern ones. The first is the relative bearing indicator (RBI), the modern displays are the radio magnetic indicator (RMI), the horizontal situation indicator (HSI), and displays fed by the flight management system (FMS). Cockpit information for the vertical element of an instrument landing system is covered in the following section on point source aids.

The relative bearing indicator was one of the earliest displays of navigational information, providing a bearing (radial) to the navigational beacon, usually a non-directional beacon (NDB). Although offering a dramatic improvement in terms of positional accuracy, it still required considerable pilot interpretation for accuracy.

The 'compass card' was fixed, with 'zero' uppermost. Needle indications displayed variance from this zero indication. Pilots had to superimpose in their minds the display message from the RBI upon the compass situated elsewhere in the cockpit.

Variations from the zeroed position indicated one of two possibilities: first, the

Navigation displays

Fig. 7.1 Controllers at the London Air Traffic Control Centre (LATCC) at West Drayton, Middlesex use a standard 23 inch plan position indicator (PPI) to monitor traffic in their sectors *Civil Aviation Authority*

aircraft was not on the desired radial to or from the beacon; second there was a planned correction for wind effect such that the aircraft nose was not pointing at the beacon, although it was physically tracking along the desired radial.

One result of this latter effect was that, if the pilot kept the needle on the zero position, the aircraft would eventually approach the beacon heading into wind, albeit having flown tighter and tighter curves towards that position!

The correction factors required to compensate for wind effects had to be mentally reversed once the aircraft had passed over the beacon and the needle was pointing behind the aircraft. The potential for inaccurately assessing the information displayed led to the development of the more 'user friendly' radio magnetic indicator (RMI).

As discussed above, the relative bearing indicator had a number of limitations,

Navigation Systems

particularly under conditions of high pilot workload. To correct many of these deficiencies, the RMI was developed.

This instrument had a rotating compass card, on which were superimposed two bearing needles, one double-barred, one single-barred. This resulted in a continuous readout of bearing to or from the navigation station to which it was tuned. The compass card was actuated by the aircraft's compass system, causing rotation of the card as the aircraft turned.

The bearing pointers displayed magnetic bearings to NDB or VOR stations. Early examples of the RMI imposed limits on the switching capability of the instrument, so that the No. 1 indicator, the double-barred one, pointed to VOR stations, while the No. 2 pointer was used for NDB stations. Later models allow full switching between the needles, so that there can be any combination of twin or single VOR or ADF displays.

Fig. 7.2 Radio direction/distance/magnetic indicator (RDDMI) *Honeywell*

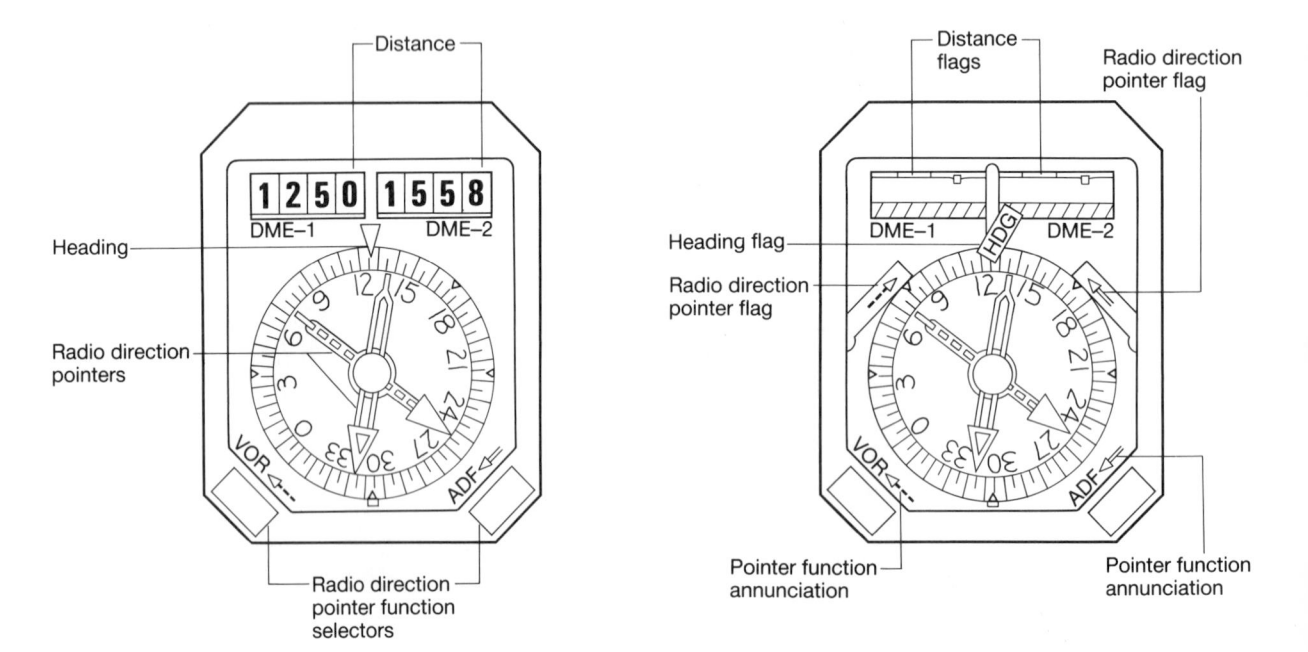

Figure 7.2 shows a typical display format, incorporating distance-measuring information on the instrument. This type of instrument is referred to as an RDDMI to indicate this additional feature.

Despite the introduction of EFIS the RDDMI remains on most flight decks as a stand-by instrument, the use of which is explained later in the chapter.

The horizontal situation indicator (HSI) is one of two primary instruments on the flight deck, the other being the attitude and direction indicator – ADI. In many modern aircraft, the electro-mechanical displays have been replaced by electronic ones for greater reliability and capability with lower weight.

Figure 7.3 shows the functions of another such electronic instrument. Many of the displays shown are not permanently activated – an advantage of electronic switching. They include the aircraft symbol which provides the visual reference of aircraft position relative to a selected course or heading. The course deviation indicator (CDI), with associated course select pointer and readouts, represents the centreline of the selected navigation input. A pictorial representation of the

Fig. 7.3 Horizontal situation indicator (HSI) *Honeywell*

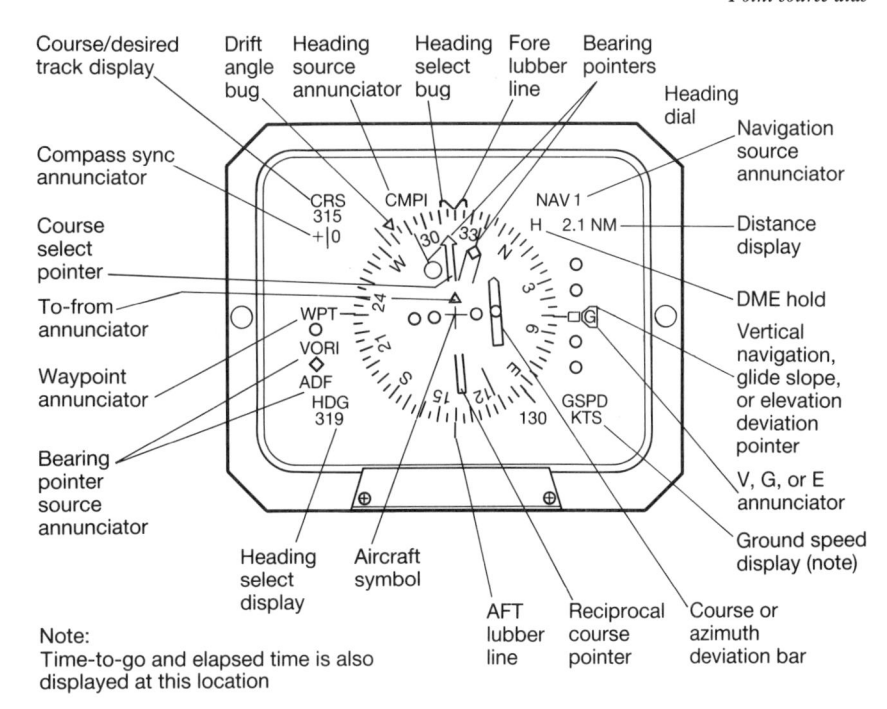

Labels (clockwise from top-left):
- Course/desired track display
- Drift angle bug
- Heading source annunciator
- Heading select bug
- Fore lubber line
- Bearing pointers
- Heading dial
- Navigation source annunciator
- Distance display
- DME hold
- Vertical navigation, glide slope, or elevation deviation pointer
- V, G, or E annunciator
- Ground speed display (note)
- Course or azimuth deviation bar
- Reciprocal course pointer
- AFT lubber line
- Aircraft symbol
- Heading select display
- Bearing pointer source annunciator
- Waypoint annunciator
- To-from annunciator
- Course select pointer
- Compass sync annunciator

Note: Time-to-go and elapsed time is also displayed at this location

aircraft shows deviation from the selected course, and provides a command indication to restore the aircraft to the centreline.

Although referred to as a horizontal situation indicator, there are some vertical functions possible in the form of part-time displays, activated for specific purposes. These include a secondary indication of glideslope deviations as well as VNAV (vertical navigation) functions for aircraft equipped with suitable navigation interfaces.

FLIGHT MANAGEMENT SYSTEMS (see Chapters Four and Six)

The navigation capabilities of flight management systems (FMS) via programmable navigation databases are displayed on the 'Map' mode of an electronic flight instrument system (EFIS). This may be displayed on the primary instruments on either side of the cockpit, or, increasingly likely, on a specialised multi-function display unit, mounted in the centre console. This MFD will act as the radar display indicator, either solely for weather or as a combination weather and map display.

Routeings between reporting points are displayed, fed either from navigation beacons such as VORs or NDBs, or from waypoints 'created' by the navigational database in a long-range navigation system.

POINT SOURCE AIDS

This section on navigation equipment covers those aids to navigation that can be considered as being derived from a source, either on the ground or in the air. This will thus cover radio beacons such as VOR stations and instrument landing systems, where interpretation is required by an aircraft's on-board receiver and radio aids like distance-measuring equipment, where there is a two way flow of data and information between the ground and the air. Lastly, some consideration will be given to the use of airborne radar for positional navigation.

For many years the primary source of navigational information was a group of low frequency devices which have now been completely superseded by the development of automatic direction finding (ADF) stations, now referred to as non-directional beacons (NDB). As will be explained later, these suffered from a number of problems, not least their tendency to be unreliable in storm activity, often the very time when they were most needed. Very high frequency devices in the aircraft followed, with very high frequency omni range (VOR), distance-measuring equipment (DME) and instrument landing systems (ILS) all contributing to greater precision and accuracy in navigation.

Future developments are concentrated upon improving these precision levels, without the inherent shortcomings of the current types. The most advanced of these is the microwave landing system (MLS), in use in the United States as a privately owned navigational aid for particular aircraft, as well as under active trial at many European airports (see p. 159). Radar technology has also improved dramatically as a result of developments in the computer industry. The results are seen in the increasing use of colour tubes, with added benefits in detecting turbulence and lightning areas.

The connecting link between all of the navigational aids considered in this section is the cockpit display for the pilot. This may indicate a deviation from the desired course, a numerical readout of distance and speed or a pictorial display of ground mapping returns. Some form of response from the ground is also relied upon, either in the form of an RF signal or response, or variations in radar signal strength.

Pilot control of the tuning of these devices has usually been accomplished by individual radios being self-contained, with separate tuning controls, bearing selectors and display indicators for each. Recent developments are reducing this level of independence, at least on the flight deck. Electronic flight displays can be switched between functions at the request of the pilot; tuning of radio stations can now be performed via the keyboard of flight management systems; radio magnetic indicators (RMI) are now common in many aircraft, with switching between ADF and VOR radios. One of the more interesting improvements in recent years has been the development of the Honeywell Primus II range of radios, whereby the tuning and selection of all navigational and communication equipment is performed via a radio management unit (RMU) that also provides a large memory capacity for common frequencies; all in a package barely an eighth of the size of the standard airline configuration.

Very high frequency omni-range (VOR)

The VOR transmitter/receiver system is the primary source of navigational guidance over populous land masses. As discussed earlier, by utilising the VHF radio wave band, certain benefits accrue, such as reduced interference, particularly static atmospheric interference; there are, of course, line-of-sight range limitations. Effective, usable range of a VOR station is thus between 40 and 300 miles, depending on the aircraft (receiver) altitude.

The current ARINC characteristic for VOR receivers is ARINC 711, developed over several years from 1978. As described in this characteristic, the function of the receiver is to receive, process and output digital signals of bearing in a frequency band from 108.00 to 117.95 MHz. (Additional functions also cater for the receiving of marker beacon signals on 75 MHz.) The receiver frequencies are used for specific functions, with relatively low powered (and hence reduced range) terminal VORs operating in the range 108.00 to 112.00, and airways transmissions between 112.00 and 117.95 MHz.

The principle of operation of VOR is based on the concept of phase differences between two alternating voltages. Detecting the difference permits the determination of positional guidance in azimuth. The two navigation signals generated by the transmitter station are called the Reference signal and the Variable signal. The first of these has a constant phase when received at any point around the transmitter station, while the second varies according to the receiver position relative to the in-phase condition at Magnetic North.

The signal from the VOR transmitter is rotated at 1800 revolutions per minute, which allows for accurate phase comparisons to determine change in azimuthal bearing, relative to the in-phase signal. The information received as a result of the phase comparison is then displayed as an indication of bearing to and from the transmitter station.

VOR displays are presented in a variety of ways. The earliest type was referred to as a course deviation indicator (CDI), which presented a visual indication of whether an aircraft was established on a pre-selected radial to or from the station. The horizontal situation indicator (HSI) shows such an indication (Fig. 7.3). The deviation displayed is known as a command indication because in order to regain the centreline of the selected radial, the aircraft must be flown towards the course deviation bar.

The radio direction distance magnetic indicator RDDMI (Fig. 7.2) shows a constant display of the radial to or from the transmitter station. The pilot may observe deviations from the desired radial in accordance with movement of the pointer.

Power output of the transmitter varies as a result of VOR classification. Terminal VORs, with their short useful range, transmit with a power output of approximately 50 watts, while *en-route* transmitters have an output of 200 watts.

Identification of the tuned station is carried out via a three letter morse code signal transmitted continuously. Failure of the station transmitter, or too weak a signal, is indicated to the pilot via failure flags, or a lack of To/From indication.

Distance-measuring equipment (DME)

Distance-measuring equipment has come a long way from the early examples. It is now an integral part of accurate navigational guidance, whether linked to a VOR facility to provide range and bearing, or by frequency scanning of a number of DME stations to provide inputs to an inertial navigation system.

The present ARINC characteristic is 709, the digital update of the earlier characteristic ARINC 568. As with all ARINC characteristics, this is primarily designed for airline transport use, with the reliability and performance requirements inherent in that market.

Distance-measuring equipment accurately measures slant range from an airborne transmitter/receiver to the ground station. (Slant range incorporates aircraft altitude as well as ground distance.) This measure is not precisely what is required by the crew, but provides an acceptable compromise in terms of reliability and complexity by removing the need for additional inputs from a radar or altimeter source. In practice, the difference between slant range and ground distance is negligible if the aircraft is one or more miles from the ground station for each thousand feet of aircraft altitude.

DME works on the basis of two sets of transmission and receiving signals, one from the aircraft, the other from the ground station. Initially, the aircraft transmits an interrogation signal, similar to that of a transponder system (see Chapter Nine).

The reception and subsequent transmission of an answering set of radio signals from the ground equipment allows for a measurement of elapsed time between transmission and reception of the signals at the aircraft, which is then converted into distance. The use of pairs of radio pulses removes the problem of ambiguity between aircraft.

The distance indicator provides a digital readout in nautical miles. This is shown embodied in the RDDMI (Fig. 7.2). Integration of DME readings with those of other navigation aids is increasingly common, although separate displays remain in production, usually for general aviation aircraft, as an 'add-on' to the original specification.

Future developments of the DME system will be the outcome of improved technology, particularly at the ground stations. This will result in greater use of the facility in high accuracy, all-weather landing requirements. The power of aircraft computers in particular, allows for rapid scanning of frequencies to ensure the integrity of display information.

Additional use of DME data is covered in the section on TACAN and area navigation.

TACAN

TACAN is now used as a civilian derivation of a military standard navigation system. Developed in the 1960s, the standard TACtical Air Navigation signal was designed to provide range and bearing guidance. The civilian development is now commonly referred to as the VORTAC system, based on VOR, TACAN and DME inputs to a navigation computer.

The TACAN system is based upon 252 different frequencies, 126 'X' and 126 'Y' channels, with interrogation frequencies ranging from 1025 MHz to 1150 MHz, and ground reply frequencies between 962 MHz and 1024 MHz, and 1151 MHz and 1213 MHz for channel X. Channel Y ground reply frequencies are from 1025 MHz to 1150 MHz. Pulse spacing varies between 12 microseconds for interrogation frequencies to 24–30 microseconds for ground reply.

By 1970, civilian approval had been granted by the Federal Aviation Administration to use short range rho-theta (bearing-distance) navigation based on the military standard. The ground facilities were based on a mixture of VOR, TACAN and DME stations, mostly co-located.

The most practical development of the merging of two guidance systems was in the field of area navigation, RNav. This allowed the on-board navigation computer to offset data received from the ground stations in order to create an imaginary 'waypoint' in space, set up by the pilot. This made possible navigational guidance to a point where no ground stations existed, through advanced computer technology, at a price and weight acceptable to operators.

A bearing and distance was established from a TACAN or VOR/DME station and the rho-theta algorithms incorporated in the RNav system calculated appropriate bearings and distance to the established waypoint, presenting them on a cockpit display precisely as if the aircraft was flying to a ground station.

With operational experience built up over several years, specific RNav routes have been established, particularly in the United States, albeit mainly for general aviation pilots.

Automatic direction finding (ADF)

Low frequency radio beacons were among the first aids to navigation. These facilities are referred to as non-directional beacons (NDBs) as they transmit a radio signal in every direction around the transmitter. Aircraft on-board equipment provides the guidance computations and display.

The present ARINC characteristic, 712, defines an ADF system as being capable of receiving and processing signals from the non-directional beacon within the frequency range 190 kHz to 1750 kHz. This is an extension from the early devices, which were restricted to 415 kHz. The new standard is based on digital techniques which allow for more accurate and more stable information being relayed to the pilot.

The principle of ADF receiver operation is based on the calculation of bearing from the aircraft to the transmitting station via two antennas, a sense antenna and a loop antenna. ARINC 712 defines the two antennas as being combined in the same unit. The vertical, sense antenna receives the signals from all directions with equal efficiency. The directional antenna extracts a portion of this energy via the loop that it is shaped into. Induced voltages in the sides of the loop and the strength of signal received are processed to determine relative bearing. The sense antenna is also utilised to determine coarse distinctions as to whether the station is ahead or behind, or to left or right of the aircraft.

Cockpit displays for ADF facilities fall into two main types. The first is the relative bearing indicator (RBI); the second the radio magnetic indicator (RMI). Both of these types of display were covered earlier in the chapter, with the varying operational benefits of each.

Voice transmission on this frequency band is often provided, subject to there being no interference with the navigation function for which the equipment was designed. The BBC transmitter near Birmingham, transmitting on 200 kHz is one of the best examples of this phenomenon. While providing radio programmes to people on the ground, it simultaneously provides high quality navigational guidance to pilots as a result of its very high power output.

The low level of quality of many stations, however, results in additional facilities being required to take advantage of any possibility of improving reception. The two devices associated with the receiver unit in the cockpit are a loop switch, which allows the rotation of the loop left and right to tune in more accurately to weak stations, and a beat frequency oscillator (BFO) which adds a signal tone to the incoming signal to produce a continuous tone of improved strength.

Automatic direction finding, by its very nature, is liable to identify the source of the strongest signal in the area. As a result, cockpit displays are often disrupted by the tendency to point towards thunderstorm activity, negating the navigational benefits of the device. Modern units are far less susceptible to these problems as a result of the digital tuning now available.

Special examples of NDBs exist as marker beacons on the approach to instrument runways, and, in the United States, as airway marker beacons (Z markers). These provide no guidance to the crew, other than an indication of passing over the appropriate beacon. Typical cockpit displays are lights of various colours and an aural tone if the receiver is switched to identify it.

Instrument landing system (ILS/MLS)

Two main types of instrument landing system are in use today: the traditional ILS, utilising a VHF localiser beam and a UHF glideslope beam (Fig. 7.4); and the relatively modern microwave landing system (MLS). This section will concentrate on the former, with some comment on what the latter offers for the future.

An instrument landing system comprises two separate components for the accurate guidance of an aircraft towards the touchdown point of a designated runway. The first of these, the VHF localiser beam is conceptually identical with a traditional VOR radial as described earlier, with one important distinction – the beam is set up along the approach path to the runway, and once identified and being

Navigation Systems

Fig. 7.4 Principles of instrument landing system (ILS) *Civil Aviation Authority*

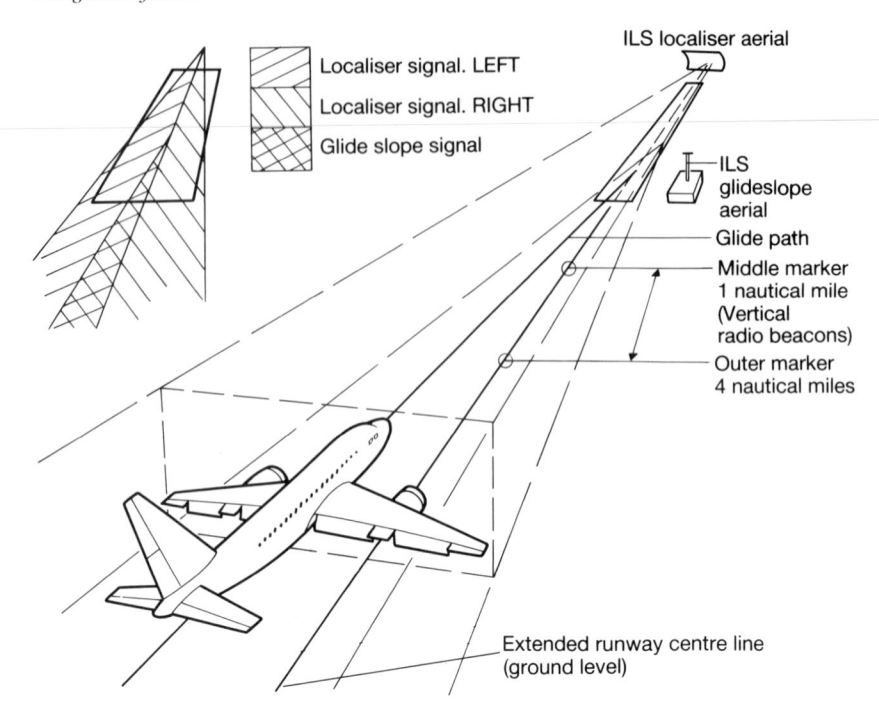

flown, is not affected by the bearing indication around the instrument. The deviation indication is always relative to the centreline of the runway, as the 'radial' is fixed.

The second beam, tuned electronically via a single tuning mechanism, is a UHF beam that provides descent slope indication, and again, is fixed at a predetermined slope angle. Typically this is 3 degrees, although certain approach paths are greater than this. The approach to the London City Airport (STOLport) for example is set at $7\frac{1}{2}$ degrees.

The localiser signal is defined as a radiated field pattern with two modulation frequencies, one left and the other right of the extended centreline. On-course signals are thus identified by signals of equal strength between the two modulated sides.

The glideslope is constructed in a similar way, with different modulated frequencies to indicate position above or below the predetermined slope.

Frequency range for ILS equipment is from 108 MHz to 112 MHz for the localiser signal, and approximately three times those figures for the UHF glide-slope signal.

Cockpit displays vary considerably. Early designs for example, the Sperry Zero Reader, had a dedicated instrument with two needles that crossed in the middle of the instrument to indicate that the aircraft was on the centreline of each radiated signal. With the needles attached at single points at the top and left-hand of the display device, there was some possibility for confusion at critical stages of the flight.

A second type of display utilised separate media, although on the same instrument (Fig. 7.5). Here, the course deviation indication was similar to the VOR generated display, whilst the glideslope deviation was shown as a vertical scale to one side of the attitude and direction indicator (ADI).

Fig. 7.5 Attitude and direction indicator (ADI) *Honeywell*

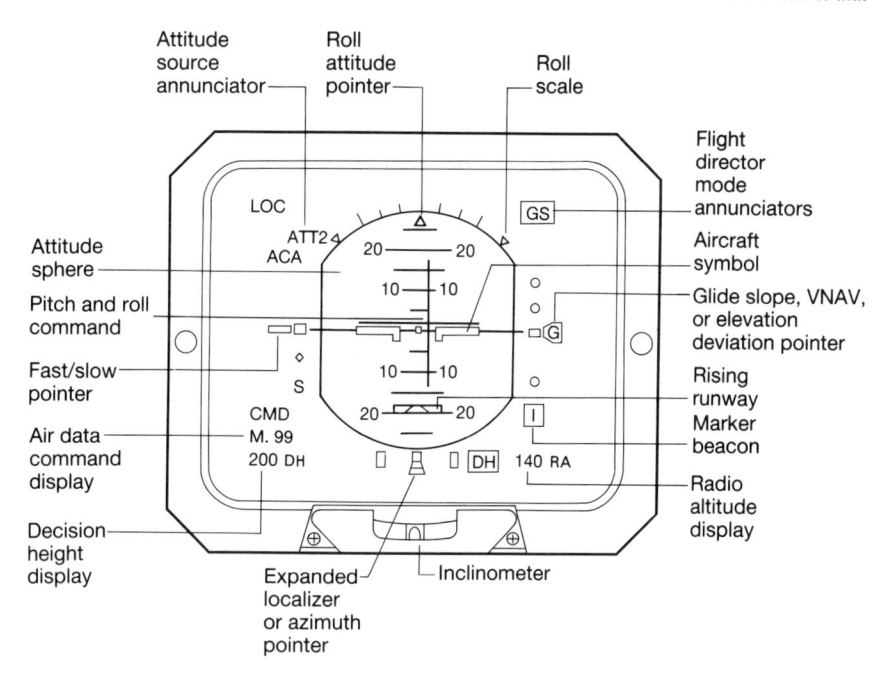

An important development in this critical navigation area was the provision of command indications that could take account of the proximity of the transmission signal, and hence distance from the touch-down point. This would allow for appropriate corrective measures to be taken, as opposed to larger than necessary measures that would require further, opposite correction as the centrelines were continually passed through.

ILS facilities are categorised by the degree of accuracy that they are capable of generating. (This has to be matched by aircraft receiver sensitivity.) These categories are designated Cat I, Cat II, Cat IIIa, IIIb, IIIc, depending on the accuracy attainable. In practical terms they define the minimum height and visual range to which a crew may descend without sight of the runway. Full blind landings were pioneered by the Royal Aircraft Establishment Blind Landing Experimental Unit, BLEU, and Smiths Industries, in association with Hawker Siddeley and British European Airways using, for the world's first blind landings, the early Trident aircraft. Such was the success of this work that it is now commonplace for the autopilots of most modern jet airliners to be capable of landing the aircraft having flown a flawless approach down an instrument landing slope to the touch-down point.

The future of runway approach guidance undoubtedly lies with microwave landing systems which provide improved capabilities in terms of reduced susceptibility to interference from buildings and hills, improved facilities for multi-aircraft approaches, even with a curved, noise abatement profile and will also provide additional guidance during overshoots.

At present, the only operational MLS stations are privately owned in the United States in mountainous areas such as the Rocky Mountains of Colorado. Operational trials are being conducted on a Boeing 757 of British Airways at London Heathrow, monitored by the Civil Aviation Authority, a body that is contributing to the adoption of European and world-wide standards of commonality.

Navigation Systems

Radar (see also Chapter Nine)

The last area to be covered in this section concerns the use of radar for navigation purposes. Although not strictly a point source aid, as mentioned in the introduction, there are areas where the effects are similar.

Two main uses can be made of radar technology for navigation. The first concerns an airport-based system for approach guidance; the second, the use of the ground mapping facility that is becoming increasingly common on airborne radar sets.

The use of air traffic controller guided approaches by airliners has been a feature of airports for several decades. The method by which it is achieved has barely changed, although microprocessor technology has improved the reliability and display for the radar controller. Typical displays require two separate radar tubes to cover azimuth and vertical aspects of an aircraft approach (Fig. 7.6). Deviations from the desired approach path are conveyed to the pilot via voice transmission on the VHF radio. Military exigency allows for approaches to very low minima, while normal civilian approaches display a greater conservatism, with lateral distances of half a mile being common for the cessation of guidance. These approaches are of value in areas where the ground returns can distort normal ILS signals, or where cost is an important consideration, particularly if the normal weather patterns do not justify a full ILS installation.

The second use of radar for navigational purposes concerns on-board weather radar. With increasing frequency, these devices have a facility referred to as 'ground-mapping'. The returns from reflected signals are analysed electronically

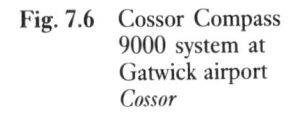

Fig. 7.6 Cossor Compass 9000 system at Gatwick airport
Cossor

and the different characteristics of cloud and ground are displayed appropriately. One result of this is to accurately map coastlines, where the water/land difference is acutely obvious. More than one major airline utilises this facility in order to maintain an appropriate distance from land that it considers hostile to aircraft overflights.

HYPERBOLIC AND GRID SYSTEMS

History and Principles of operation

This section is devoted to those navigation systems that utilise a number of earth or space stations to provide a continuous readout of position for aircraft. The systems that will be covered are Loran C, Omega, and Global Positioning System. The latter relies on satellite coverage, and at the time of publication, is yet to be formally commissioned with the full number of satellites.

The ground based hyperbolic systems have been in general aviation use since the 1950s, with the development of Loran and the Decca Navigator. Early methods of judging the position of an object had been developed from the First World War, with ranges for artillery being gauged from the differences in arrival times of the sound of gunfire at a variety of microphone listening points.

A number of variations on the basic theme of phase differences for the creation of lines of position evolved between those early days and the development of hyperbolic systems that developed at much the same time in the United States and Great Britain in the early 1940s. These developments were to utilise pulse measurements, and relied on the invention of radar technology for their advancement.

Hyperbolic systems differ from the more common range and bearing techniques discussed in the previous section as they produce lines of position in a hyperbolic form rather than the circles and radials approach of DME and VOR (Fig. 7.7). The transmission of radio signals is confined to the non-aircraft stations, removing the possibility of saturating the frequencies that can occur in two-way transmission.

The method used to create the hyperbolic lines of position is to establish a group of master and slave stations suitably distant from each other. Considering a two station pairing, a single pulse issued simultaneously will travel in concentric rings from each station, each with a measurable time interval at each 'ring'. The relative position of a device capable of measuring the time elapsed since trans-

Fig. 7.7 Hyperbolic grid pattern

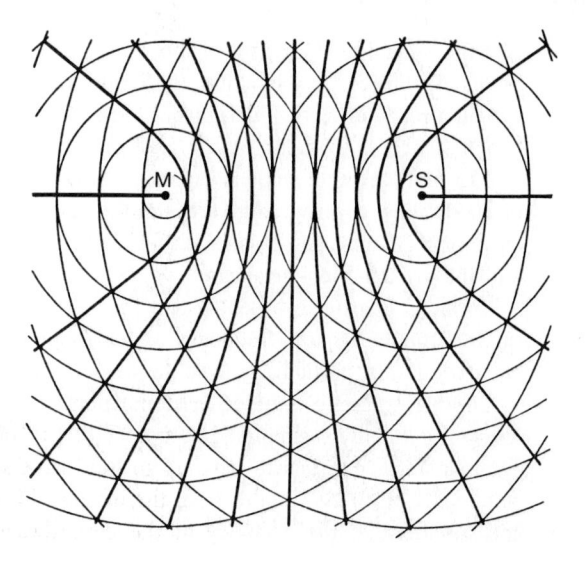

M – Master station
S – Slave station

mission will be located along a hyperbola located on possible points that can be considered as lines of constant time difference. Three possible straight lines exist: the first is when a measuring device is exactly on the line whereby the time difference between the two pulses is identical; the second and third where the device lies on the extensions behind each station but is on the imaginary line drawn between the two.

The description above implies that a single pulse is used as a measure of time difference. In practice, of course, there is a steady stream of such pulses synchronised between the stations. An alternative philosophy relies on the transmission of a synchronised continuous wave (CW), with measurement relying on variations in phase detected at a common frequency.

In the latter case, the concentric circles created are spaced at intervals of one wavelength. The pulse method utilises time differences. Each system has properties that render it most suitable for particular applications; Loran C, discussed in greater detail later, uses both techniques. The time variation element is used to derive an unambiguous position line, with the CW technique being used to generate greater accuracy along that line.

The discussion so far has referred to a pair of stations, one master and one slave. In reality, it is necessary to obtain an intersection of position lines in order to determine actual position of the receiver station. This is done by using a second pair of stations. Often, the master station is common to the two pairs, enabling position to be determined from a minimum of three stations, a master and two synchronised slaves.

The physical display of the resulting lattice pattern depends on the use for which the hyperbolic system is intended. Relatively slow moving craft such as ships will have charts drawn with the lattice superimposed; aircraft on the other hand rely on computational techniques to drive a more suitable display.

As late as the early 1970s, there was a general belief that hyperbolic systems would form the basis of all long-range navigation as well as supplying the localiser element of an Instrument Landing System. The development of inertial devices, self-contained within the aircraft has amended that philosophy. However, the development of satellite-based systems that can provide positional fixing for ground based personnel as well as navigational capability for the crew represents a further advance.

Loran C   Loran C was a development of the work carried out in the 1940s by the United States Air Force. It was a low frequency variant of the Loran A, or Standard Loran, already in service, designed, ultimately to replace the older system, offering benefits such as 24-hour service and improved range data over land.

The first 'chain' of stations came on stream in 1957, and by 1970 had expanded to 30 stations, covering large areas in the Northern Hemisphere. These included the North Atlantic, the eastern seaboard of the United States, areas of the Pacific, and the central Mediterranean.

Loran C is a pulse-based system that utilises phase comparisons to fine-tune measurements. Its low frequency and high power combine to provide excellent accuracy levels, particularly over the sea, but also over land.

Loran C utilises a common master station for the two pairs of transmitters. This allows for relatively simplified computer calculation, making airborne equipment feasible. Each master can have up to four slave stations associated with it, each potentially up to 1000 miles from the master. These distances are determined by measurement of the accuracy of the groundwave generated from the stations.

However, especially at night, there are skywaves generated as well, and these travel by a different path to the groundwave, taking longer to arrive and exhibiting a degradation in pulse duration. Although groundwave measurement is more accurate within its operational range, it is feasible to utilise the skywave at ranges, occasionally up to 4000 miles, at night.

The accuracy of the Loran C system is very high at the points of greatest signal strength and differentiation. Typical values are better than a tenth of a mile at 350 miles range from the master station reducing in accuracy to a fifth of a mile at 750 miles range. It has been widely adopted in areas of traffic congestion and has been particularly valuable for its high degree of accuracy achieved in helicopters operating to off-shore oil rigs, although the new generation of radar is taking over this particular function.

Omega The range of techniques that evolved from hyperbolic principles led to the development of the Omega system. This is based upon the continuous wave method utilising phase comparison to determine lines of position. The very low frequency band is used, specifically in the 10 to 14 kHz range.

The reasons for the choice of such a low frequency stemmed from the requirement for a low cost system. Operating in this wave band would allow very long base lines between stations. This, in turn, implied a reduced number of stations for global coverage. The total required reduces to eight with base lines of 5000 to 6000 miles between stations.

Other benefits from base lines as long as this include the ability of navigational aids to utilise more than the minimum number of fixes for position identification. This improves the accuracy of the resultant position.

Modern Omega systems are now available for business aircraft as well as conventional large transport aeroplanes for which they were originally conceived. This is a result of the considerable advances made in computer technology, allowing fast processing of the transmitted information at a cost that is acceptable to a growing group of users.

The sensors consist of a receiver processing unit RPU, and an antenna coupler unit ACU. The first of these provides the interface with the aircraft's flight management system FMS to determine present position and speed. The second unit receives the signals and converts them for processing by the RPU. A number of antenna types is available, each with specific characteristics. This area was one of the most difficult to develop in the early days of Omega design.

The positions of the eight transmitters are divided equally between the Northern and Southern Hemispheres; the first four having been located in Norway, Trinidad, Hawaii and on the east coast of the United States.

Global positioning system (GPS) GPS is a satellite navigation system that will provide continuous worldwide coverage, once all the planned satellites are in space and operational. When fully commissioned, 18 satellites will be required to provide the most accurate level of positioning fixing so far developed.

Originally designated NavStar, the system was conceived in the mid-1970s to provide highly accurate position and velocity information, primarily to military users, on a continuous basis. The inherent benefits of the system have encouraged civilian use to be explored as well, although as a result of military pressure, probably at a reduced level of accuracy.

GPS will comprise three distinct components: a ground based control segment comprising a master station and four other monitors; the space segment comprising the 18 satellites; and the user segment installed in the vehicles utilising the system. The control master station will be positioned in the continental United States, with the four monitoring stations on islands in the Atlantic, Pacific and Indian Oceans.

The user segments will only receive data, and hence the signals can be utilised by any number of users in the same area. The on-board computations will include position in three dimensional space. (An interesting use of the GPS while in test mode was to verify the height of several mountains, confirming Mount Everest as the highest point on Earth.)

The accuracy of the civilian system has an inherent value approaching 20 metres, but it will be deliberately downgraded to approximately 100 metres in operational use. The system is not affected by terrain or geographical interference, and, as such can be utilised to a far greater extent than Earth-based systems.

A number of companies are intending to produce receivers for civil use, both on aircraft and ships. The planned cost of the system should allow other uses for it on land-based vehicles, particularly those used in harsh environments.

The method of operation is for an estimate of present aircraft position, speed and time to be entered into the system automatically by the normal navigational system. This will initiate a search for the satellites and a determination of their tracks. From three of the satellites data is picked up by the GPS receiver, the range of each being calculated by multiplying the time taken to receive the signal by the speed of light. A fourth satellite is interrogated and its range established to correct any error which may have arisen due to a time difference between the atomic clock in the satellite and the crystal clock in the GPS. Speed is established by measuring the Doppler shift from the frequency of the GPS signal.

Production receivers are planned to be five channel; three are required for position determination, one for precise time, and a fifth to capture a new satellite coming into view as one of the three used for position is disappearing.

Although 18 satellites are needed for global coverage of all functions, a further three will be in position. It is anticipated that the full range of satellites will be in orbit by 1991, several years late as a result of the Space Shuttle accident that stopped satellite launch flights by this type of vehicle in 1986.

DOPPLER NAVIGATION

This section details Doppler sensors of aircraft speed and their use in self-contained navigation systems. It is included in the section on grid systems as a link between the specific navigators described above and the inertial systems.

The concept of Doppler navigation stems from the observation, in the last century, of an effect caused by relative motion between a wave source and an observer. This effect results in a difference between the transmitted and received frequencies. A common example of this occurs with the passage of a high speed train, when the frequency of the sound perceived by an observer on the platform varies as it approaches and departs.

Doppler navigation requires the measurement of this frequency shift between a transmitted signal and the subsequent reception of the same signal reflected back from the Earth's surface, be it terrain or water. If the transmission wavelength is known, and the angle of depression of the transmitted beam, and both are constant, then the detected Doppler shift is a direct measure of the aircraft's speed (Beck 1971).

The velocity vector is resolved into the three basic components, horizontal, lateral and vertical. As a result it is possible to provide a combination of speed and distance information, true height, angle of attack, wind velocity with climb and track angles. These are presented in a manner suitable for crew interpretation as well as providing direct interfaces to a flight control system and flight management system for navigation.

The accuracy of the system depends on a number of factors, some mechanical and some relating to the environmental conditions in which the aircraft is flying. Antenna design is of crucial importance, as it is desirable to produce narrow beams for accurate tracking of the reflected signal. Alignment of the antennas also plays a crucial role in determining accuracy (cf. strapdown inertial platforms).

The largest area of inaccuracy lies in the relative reflectivity of the surface over which the aircraft is flying. Land and water masses display different levels of attenuation that distort the returning signal. A number of potential compensators exist to reduce the level of this effect. First a measured factor can be installed based on the probability of operating over land and sea. Second, a crew initiated device is possible, operated as the terrain changes from land to water, or third, an autocorrect system is possible, with additional expense and complication, by using twin-, split- or switched beam antennas that recognise the variation in reception characteristics and compensate accordingly. Lastly, the movement of water droplets, either in the air or due to wave action over bodies of water causes errors in measurement. These factors produce inaccuracy levels of between one and two per cent in total position.

The development of Doppler navigation systems, still in active airline use in the 1980s, brought a number of significant benefits to airline operations. These included a reduction in crew workload, and in some cases the reduction in crew complement as navigation specialists were no longer needed and more accurate navigation resulted from better information about winds. This had a consequential effect in that fuel conservation policies became more realistic and grew in importance as the price of fuel trebled in the early 1970s, when aircraft were growing larger with the introduction of the wide-bodied types such as the Boeing 747.

A disadvantage of Doppler, at least for military applications, is that the transmitted beams can be received by ground-based missile detection and tracking systems.

SELF-CONTAINED SYSTEMS

Principles of Gyros/Flux valves/Coupled compass

Gyroscopic behaviour can be explained by the application of Newton's Laws of Motion. By reason of rapid rotation, the axis of a spinning gyroscopic rotor mechanism maintains its direction in space.

A gyro is a mechanical system, based on a rotor, universally mounted, to give three axes of freedom. The two fundamental properties, gyroscopic inertia and precession, are utilised in gyro instruments to establish a reference in space independently of the supporting body and to control the effects of the Earth's rotation, its own bearing friction and unbalance. This latter property can also be referred to as gyroscopic drift, made up of mechanical drift, caused by physical elements during manufacture and operation, and apparent drift, caused by the Earth's rotation.

Directional gyroscopes are maintained in their level operating position via torque motors and tilt switches. However, they will, due to deficiencies in balance and friction, drift in yaw. This tendency is controlled by a stabilisation process accomplished by the use of a flux valve, sometimes known as a fluxgate, which helps to slave the gyro to the Earth's magnetic field.

Surrounding the Earth are magnetic lines of force which can be used to establish the azimuth reference of an aircraft. Mounted pendulously, the flux valve is maintained parallel to the horizontal component of the Earth's field.

Components of the valve include a sensing element, mounted in a damping fluid, secondary coils, and an excitation coil. Output signals from the flux valve vary, depending on the position of the secondary windings relative to magnetic North. Sensing the direction of the lines of force of the Earth's magnetic field results in a transmission of this information, electrically, to a suitable heading device.

The valve is usually mounted in the wing tip where magnetic interference from the aeroplane is at a minimum. Errors in data transmission can be induced from a variety of sources. Examples include misalignment of the unit, permanent magnetism in the aircraft and acceleration errors during prolonged turn manoeuvres.

As the normal motions of aircraft in flight cause errors and fluctuations in ordinary magnetic compasses, the addition of synchronised directional information allowed the introduction of slaved compasses (such as the pioneering Sperry Gyrosyn Compass, developed in 1946) giving extremely accurate and stable magnetic heading information.

Inertial navigation The modern inertial navigation system is the only self-contained single source for all navigation data. Once the device is supplied with initial position information, it is capable of continuously updating extremely accurate displays of aircraft position, ground speed, attitude and heading. It also provides guidance to aircraft sub-systems such as the auto-pilot and flight instrumentation. Following a US Department of Defense requirement, the first inertial navigation system was developed by the Massachusetts Institute of Technology in the 1950s, but not introduced operationally until the early 1960s.

By applying the principles of Newton's Laws it is possible to produce a device that is capable of detecting the minute changes in acceleration and velocities that occur. This forms the heart of an understanding of an inertial reference system (IRS).

The basic measuring instrument in an inertial navigation system is the accelerometer. Inertial accelerometers contain a pendulum that swings off its null position when affected by an acceleration or deceleration. The extent of the swing is measured electrically by a suitable pick-off device.

Two accelerometers are mounted in the system, one to measure movement in a North–South direction and the other to measure movement in an East–West direction. An amplifier circuit converts the electrical signals into a current that drives a torque motor that returns the pendulum to its null position.

The acceleration signal from the amplifier is also sent to an integrator, which is a time multiplication device. The acceleration is in feet per second squared. In the integration, this acceleration is multiplied by time, with a resultant velocity in feet per second. A second integration, in which the input of feet per second is multiplied by time, gives an answer in feet or in miles.

As the computer associated with the inertial system was given the original position of the aircraft, in latitude and longitude units, and has now calculated distance travelled in the two planes, it is a simple calculation for the digital computer to continuously update the present position of the aircraft.

The simplified explanation given above implies that the accelerometers are hard mounted in the aircraft. However, this may result in acceleration errors from a number of sources. The most prevalent are errors that would be induced by gravity. As the aircraft moves in attitude, for example pitch, the pendulum of the accelerometer would tend to swing off its null position as a consequence of gravity. This would induce a false signal as to distance travelled. In order to keep the accelerometer level at all times, a gimballed device has been necessary. This is commonly referred to as the 'inertial platform', usually known as a 'single-axis platform'. In reality, there are similar arrangements in all three axes – pitch, roll and yaw.

A second source of error may occur as a result of Earth rotation. As the gyro-stabilised platform tends to remain fixed in space, but the aircraft is operating relative to the Earth, it is necessary to keep the accelerometers level with the Earth. The apparent tilting of the Earth relative to a space reference is corrected by forcing the platform to tilt in proportion to the Earth's own rotation. The required earth rate compensation is a function of latitude, as the compensation is being applied to the horizontal component of the Earth rate felt by the gyros. At the Equator, this value is 15.04 degrees per hour, reducing to zero at the Poles.

Strapdown systems Accelerometers have shown a high degree of reliability in service so little design development work has been carried out upon them. The problems inherent upon gimbal mounting of the gyro have, however, been investigated in depth and the outcome has been the introduction of a principle which has achieved almost universal acceptance – the strapdown system.

In a conventional platform type of INS the gyros, with $2\frac{1}{2}$–3 degrees of freedom, are gimballed inside the inertial device. The spinning mass is thus isolated from the airframe, and the output signals from it are displacement sensitive.

In a strapdown system, gimbals do not exist, therefore the spinning mass follows the airframe, and the output signals are rate sensitive. The gyros and accelerometers of the system are connected solidly to the aircraft through a chassis arrangement, with the accelerometer's axes aligned with the aircraft's longitudinal, lateral and vertical axes. The gyros are still mounted to detect aircraft yaw, pitch and roll.

Strapdown differs from the gimballed system in the highly sophisticated microprocessor which is used to measure the angular rates of displacement defined by the three gyros and resolve the acceleration signals into Earth-related accelerations, followed by a calculation of the horizontal and vertical navigation components.

Additional compensations have to be performed to take account of the Earth's gravity, rotation and shape, exactly as in a gimballed system. However, these calculations are now performed inside the microprocessor's software.

There are a number of inherent benefits which accrue as a result of utilising a strapdown configuration. These can be summarised as follows:

(a) the gimbals of a conventional system are replaced by a mathematical platform;
(b) the measurement of rate data rather than displacement information has important effects on flight control systems;
(c) as a second order system, manoeuvre-induced errors are minimised;
(d) a variety of function extensions to the system is possible through the use of computer technology.

It also displays improved simplicity, with fewer units, offers improved performance and greater ease of maintenance.

Navigation Systems

AIR DATA ACQUISITION AND AIR DATA COMPUTERS

In aviation terminology, 'air data' refers to the measurements of atmospheric conditions through which the aircraft is flying. Pressure and temperature are measured to provide vital data for the operating crew. Examples of the type of information provided include: barometric altitude, vertical speed, speed relative to the local speed of sound (Mach number), indicated airspeed, true air speed and a range of temperatures, such as static air and total air temperature.

The requirements for improved acquisition of air data information have increased in recent times as a result of the larger flight envelopes available to modern jet aircraft and the proliferation of on-board systems such as auto-pilots and long-range navigation units requiring accurate atmospheric data.

However simple, an air data system requires some form of computational capability in addition to the basic sensing and display functions. This is because only the basic parameters of barometric pressure and temperature can be measured; all other elements have to be calculated.

There are two types of pressure that need to be measured: static (atmospheric) pressure and pitot pressure. The first measures the 'weight' of the atmosphere in a static environment, undisturbed by the aircraft, but varying with changes in altitude and general atmospheric conditions.

Pitot pressure measures impact pressure, caused by the passage of the aircraft through the air. Pitot pressure is higher than static as a result of the ram effect of aircraft motion. The differential between the two measurements is referred to as dynamic pressure, and is sensed by connecting both pressure lines to the sensors.

The practical application of this system requires a pitot tube to be aligned with the longitudinal axis of the aircraft, and static ports attached to the outside skin of it in an area of undisturbed airflow (see Fig. 5.7).

Temperature is measured in a similar way, with a suitable probe extending into the airflow.

Computations are carried out in a variety of ways so that the raw data measured by the systems can be converted into a more useful form. Examples include: static pressure converted into aircraft altitude; a combination of airspeed and altitude to obtain Mach number; a combination of airspeed, altitude and temperature to obtain true air speed. In addition, error correction is possible, either via a manual system, such as corrections in the aircraft's Flight Manual, or through more sophisticated computational processes in an air data computer (ADC).

The development of air data systems began with an elementary pneumatic system, consisting of aneroid bellows, mechanically coupled to the display pointers. To some extent, they are still used as back-up devices, despite inherent accuracy problems.

The ADC has benefited from rapid advances in technology and now has improved performance and reliability, lower power consumption, the capability of performing vertical navigation calculations with reduced weight and increased simplicity leading to shorter repair times and consequent reduction in cost of ownership.

A digital air data computer offers a centralised source for all the air data information required by the crew as well as the automatic flight control systems aboard. Typically, two pressure sensors, one for static, one for pitot, plus a total air temperature probe provide the basic inputs to the computer; controlled by the central processing unit, these signals are converted, if necessary, from analogue to digital data. One feature that becomes feasible as a result of the computational capabilities is the capability of compensation factors for calibration and temperature to be built in via programmable read-only memory (PROM) chips, allowing complete interchangeability between aircraft types.

One important error associated with air data systems results from disturbed air flow around the static ports. This causes differences between the measured static pressure and the true measurement. At low air speeds, the effect is minimal, but with modern jet aircraft, it may be considerable, resulting in variations in altitude and true air speed. Static source error correction (SSEC) is a function of Mach number and it is possible, with the computational facilities of modern air data computers, to make adequate error correction of those aberrations which are long-term and measurable.

Developments in ADC technology now concentrate upon the sensor elements. The most interesting changes are currently occurring on the Airbus A320, where air data modules (ADMs) are positioned very close to the pressure sources. The ADMs convert the pneumatic pressure measured into a linear ARINC 429 databus format. This in turn is fed to the combined air data computer and inertial reference unit (ADIRU) which converts the output into compatible formats for flight guidance functions. The main benefit of these modules is the high level of simplicity which will improve failure rates and reduce maintenance costs.

Lastly, the digital nature of the ADCs and associated data inputs make extensive test and condition monitoring functions more readily available. This includes both continuous health monitoring by the components as well as fault logging in the non-volatile memory. Further discussion of these functions appears in Chapters Two and Five.

ATTITUDE HEADING REFERENCE SYSTEM (AHRS)

It has already been seen that an attitude and heading reference system (AHRS), being an inertial system, provides aircraft attitude, heading and flight dynamics information for cockpit displays, flight controls, weather radar antenna platforms and other aircraft systems and instruments.

The gyros are strapped down and a digital computer integrates the rate data to obtain heading, pitch and roll information, while a flux valve and three accelerometers provide the aircraft computers with the level of dynamic information demanded by high-performance flight control systems.

Typically, three line replaceable units (LRUs) make up each AHRS installation: the flux valve, the compensator/controller, and the attitude and heading reference unit (AHRU). As previously stated, the flux valve detects the relative bearing of the Earth's magnetic field, and the compensator/controller provides the single cycle error correction for the flux valve.

The AHRU is the major component of the system, and, in the Honeywell–Sperry AHZ-600, is made up of four major sub-systems (Fig. 7.8). The inertial measurement unit (IMU) senses the aircraft's body dynamics. It contains the rate gyros, accelerometers and support electronics. The central processor unit (CPU) performs the numerical computations necessary to extract the attitude and heading information. It contains the microprocessor, arithmetic logic unit, programme memory and scratchpad memory. In addition, this unit controls and monitors the operation of the entire system. The input/output (I/O) Unit supervises the analogue-to-digital and digital-to-analogue conversions. The flux valve is connected to the I/O unit through the compensator using the current

Fig. 7.8 AHZ-600 simplified block diagram *Honeywell*

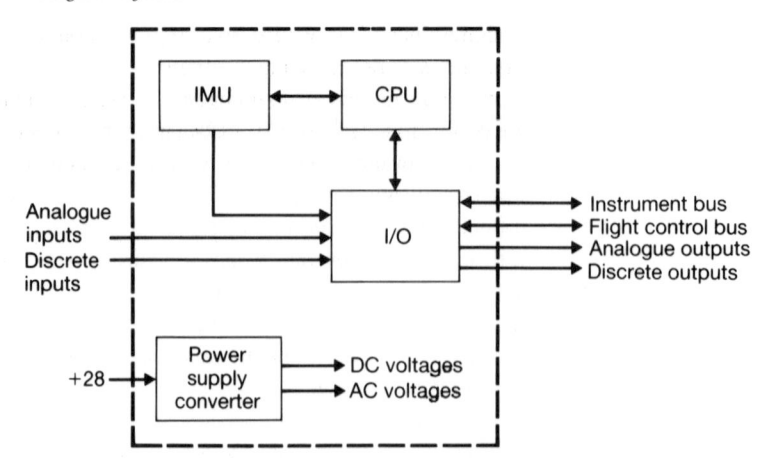

servo approach which provides increased accuracy over conventional open loop connections. The power supply converts aircraft power to the regulated d.c. voltages and a.c. power signals required by the system. The AHRU should be mounted at the centre of gravity of the aircraft.

The AHRS offers several advantages over traditional vertical and directional gyros. Principal among these is in the area of performance. In a conventional vertical gyro, automatic vertical erection is 'cut-off' when the roll angle, and in some cases the pitch angle as well, exceeds a certain value, typically 5 to 10 degrees. This causes the gyro to produce a vertical error proportional to its free drift during extended large bank angle manoeuvres. A related problem concerns shallow bank angles just below cut-off, where the automatic erection loop causes the gyro to erect to a false vertical induced by the turning acceleration. In order to overcome these problems, the AHRS uses velocity-damped vertical erection loops proven in advanced military applications.

Conventional gyros are susceptible to 'gimbal lock' under certain conditions. As the AHRS dispenses with gimbal mounted gyros this circumstance cannot occur.

As modern flight control systems are able to make effective use of rate and acceleration feedback terms, which are not directly available from simple directional and vertical gyros, the use of an AHRS has reduced the number of components such as additional rate gyros and accelerometers. This is possible as the device provides direct measurement of these quantities and supplies all required data for the DAFCS in both Earth-based and aircraft-body axis coordinate reference frames. Angle of attack is computed for the flight control system as a function of true airspeed, altitude rate and pitch angle.

In addition to the special outputs, the AHRS provides the standard analogue signals such as pitch, roll and heading synchros, slaving error, true rate of turn and normal acceleration. Also, analogue outputs are generally available for weather radar antenna stabilisation in pitch and roll.

AHRS systems offer improved levels of self monitoring over earlier devices. The central processor in the AHRU performs continuous self-checking of data and computations. A pre-flight test provides pilot verification of system operation.

The AHZ-600 has two modes provided for routine operation, the normal mode in the attitude channel and the slaved mode in the heading channel. An air data computer provides true air speed allowing compensation for acceleration-induced attitude errors. The slaved mode for heading utilises the flux valve to align the heading outputs.

In-built safety devices compensate for the failure of any elements in the system, allowing safe continued flight in the event of failed inputs such as true air speed, or a slaving loop failure.

Several test modes are routinely incorporated in AHRS systems. These can be automatic, on power-up, or initiated by crew command, and are in addition to continuous monitoring of function and output.

LASER TECHNOLOGY

In recent years, more and more extensive use has been made of laser technology. The medical field was among the first to make radical use of the special properties associated with light. However, the laser gyro is among the more remarkable of a growing list of uses.

As discussed earlier, strapdown inertial technology measures rotation rate. The laser gyro measures this rate by using the properties of two laser beams rotating in opposite directions within a cavity.

The principles of laser generation go beyond the requirements of this book. However, a brief description will prove useful. A d.c. voltage is applied across a laser cavity, establishing an electrical discharge in the mixture of helium and neon gases present in the cavity. The discharge that develops is very similar to that produced by a neon sign.

A complex physical activity now takes place, whereby the neon atoms are 'excited' by collision with the helium atoms. The high energy neon atoms cause the generation of additional photons, resulting in the amplification of light, or 'lasing'. The repeated amplification of the light that is forced to reflect between mirrors results in a steady state oscillation – referred to as a 'laser beam'.

A laser gyro measures attitudes, as does a traditional gyro, but it operates in an entirely different way. It contains at least three mirrors so that the laser beams travel around an enclosed area. Such a configuration allows the generation of two distinct laser beams occupying the same space; one travelling in a clockwise direction, the other anti-clockwise. The operation of the ring laser gyro is based on the effect that rotational movement has on the two beams.

The effect utilised is called the Sagnac effect, which, put simply, is that variations in the laser beam's length, caused by any rotational effects cause a change in frequency of the beam's wavelength. These frequency changes are extremely small, and would be impossible to measure to the required accuracy. However, the difference in the frequencies of two beams travelling in opposite directions within the same cavity can be measured, because of the interference or fringe pattern formed. The signal generated is inherently digital, and hence needs no processing prior to transmission to the microprocessor element of the inertial unit.

Since its invention in 1960 the device which, strictly speaking, is neither a ring nor a gyro, but an accurate and extremely sensitive sensor of rotation, has shown remarkable reliability with mean time between failures measured in thousands rather than hundreds of hours. On tests the laser gyros have run for 60,000 hours without failure. The British Aerospace facility at Bracknell tested their first system in a Royal Aircraft Establishment Comet in October 1981. In two years of flight test the system exceeded its design accuracy, achieving two nautical miles per hour in flights of over ten hours' duration. BAe is currently developing 'Triad', a three-axis monoblock laser gyro in which rotations in all three axes are measured

Navigation Systems

by the same gyro. This is expected to be less expensive than the three single-axis units which it will replace.

The Laser gyro Using the Honeywell Ring Laser Gyro as the basis for description, each gyro is a triangular shaped laser that produces two light beams travelling in opposite directions. Laser radiation from a central source is reflected around the triangle by mirrors at each corner. One of the three corners has the prism that allows the two beams to mix together to form the interference fringe pattern on the detector (Fig. 7.9).

Fig. 7.9 Ring laser gyro – schematic drawing *Honeywell*

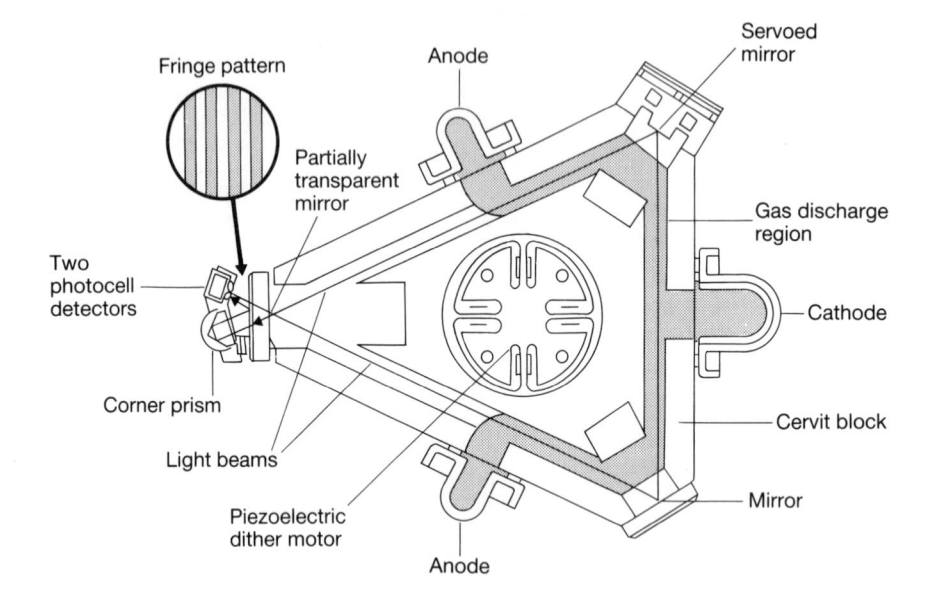

While the gyro is stationary, the fringe pattern will also remain stationary since the frequencies of each of the beams is the same. Rotational movement of the gyro will cause a shift in the fringe pattern to compensate for the variation in frequency now measured.

At low rotation rates, the two light beams can get 'locked-in' together. To prevent this happening, resulting in a loss of information at these rates, a dither motor is installed to vibrate the assembly thoughout the lock-in region, and the beams are then optically decoupled from the gyro output.

Accuracy of a laser gyro is a function of the length of the optical path, with changes in accuracy improving greatly for a relatively small increase in length. Considerable research has taken place on the question of four-versus three-sided gyros. Honeywell opted for the triangular shape as a means of obtaining acceptable accuracy from the smallest possible space.

The outstanding level of reliability and accuracy which has been achieved with inertial guidance systems has led to their introduction in most major jet transport aircraft, a trend which is likely to continue for, at least, the rest of this century. The greatly superior reliability of electronic components over the simple mechanical systems which they have replaced has been proved conclusively in well over half a million flight hours.

Figure 7.10 demonstrates the extent to which inertial reference units provide vital data to all of a modern jet aircraft's sub-systems, hence the need for the levels of reliability being achieved.

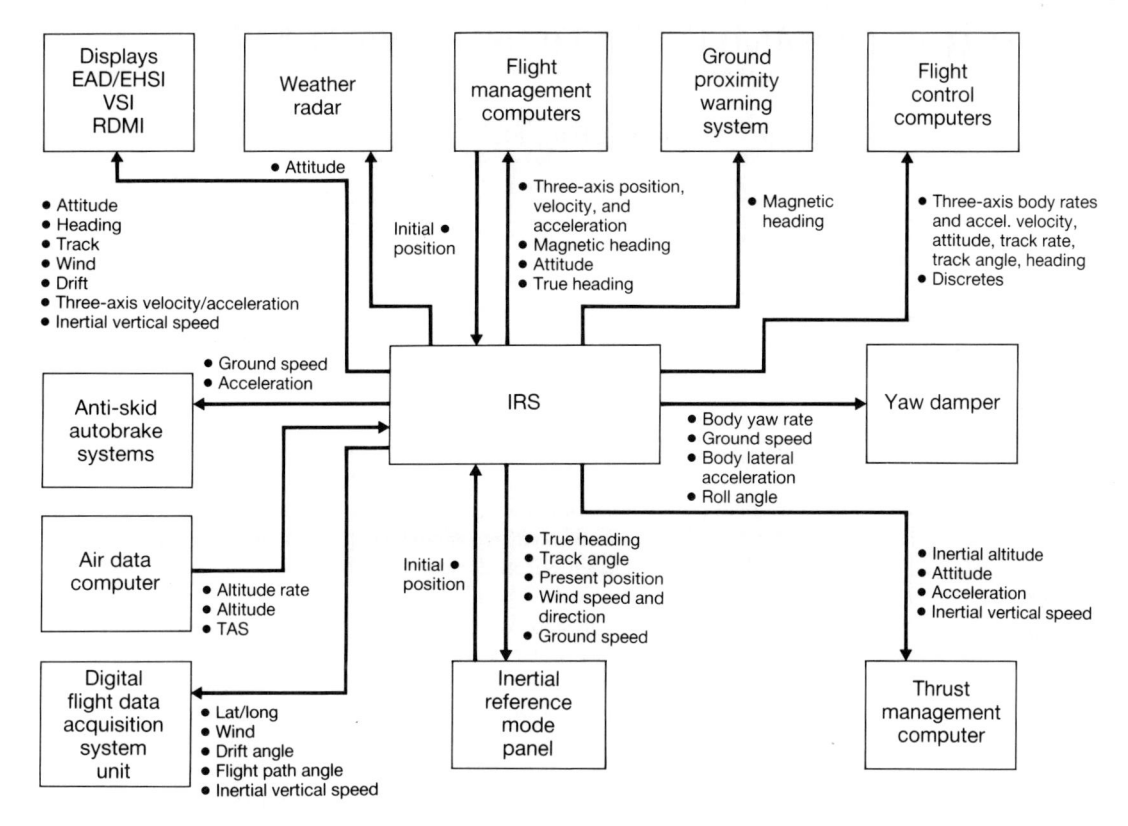

Fig. 7.10 Boeing 757/767 inertial reference unit interface diagram *Honeywell*

The future offers a new generation of inertial references based on fibre optic technology, still generally at a concept stage. This will allow for improvements in data transmission due to the virtually 'noise'-free environment that would then exist.

BIBLIOGRAPHY

Beck, G. (ed.), *Navigation System*, van Nostrand Reinhold, 1971.
F A A, *Instrument Flying Handbook*, Various.
Flight International magazines, Various.
Honeywell Avionics, Training Manuals, Various.
Sperry Commercial Flight Systems, Training Manuals, Various.
Stensland, R. A. *Principles of Strapdown Laser Inertial Navigation*, 1984.

CHAPTER 8 COMMUNICATIONS SYSTEMS

THOMAS T. BROWN CEng, FIEE

Mr Brown has recently retired from GEC Avionics Ltd. Basildon where, since 1960, he has been involved in major research and development programmes, such as VHF, high power SSB equipment, Selcal, Area navigation, TACAN and UHF. Since 1976 he was worked on computerised AF systems, secure speech, introduced ATE and carried out investigations into JTIDS.

In 1979 he was seconded to The Hague to join the MIDS team, an international body on Command, Control and Communications, while from 1981 to 1984 he was involved in JITDS frequency clearance with the CAA and RAE.

In 1939 Mr Brown qualified with a Higher N.Dip 1st Class and the Polytechnic Bronze Medal in Radio Engineering at the Polytechnic of Central London. From 1940 to 1946 he was a Royal Air Force Signals Officer, the last two years of which were spent in the Interservices Research Bureau on clandestine radio design for the Special Operations Executive.

After the war he joined Marconi W. T. Ltd. in the Air Transmitter Dept. Writtle. Later, as head of the Air Receiver Dept. he designed, among other equipment, the Comm/Nav VHF receiver used for the original Autoland system.

Mr Brown holds 16 patents and, excluding restricted reports, has published seven technical papers.

All branches of electronics have sprung from and are based on classic radio theory and use components originally developed in this field. Later, specialised branches evolved in radio communication, such as propagation and antenna theory, navigational aids, radar, (and television) servo control, computers, digital techniques, and information theory.

Avionics makes use of all these disciplines and, while radio communication itself has been in use since the early part of the century, its functioning is being continuously transformed by re-incorporating devices and systems drawn from subsequent development in the above branches in a somewhat 'chicken and egg' process. It has been suggested that major advances in science occur in areas in which two disciplines overlap. In the case of avionic communications, examples are frequent. The necessity of miniaturisation of components plus the development of transistor technology led to the advent of the integrated circuit, this in turn was incorporated in the computer field leading to the microprocessor which again has transformed the way in which signal processing and data handling are carried out. For example, the availability of powerful computing facilities has realised mathematical processing with information theory for signal coding and error control, real time channel evaluation (RTCE), voice synthesis (Vocoder) and voice recognition (probability theory).

Servo controls originally developed for operation by remote sensor or manual control are now operated by microprocessors utilising stored data. Such systems automatically tune receivers and transmitters and match antennas, if necessary, according to the aircraft location.

In conjunction with appropriate flight control systems, aircraft on scheduled flights have been landing automatically utilising the signals from the VHF/UHF ILS ground transmitters for about the last two decades, showing among other things, the reliability and stability of this generation of receivers.

In the succeeding sections, only a brief aspect of certain avionic communications can be covered in the space. Further coverage of related subjects as above can be found in other chapters of this book and more detailed information may be obtained from the bibliography at the end of this chapter. However, at this stage a summary of the principles of engineering design of avionic units may be helpful.

DESIGN PRACTICE

The specific design requirements of avionics communication equipment is not apparently dissimiliar to that of normal high-grade communication equipment except that it usually covers greater frequency ranges and is intermittently being channelled by remote control. In its civil or military aeronautical environment, equipment differs significantly both in performance and reliability from equipment designed for working in a conventional environment. Also, safety and lives may depend on its functioning at all time and in all conditions.

A prime requirement is reliability, and a high proportion of the cost results from the extra circuit and mechanical design needed to maintain a full operational performance in an environment far more severe than in office or factory. Reliability depends on both circuit and mechanical design. Circuit design comes initially and depends on concepts in part original, but which must have a sound engineering basis. Much of this depends on the experience gained with earlier designs in practice, and may not necessarily be obvious to those coming in from other fields of engineering.

Much may be summarised as 'tolerancing'. The overall equipment must perform with only a small given percentage degradation under climatic variations of temperature from $-60\ °C$ to $+70\ °C$, atmospheric pressure from that equivalent to sea-level to 80,000 ft, humidity to 100 per cent, effectively from steam to freezing water. These conditions are realised practically by testing in environmental chambers in which the conditions are cycled over the extremes for periods of more than a week. In addition, mechanical tests of vibration using frequency sweeps from 0 to 2000 Hz acceleration and bump tests are also carried out in this period. Various selected circuit parameters will be monitored by instruments and recorded throughout all these cycles. Later the equipment will be tested for power supply variations including voltage surges, spikes and ripple. EM testing including susceptibility to external electrical interference and production of unwanted harmonics and spurious radiation products is also carried out.

The reason for the 'tolerancing' concept in design thus becomes apparent, as there is no point in designing a 100 MHz oscillator using a ceramic capacitor with temperature coefficient of -750 ppm as tuning element, since the frequency of the oscillator can vary by about 3.5 MHz between the operating temperature extremes, although this design might be acceptable for domestic equipment at room temperatures. Similarly, transistor parameters vary greatly with temperature; until recently the design of high gain d.c. amplifiers to function over such temperature ranges was difficult, mainly because of temperature differentials between the transistors and compensating components. Later, when i.c. operational amplifiers on one chip became available, this and any other compensation were already incorporated in the chip, so the difficult part of the design had already been pre-packaged.

As a consequence of these environmental conditions, it is essential to 'know your components', and the effects of variation of supply voltage, temperature, humidity and drive level on them and also the effects of ageing and overloading on overall performance characteristics.

The function of analogue circuits in particular is dependent on their transfer characteristics, which are only approximately linear with environment and ageing. Variation of components causes drift in overall system parameters such as gain, tuning, power output, feedback, etc.

There is thus a very large number of possible combinations of characteristics drifting in different ways, the sum of which must remain within an overall performance tolerance. For this reason it is usual and necessary to produce several models of the same equipment to check that these drifts are minimal, and that the manufacturing tolerances are acceptable to give the required performance. Of course in the initial design all these limits should have been calculated and/or measured in detail. Today it is practicable to do these calculations both faster and to a greater extent by computer simulation, but nevertheless it is still essential to prove the design by making such models. These normally consist of a 'breadboard' produced in the laboratory to devise the circuits experimentally, followed by 'A' models to approximate the mechanical layout from preliminary DO drawings from engineer's sketches, and then after preliminary testing, a series of 'B' models made to prove the DO mechanical and assembly production drawings incorporating modifications found necessary from the 'A' models. Apart from proving the production drawings, 'B' models also check the ordering system. One of the models is used for Type Approval testing in the environmental chambers, while the others are used with the customer for system and flight trials, checking Ministry of Defence specifications etc. The number of models made depends on the complexity of design and size of order, a sequence of 1, 4 and 10 models being typical. Production then follows these. Some companies term 'B' models, 'pre-production' models, and in other cases, the 'A' and 'B' models are telescoped into 'AB' models when printed-circuit board (PCB) design has been used initially.

The performance of current digital type systems shows a considerable improvement over the older analogue systems in both manufacturing consistency and reliability in the field. This appears inherent, as each element in a digital system has only two states, whereas there is continuous gradation in parameters in analogue systems. Good design in the past has often made use of this knowledge where appropriate – as the old transmitter design adage says, 'Bias hard and drive hard'. This was evinced in another aspect by the speed and consistency in which ranges of digital i.c.s appeared on the market compared with analogue i.c.s, which tended to be both late in availability and variable in performance, indicating greater design and manufacturing difficulties. One might safely conclude that the steady replacement of analogue by digital systems has already resulted in improved communications reliability and will continue to do so.

FREQUENCY BANDS

Allocation of frequencies is by international treaty the responsibility of the International Telecommunications Union (ITU). Geneva. Frequency band nomenclature is defined in the ITU Radio Regulations, sect. 2.1. in decade steps of 3 to 30 Hz upwards. (See Chapter Seven.)

Aircraft communication is carried out by radio in the following frequency bands:

VLF, Very low frequencies.	3 to 30 kHz
MF medium frequencies	200 to 1600 kHz
HF high frequencies	2 to 30 MHz
VHF very high frequencies	100 to 200 MHz
UHF ultra high frequencies	200 to 400 MHz.
” ” ” ”	1540 to 1660 MHz

These adjectival designations are convenient in referring to transmissions in those bands allocated for radiocommunication by aircraft. Other designations, such as C-band and X-band etc, used for radar and ground/satellite transmissions at microwave frequencies,. are only semi-official, and not being directly involved in airborne radio communication installations are not quoted.

VLF These frequencies are receivable world-wide owing to their low propagation attenuation and the high power of the ground transmitters used, e.g. Rugby at 5 megawatt.

Owing to an appreciable ground penetration, they can be used for communication with submerged submarines and they have some use for this purpose aboard aircraft. Otherwise the main aircraft use is for Omega navigation signals.

MF This band is mainly used for nondirectional navigation beacons (NDBs), for use with automatic direction finding receivers (ADF).

HF This is the basic band for long-range communication, mainly because its transmissions are reflected from the ionosphere. These reflections cause the 'skip distance' effect with areas of non-reception depending on frequency, day/night ionisation state over the path-length, number of sunspots, etc., and require sophisticated estimation of choice of channel to ensure reliable communication.

VHF This is the standard air traffic control band using speech modulation. It is limited to LOS, 'line-of-sight' range, normally about 200 nm. The lower end, 108–118 MHz, is restricted to VOR/ILS functions, the remainder is used for civil aviation.

UHF This is similar to VHF, but is restricted mainly for military aviation use. Although in the navigation band with Tactical Air Navigation TACAN, and distance-measuring equipment (DME). JTIDS on 969–1215 MHz will be used for communications as well as navigation. 1540–1660 MHz will be available for air/satellite up/down links, the corresponding ground/satellite links being around 4.2 and 6.4 GHz.

HF TRANSMISSION

High frequency (HF) transmission nominally in the range 2–30 MHz for avionics is the original frequency band for aircraft communication dating from the earliest days. Its use falls into civil and military areas. In the civil area it is expected that it will be replaced by satellite communication by means of geocentric satellites at a distance of 23,000 miles above the Earth. However, the vulnerability of the satellite in wartime makes this link highly suspect for military use, and the flexibility of HF transmission for this purpose makes this still one of the most favoured communication systems. Originally limited by the cost and weight of higher-powered airborne transmitters, these limitations have been greatly reduced by the use of single-sideband (SSB) and all solid-state output stages enabling output powers up to 500/1000 watts to be practical. However, it should also be remembered that amateur transmitters using low powers of only up to 20 watts, for decades, and clandestine portable transmitters during the Second World War, gave remarkable ranges of many hundreds and thousands of miles, by using wireless telegraphy, the oldest and still possibly, the most effective system, which involves keying the continuous wave (CW) carrier of a transmitter in Morse code. and receiving the message by heterodyning the carrier with an oscillator in the receiver to give an audible note. Such synchronous or coherent demodulation gives a maximum signal/noise (s/n) ratio, as opposed to the unsynchronised envelope demodulation of amplitude modulation (AM), which is normally at least 9 dB inferior. (although this is less of a problem with the introduction of single-side band (SSB) systems – see Transmitter modulation).

Owing to the range of HF propagation, all kinds of electrical interference caused by ionospheric disturbances, thunderstorms ('static') electrical apparatus and other transmissions are continuously receivable from great distances all the time. This provides the typical radio noise background, the absence of which accounts in part of the popularity of VHF.

Fading is another characteristic caused by a combination of signal strength variation due, in turn, to skip distance effects leading to multi-path reception. This results from long term and instantaneous changes in the ionosphere and there is a critical frequency at which the ionosphere reflects a vertically incident wave; frequencies higher than this pass through the ionosphere and are not reflected. For given distances and hence oblique angles of incidence, the maximum usable frequency (MUF) is given by the critical frequency times the secant of the angle. Similarly there is a lowest usable frequency (LUF) in which noise level and ionospheric absorption limit transmission. These two limits change continuously throughout the 24 hours, so that an operating frequency is chosen which stays between them as long as possible.

Even using special charts, the calculation and forecasting of these frequencies can be a formidable task requiring a degree of experience of local conditions (Fig. 8.1). It is important to note that the difficulties and technique of HF were

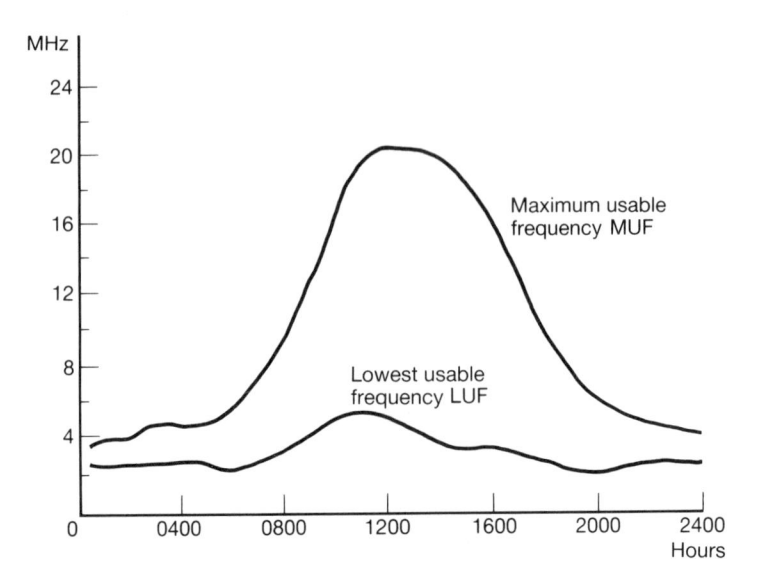

Fig. 8.1 Variations of HF transmission with time of day

such as to require a dedicated crew member on long range commercial and military aircraft. With the advent of microprocessors able to search for, and automatically lock on to, the best HF frequencies, many companies consider HF is making a comeback, and some US electronics industries are reputedly re-training their engineers in HF technology. In this context techniques are being developed to assist the problem of unattended HF sets with automatic continuous propagation and similar measurements – see Real Time Channel Evaluation.

To mechanise transmission and obviate the need for trained Morse operators, digital data links have appeared. These tend to be mostly military, such as Link 11, etc., and, apart from speeding message rates, enable sophisticated signal processing systems to be developed. A comparatively straight-forward system is RTTY or radio teletype.

A predecessor of data links was machine Morse, an interesting and very simple military system which pre-recorded the message in Morse on alternate metal and insulating rings on a rod. Brushing the rod with a metal input contactor gave a data rate of several hundreds of words/minute compared with 10/25 w.p.m. of normal manual Morse. This technique, called 'squirt' was evolved during the Second World War for clandestine agents to manage without trained operators and minimise the danger of locating the transmitter by radio direction-finding.

UHF/VHF TRANSMISSION

The VHF and UHF bands, nominally 100 to 200 MHz and 200 to 400 MHz respectively are line-of-sight (LOS) propagation, aircraft transmitters are normally rated at about 25 (watts output power maximum which is sufficient to give a range just beyond the visible horizon from an aircraft (Fig. 8.2). The effective Earth radius for VHF is about 1.33 times the actual Earth radius. A convenient formula is that maximum range in miles is: $\sqrt{2 H_t} + \sqrt{2 H_r}$ where H_t is the height of the transmitter and H_r is the height of the receiver in feet.

Fig. 8.2 UHF/VHF transmitter/receiver. AD 3500. Frequency range 108–173, 975, 225–399.975 MHz. Channel spacing 25 kHz. Power output 15 watt FM, 10 watt AM *GEC Sensors*

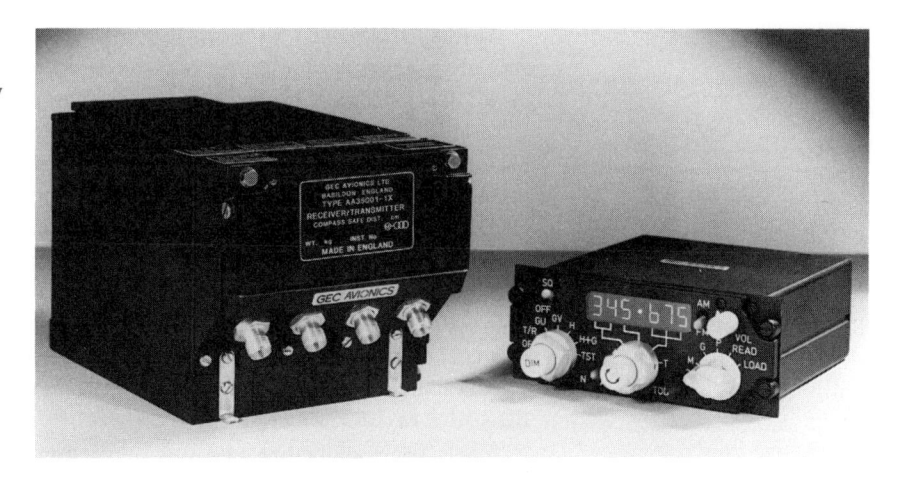

Mostly these bands are used for air navigation – VOR/VHF omnirange and ILS at the lower end of the VHF band from 108 to 117.975 MHz while the glideslope functions of ILS operates between 328.6 and 335.4 MHz. Otherwise the VHF band is used by civil aviation for voice communication such as ATC air traffic control, position reporting, etc. The UHF band is used similarly by military aircraft for ground/air, air/air operational purposes. There is also a semi-VHF band between 30 and 88 MHz that is often provided on VHF/UHF sets for close ground support in military operations by helicopters, etc.

In the case of long-range civil routes there is a tendency to replace HF by VHF, resulting from the increase of ground stations, and the better quality of voice transmission. However there is the disadvantage of having to change channel frequently owing to VHF short range and the high speed of modern aircraft. This has been met by two developments; in the receiver, by area navigation (AN) in which VOR and DME information is processed by a computer with the appropriate channel being selected automatically, and on the ground by the installation in ground stations of the offset carrier 'Climax' system. Here three stations equidistantly separated by just under twice the line-of-sight are distributed on the circumference of a circle. The three transmitters are on a common channel but each is offset in frequency from its neighbour by 7 kHz, so that all three transmissions fall within the 25 kHz bandwidth of a receiver. With a voice modulation common to all three, normal voice modulation is receivable over an area approximately four times that of a single transmitter without needing to change channel at the receiver. A disadvantage is that the modulation is accompanied by a 7 kHz beat note which approaches 100 modulation when two signal strengths are equal, i.e. on a line equidistant and normal to the line joining two transmitters. This note is rejected by the AF bandwidth of the receiver cutting off at 3/3.5 kHz. Problems may occur however, with receiver 'squelch' systems which are intended to mute the voice signal when the signal/noise ratio falls below a pre-set level, as the beat note appears as a noise level. This requires additional circuit design to overcome the problem.

REAL-TIME CHANNEL EVALUATION

Real-time channel evaluation is a systematic continuous method of selecting radio channels to obtain those with optimum signal/noise (s/n) ratio using ancillary search equipment.

Basically it amounts to measuring the attenuation between base station and aircraft and also the noise on each of a selection of channels which are continuously scanned to enable best choice for communication.

Various parameters have been found to correlate with these characteristics. A low-level pilot tone transmitted from base is phase compared with a high stability reference oscillator in the aircraft, the phase difference being measured every 10 msec and the errors counted in every 200 sec. The phase error rate or phase stability gives a very good measure of bit error rate (BER) to be expected if this channel is used for data or digitised speech.

Similarly, a probe bit-stream signal can be used to evaluate a channel making use of a count of the BER from existing error-correcting code equipment. A BER of 10^{-3} is a minimum for digitised speech, whereas machine data requires BER better than 10^{-6} or 10^{-8}, according to the system.

A selectively addressed data test signal from base to aircraft on each of the available channels gives the value of the noise measured on that channel at the base together with a reference tone. These channels are scanned by the aircraft receiver controlled by a microprocessor which calculates the s/n ratio from the level of the received tones and noise values transmitted. Assuming reciprocity of propagation, the optimum channel is selected by the microprocessor algorithm allowing for antenna characteristics and power levels.

For 'single-hop' radio transmission, the amount of back scatter from the ionosphere from a pilot transmission is an indication of the amount of attenuation on that channel. The transmission can be a frequency sweep or 'chirp' pulse with the reflections or ionograms displayed on a cathode-ray oscilloscope (CRO) screen against a frequency base rather after the fashion of the pre-radar experiments of Appleton or Breit and Tuve.

The first three methods lend themselves to continuous automatic selection and tuning of microprocessor controlled aircraft equipment.

AIRCRAFT ANTENNAS

Aircraft antennas fall into five categories – blades, slots, bowls, loops and long wires – and they are also grouped into frequency bands. An essential requirement is minimal wind resistance, structures being either streamlined as fins (blades) or suppressed as slots in the empennage.

VHF and UHF antennas are usually rods or fins, acting as monopoles, which are effective as the wavelength is comparable with practical physical dimensions.

At HF, the notch antenna is particularly efficient as it excites the airframe as a radiator for only small changes in the aircraft structure with virtually no aerodynamic losses. It is usually regarded as an open ended slot and used as a broadband radiator.

Slot antennas are based on Babinet's principle that the slot in a conducting surface behaves as a radiator of comparable dimensions to a conductor in a dielectric medium. The impedance of the slot is $(60\,\pi)^2$/(impedance of the complementary conductive antenna). For example, for a slot $\frac{1}{2}$ wavelength long, the impedance is 496 ohm compared with the 73 ohm of the dipole.

The siting of antennas on the fuselage to provide polar diagrams for required coverage, normally requires measurements from scale models of the aircraft. Adequate attenuation between transmit and receiver antennas is necessary, particularly at VHF, and between communication transmitters and ILS or other navigation receivers. With an antenna above and one below the fuselage, about 40 dB is possible, the metallic aircraft structure acting as a screen. It is also possible to design the vertical polar diagram of the ILS receiver to be complementary with that of the ground transmitter so that almost constant signal strength is received at distances from near LOS to touchdown.

For helicopters, a successful antenna is a long rod spaced off and running almost the length of the fuselage. Providing the fuselage has a conducting skin, the antenna forms a large loop with the aircraft body.

VLF antennas present special difficulties; with a frequency between 17 and 60 kHz, wavelengths are 17,647 and 5000 metres. With an aircraft length of 70 metres, this is much smaller than even 1/8 wavelength. Trailing wire antennas have been used on subsonic aircraft consisting of either a single wire from 2.5 to 8.5 km, or a wire and counterpoise wire of 1.2 and 7.2 km respectively trailing with an angle of 3 degree/separation.

Receiver antennas do not need high radiation efficiency as do those of transmitters. Here s/n ratio is the criterion and if the atmospheric noise input is very much greater than the receiver noise figure, there is no advantage in using large antennas to pick up both more signal and noise. At MF and below, atmospheric noise increases with decrease in frequency and at VLF, galactic noise, atmospherics and man-made static are very high. Precipitation and local man-made static tend to be electrostatic and can be reduced by electrostatic screening of a loop antenna. Loop dimensions can be reduced with a ferrite core and a pre-amplifier to match to a transmission line, when it is termed an active antenna.

Another form of active antenna uses a short rod monopole with a remote amplifier at the base; such a system is very broad-band, from 16 kHz to 30 MHz. These arrangements are subject to intermodulation by interfering signals, requiring special type FET input circuit design. By raising the antenna and amplifier above the fuselage, out of the static area by a grounded metal mast of height M as a monopole, the effective height of the antenna of height A is A/2 when M = 0, but when raised, is A/2+M. So ratio of effective heights is: (M+A/2)/(A/2) If A = 1 metre and M = 2 metre ratio = 2.5/.5 = 5 (15 dB).

Antennas for satellite operation at 1.6 GHz must be highly directive to ensure sufficient gain to overcome the space attenuation of 187 dB due to the distance of about 36,000 km to a geocentric satellite. Such an antenna must be steerable, either mechanically, or by a phased array in which the steering is done by differentially shifting the phase angle of each of the elements of the antenna. Such a design could consist of an array of discs plated on to a dielectric base sandwiched on a metal baseplate. These circular plates of quarter-wavelength diameter, are each fed through a variable phase-shifting network controlled electronically. By programming the phases correctly, the polar diagram of the assembly can be focussed onto the satellite to give antenna gains of the order of 10 dBi, which, taken with the satellite antenna gain, is adequate to provide sufficient signal strength.

Antennas: matching systems

The basic antenna can be considered as a twin wire transmission line opened up so that the parallel wires now fan out 90 degrees apart from the input. Fundamental transmission line equations hold, the real part of the input impedance being composed of the radiation resistance and series and shunt losses of the antenna.

At VHF and UHF, antennas are designed to provide relatively constant input resistance across the working frequency band matching the characteristic impedance, Zo, of the coaxial cable connecting to either the transmitter or receiver, usually 50 or 70 ohm. When matched, there is minimum loss in the cable, the transmitter is correctly loaded and most of the output is radiated.

Should these impedances differ, the cable is mismatched and energy is reflected back from the cable/antenna junction, interfering with the forward waves to provide a standing-wave pattern of alternate voltage maxima and minima along the cable occurring at physical lengths corresponding to about a quarter the wavelength of the frequency transmitted. The ratio of the amplitudes, V_{max}/V_{min}, is the voltage standing-wave-ratio (VSWR) and is a measure of the mismatch. Apart from power loss, the cable itself radiates and this spurious radiation can cause interference with adjacent receiver circuits, or if a receiver cable, allow pick-up of interference from local sources in the aircraft.

The voltage SWR should not be greater than about 1.3 to 2.0 for transmitter cables, but in the case of receivers, SWR may be larger, say up to 10, since the power levels are orders lower than the transmitter and the receiver a.g.c. will mask input level variation, the main effect here being a loss in signal/noise ratio. As SWR effects depend on wavelength and cable length, the effect on loss of short cable runs at lower frequencies is not great. However at higher frequencies and/or with longer lengths, the effects become more critical, until near the GHz band even the effects of plug and socket connectors have to be considered and generally the antenna connections predicate waveguides rather than coaxial cables.

Measurement of SWR from which antenna impedance could be calculated is made by inserting a section of tapped or slotted line in series with the cable and antenna and measuring the difference of voltage level at successive intervals. Alternatively the resistive and reactive components of the input impedance may be measured at discrete frequency intervals with an r.f. bridge, and in either case the readings plotted on a Smith chart. More conveniently, this is now done directly with a swept frequency device such as the ZG Diagraph or HP Network Analyser giving real and imaginary components and phase angle of the input impedance (or admittance) as a continuous characteristic across the frequency bandwidth. These devices have the advantage that matching stubs (sections of line to simulate desired reactance characteristics) may be designed to bring the input impedance of the antenna to match the transmission line across the band.

With HF antennas, the case is quite different. The frequency band ratio is about 15:1 (2–30 MHz) compared with only about a 2:1 ratio for the VHF and UHF bands of 100–150 and 200–400 MHz.

The antenna length is normally small compared with the wavelength, so, together with the variety of designs encountered on aircraft, the problem arises of matching a 'random antenna' throughout this band to 50 ohm.

Input impedance/frequency characteristics of such antennas show fairly violent changes in series reactance across the band changing sign every quarter wave fractions of the wavelength.

The typical 'short antenna' has a very low series radiation resistance in series with a large capacitive reactance, say a few ohm resistance and up to 500 ohm or more reactance at any one frequency. To obtain maximum efficiency it is necessary to resonate out the reactance with another of opposite sign and transform the series resistance into an equivalent shunt resistance to match the transmission line. Earlier designs of transmitters coupled the antenna directly into the output stage using a simple parallel LC circuit in which the antenna was tapped up the power amplifier PA tank inductance, while manually maintaining tuning with a variable capacitor until an r.f. ammeter in series with the antenna read maximum antenna current. This necessitated a series of interrelated adjustments for tuning and matching at each frequency change.

The next generation design used in the Second World War was to preset controls for a small given number of frequencies by mechanical click-stops that were set up before flight. After the War, these click-stops were replaced by designs using remotely controlled stepping motors. These systems were limited to not more than a dozen or so preselected frequency channels, so with the growth of aircraft communication traffic and the concurrent appearance of frequency synthesiser systems permitting almost limitless choice of frequency channels at will, limitations in operation with such types of antenna tuning units (ATUs) led to the design of more complicated automatically tuned matching units using pi-filter networks in which the elements were driven by servo motors controlled with sensing devices such as phase-detectors and resistance magnitude detectors. The matching network variables were rolling coil inductors and vacuum sealed variable capacitors (to deal with the high r.f. voltages). The sensors measured the resistive and reactive components at the antenna base and servoed the output of the antenna tuning unit to 50 ohm to match the transmission line (Fig. 8.3). In order to reduce losses, later ATUs were designed in sealed containers mounted directly at the aircraft antenna base, since, with short antennas, particularly notches, having fractional or only a few ohm series resistance, most of the power could be lost in the series resistance of the connecting wire and earth return.[1] Similarly, the voltage across the high antenna series reactance will be many hundreds of volts to force sufficient current through the antenna resistance. In fact in one initial

Fig. 8.3 Automatic antenna tuning unit (ATU) – basic circuit

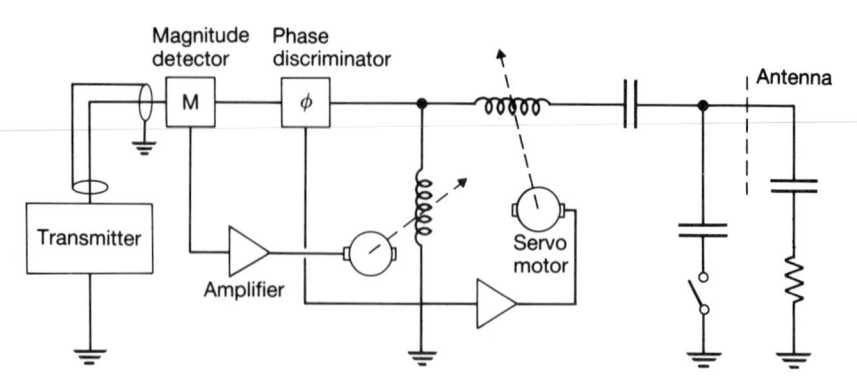

installation, due to partial mismatch, keying the transmitter turned on some of the aircraft lights and interfered with the flight control servo system.

These ATUs are relatively bulky and being mechanically driven suffer normal mechanical maintenance problems. For some later applications, they are somewhat slow to tune, 20–40 seconds being typical. The latest designs use fixed inductors and capacitors switched by vacuum reed-relays in binary banks after the fashion of digital-to-analogue converters, being controlled by microprocessors from the antenna sensor information. Such ATUs can tune in five seconds or possibly less and have virtually no moving parts. Network combinations can be set up from programmes in the computer read-only memory (ROM) so that with given antennas, operation in say a frequency-hopping mode, need only depend on very few relays with an order of milliseconds operating time. By using heavy duty PIN diodes, this can be reduced still further.

RECEIVERS

Current airborne receivers are virtually free of mechanical moving parts such as tuning capacitors and waveband switches, these functions now being carried out by remotely controlled varactor tuning diodes and diode switches.

The replacement of RF transistors by MOSFETs has greatly improved cross-modulation and spurious responses due to unwanted high level transmissions. Figures of −80 dB spurious rejection and comparable cross-modulation are now available. These characteristics are particularly important when receivers are used in conjunction with navigation processing units or are self-contained navigation receivers for ILS or ADF use, as spurious inputs can lead to navigational errors.

Adjacent channel selectivity obtained by ceramic IF filters give flat frequency passbands varying between about 8 and 30 kHz for 6 dB cutoff with about twice these bandwidths at 60 dB or 80 dB down according to the operational requirements.

Automatic gain control (AGC) line filtering is important, as 30 Hz VOR or 90/150 Hz ILS signal traces can remodulate the incoming RF signal causing errors in the demodulated navigation output. Similarly it is necessary that phase-shift through the RF and detector stages is zero at these frequencies. Phase-shift of the 480 Hz deviation of the 9960 Hz VOR FM sub-carrier may be caused by ripples in the ceramic IF filter passband causing frequency discrimination and resultant amplitude modulation conversion of the 30 Hz FM modulation which will give an error when added to 30 Hz AM modulation.

The shift from analogue to digital signalling has resulted in new concepts in the design and evaluation of receivers, the probability of bit error rate (BER) is now a measure of sensitivity rather than minimum signal for specified output in a given s/n environment. Pre and post digital processing of the signal by error-coding systems is now practicable using specialised very-large-scale integrated-circuit (VLSI) packs to implement functions like Reed-Solomon coding, hitherto impracticable owing to the amount of components required.

Tuned circuit filters are being replaced by digital filters, at first at LF, but increasingly at higher frequencies as LSI technology improves. This obviates one of the main design problems in airborne equipment, namely, temperature and ageing changes in tuned circuits due to the severe environmental conditions. Miniaturisation and cooling problems are also reduced by such replacement, since tuning inductors and capacitors are both bulky and susceptible to heat differentials.

Automatic gain control, (AGC) functions normally requiring amplifying stages biased back by a d.c. control voltage, are being replaced by an integrated circuit IC digital/analogue converter driving an attenuator. AGC filtering problems noted above do not occur in this system and time constants may be adjusted to any value up to infinity, i.e. as a limit, the gain of an amplifier may be set to be constant at a desired level after the incoming signal has momentarily disappeared.

The use of local microprocessors enables these functions and others like automatic channel selection with time or geographical location to be carried out by digital signals either locally or even from the ground. Such cases may arise in the use of military frequency-hopping receivers, where both tuning and frequency synthesisers must be controlled in operating time in the order of milliseconds.

Receiver sensitivity The ultimate sensitivity of a receiver is set by the thermal noise in the input circuit. The sensitivity of broad band receivers used for radar is usually qualified by noise factor, i.e., the input level from a noise generator to raise the noise power output of a receiver by 3 dB in the absence of a signal. A perfect receiver would have a noise factor of 0 dB.

For practical measurement of HF and VHF/UHF communication receivers using voice modulation it is customary to use an RF signal generator with AF modulation depth set to 30 per cent at 1 kHz, and an AF output power meter on the receiver output. After tuning the receiver, the signal generator level is reduced until the AF output changes by 6 dB when the modulation is switched on and off. The signal generator level is noted, say 3 microvolts EMF, and the receiver is said to have a sensitivity of 3 microvolts for a signal plus noise to noise ratio of 6 dB.

Some generators are calibrated in 'soft microvolts' or PD across the matched 50 ohm output, and this gives an apparent receiver sensitivity optimistic by 6 dB.

Similiarly the AF bandwidth of a receiver affects the noise on the basis of:

$N^2 = 4kTRB$ where $k = 1.37 \times 10^{-23}$ J/K
                  T = temperature, Kelvin
                  R = resistance, ohm
                  B = bandwidth, Hz

For a normal communication receiver in voice mode, the AF bandwidth response is typically 300–3500 Hz. Values of 1–3 microvolts EMF for 6 dB s+n/n ratio would be typical values for such a receiver. A receiver with a narrower response would give an apparently more sensitive figure using this method, although in prac-

FREQUENCY SYNTHESIS

A measure of progress in avionic communications has been the steady improvement in accuracy and stability of frequency generation. It is fundamental in information theory that the amount of information that can be transmitted is proportional to the bandwidth. The continual growth in air traffic has necessitated allocation of more and more channels in given frequency band limits, as well as the need for information to be transmitted in narrower channels to improve the signal/noise ratio, noise also being proportional to bandwidth.

In earlier days, frequency control was by LC oscillators tuned by hand-calibrated variable capacitors. As receivers became more stable and easy to check against broadcast, base stations or other standards, it was necessary to back-tune (or net) the transmitter to the receiver so that both would be on the same channel. This applied in the early part of the Second World War. During this period, plug-in crystals became available for transmitters, but as such large numbers of crystals were required for the thousands of aircraft in service together with the number of frequencies involved, production became a problem. This was alleviated by the discovery that a crystal's frequency could be permanently shifted a small percentage by X-ray radiation and also that almost all operational flights were in 'radio silence'. Even so, each airfield had to maintain very large crystal stores.

Towards the end of the war, an HF military receiver was designed with a crystal calibrator to give harmonics for calibration 'pips' at 100 kHz intervals (the R1475), and the first multichannel VHF transmitter/receiver (T/R) used a telephone uniselector to count harmonics from a crystal derived harmonic comb to select the channel (the TR1407). These sets were developed by Marconi and GEC respectively.

Civil aviation radio was normally on HF after this period. After the Atlantic City Convention laid down that frequency tolerance of transmissions should be less than 0.02 per cent from all causes, HF T/R combinations using banks of remotely switched crystals of up to 200 plug-in units; were designed.

Subsequently, frequency synthesis was originated and concentrated on VHF equipment, where direct mixing from banks of crystals switched in decade steps of frequency gave output sum frequencies in 1 MHz, 100 kHz and 50 kHz steps. One HF drive was developed, however, using sum and difference frequencies from a bank of 10 selectable crystals between 2.5 and 3.5 MHz, directly mixed with frequency from a stable variable oscillator of 400–500 kHz. Continuously variable frequency output between 2.0 and 4.0 MHz could then be frequency multiplied to cover the remainder of the HF band, the output frequency having been stabilised in the ratio of the crystal to variable oscillator frequencies to give overall stability approaching that of a single crystal (Fig. 8.4).

Synthesisers using crystals banks were the norm until the mid-1970s, the miniature metal or glass cans being physically convenient to use. Circuit configurations were direct mixing, described above, and so-called indirect mixing, where a continuously tuned VHF variable oscillator was used as the input to a multiple superheterodyne chain of mixers and decade crystal oscillators to give a 5 MHz output. This was phase compared with a 5 MHz crystal in a phase discriminator whose d.c. output was used as a control voltage to lock the oscillator via a varactor diode acting as a voltage controlled capacitance. Preliminary coarse tuning was

Frequency synthesis

Fig. 8.4 Direct mixing frequency synthesiser

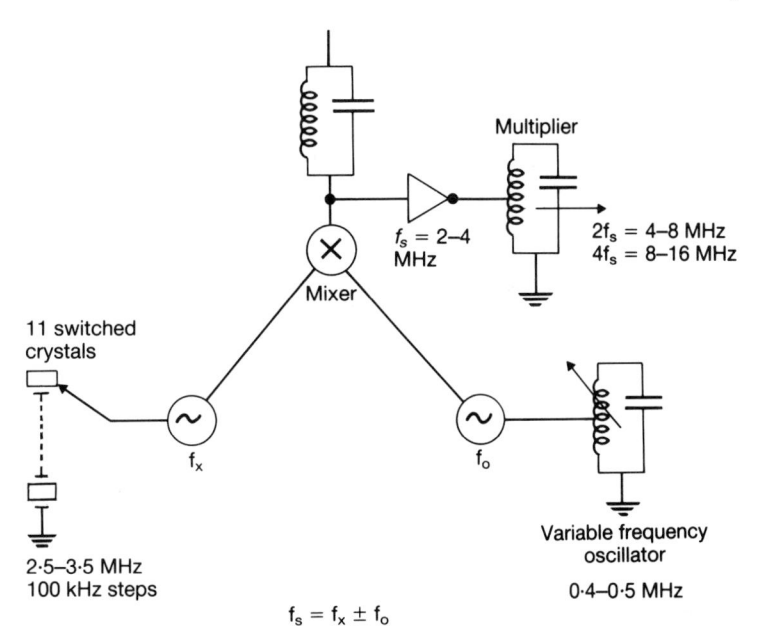

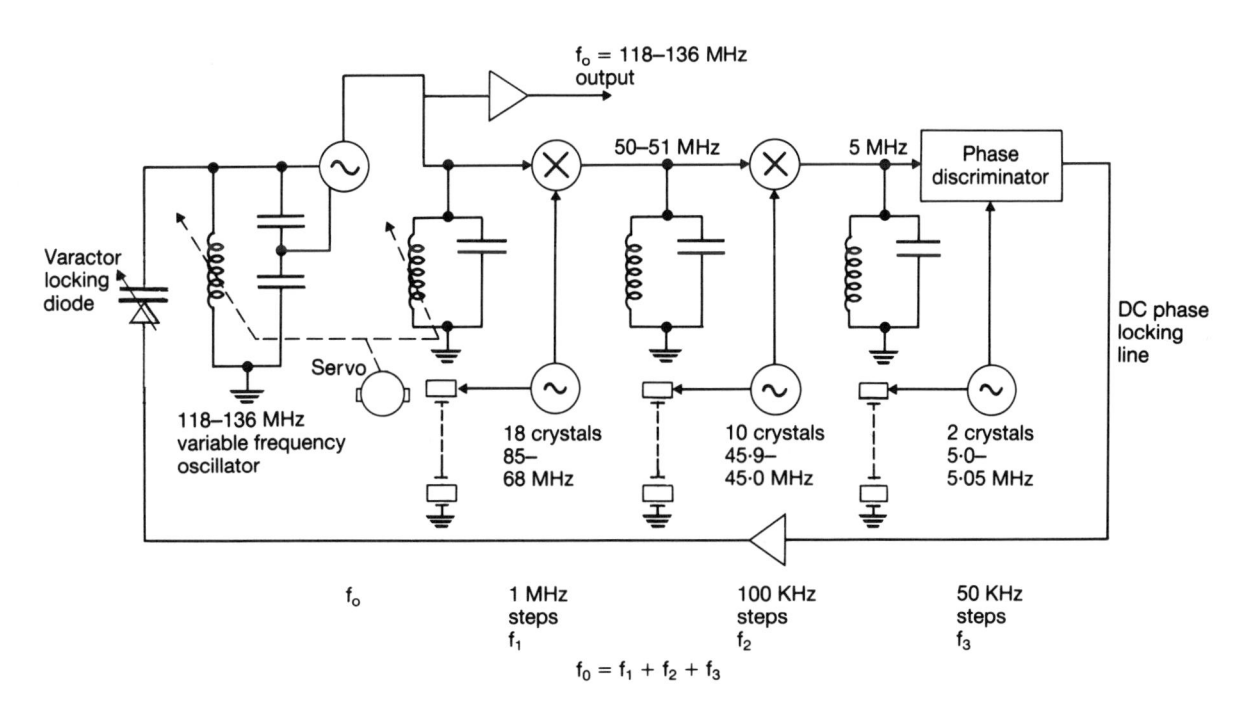

Fig. 8.5 Indirect mixing frequency synthesiser for 118–136 MHz transmitter drive

by motor drive and the final locked frequency was the sum of the crystal frequencies selected (Fig. 8.5). It will be seen that this is basically the same system as current digital synthesisers, the successive heterodyning down with crystals being replaced by successive dividing down with digital dividers (Fig. 8.6).

The reason for developing indirect mixing is to minimise the production of spurious intermodulation terms between harmonics of the two oscillator frequencies which is inherent in the mixing process. Sum and difference products of up to the fifth and sixth harmonics are troublesome and if close to the wanted carrier

187

Communitations Systems

Fig. 8.6 Comparison of indirect mixing frequency synthesiser (a) with digital frequency synthesiser (b)

(a) Variable controlled oscillator VCO is heterodyned down to the h base discriminator ø
(b) VCO is divided down to discriminator. 6.4 MHz frequency standard is oven controlled.

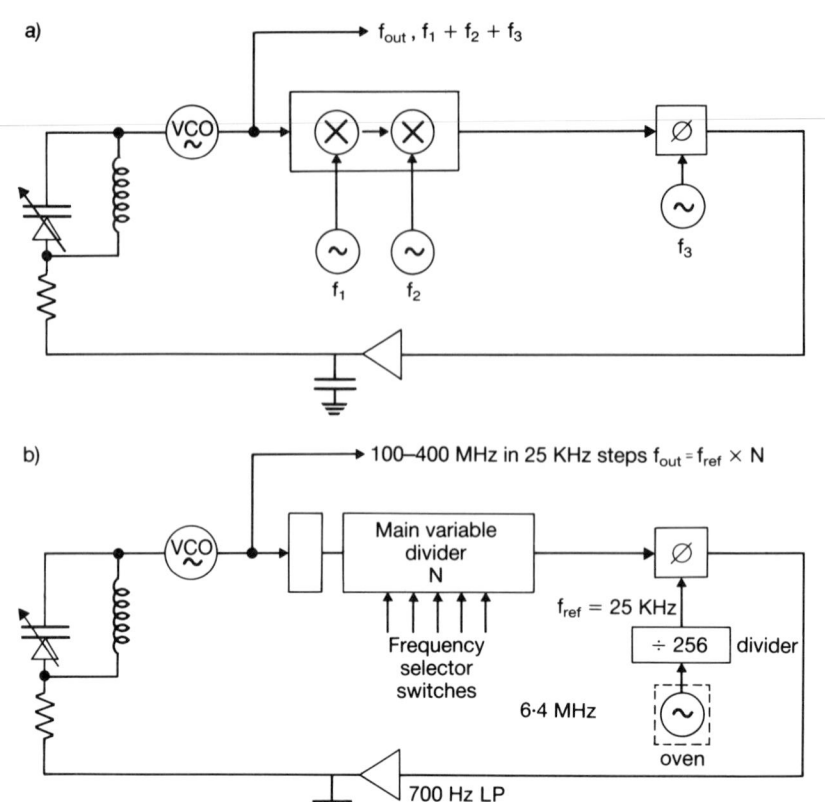

a) $f_{out}, f_1 + f_2 + f_3$

b) 100–400 MHz in 25 KHz steps $f_{out} = f_{ref} \times N$

Main variable divider N

$f_{ref} = 25$ KHz

÷ 256 divider

6·4 MHz

oven

700 Hz LP (loop filter)

Frequency selector switches

radiated, will cause interference in other channels. Specifications call for all spurious radiation to be below 60 dB of the carrier, and since VHF amplifiers tend to be broad band, these are radiated almost unattenuated.

With the introduction of single sideband transmission (SSB), the need for more channels and greater frequency stability on HF arose, as the suppressed carrier of the transmission had to be restored in the receiver with less than 20 Hz frequency error. Several synthesisers were developed based on the superheterodyne technique of direct mixing with crystal banks, and in one case the inaccuracy of the crystal oscillators was taken out by pulling them to harmonics of an oven-controlled reference oscillator of 10^{-8} accuracy. In another, all oscillators were oven-controlled. Small AT cut crystals have a temperature coefficient of about plus or minus 2 parts per million (p.p.m) (2 Hz per MHz) per degree C. Military climatic requirements span −60 to +70 °C, so that at 20 MHz the frequency error could be 5.2 kHz either way plus a grinding tolerance of 50 p.p.m. or 1 kHz. It was therefore necessary to both oven-control the crystals and trim them individually for tolerance.

Although improvements in crystal units have been made such as temperature compensating circuits within the glass containers, there is clearly an advantage in using oscillators whose frequency is locked in an indirect mixing circuit to one oven-controlled reference oscillator.

Such circuits became practicable with the availability of high frequency integrated circuit (IC) digital dividers. These take inputs at any frequency and divide this by an integer factor, often selectable.

Current synthesisers use a variable frequency oscillator coarsely tuned by diode-switched capacitors to approximate the required frequency. The frequency of an oven-controlled crystal reference oscillator of 6.4 MHz is divided down to 25 kHz by a fixed digital divider of 256 times and applied to a phase discriminator. The variable oscillator frequency is divided by a variable divider chain whose division ratio can be set to give a nominal 25 kHz from 108 to 400 MHz by the remote frequency selector which is selecting the coarse tuning capacitors, and this nominal 25 kHz then goes as the other input to the discriminator. A beat note produced between the two frequencies appears at the discriminator output and is applied via the control line to a varactor diode across the variable oscillator tuning circuit. This voltage sweeps the variable frequency across the sub-band range. As coincidence of frequencies occurs in the discriminator the beat note falls to zero, and the resultant d.c. control voltage, now proportional to the phase error between the two signals, phase locks the variable oscillator to the reference via the varactor. The variable oscillator may thus be set in 25 kHz steps to the desired UHF/VHF frequency having the frequency stability of the reference oscillator (Fig. 8.6).

VOICE SYSTEMS

Aircraft voice transmissions operate in a high acoustic noise environment. Unpressurised military aircraft incorporate microphones in the oxygen mask or use throat microphones. Noise-cancelling microphones may be used in pressurised aircraft.

Some improvement in s/n ratio is attempted by restricting the AF frequency response of microphones, amplifiers and headphones to between 300 Hz and 3.5 kHz.

Earlier microphones of the carbon granule type gave a high voltage output (up to 0.5 v) simplifying electrical noise pick-up in the pre-amplifier, but had a peaky frequency response, were electrically noisy and suffered from non-linearity with resultant distortion and acoustic noise intermodulation. Later electromagnetic microphones, although of lower output, typically 10 mV, are linear in transfer response.

Many microphones and headsets response peaked around 1 kHz, to which the human ear was thought most sensitive. Historically, this is a relic from the days of continuous wave CW reception when Morse code using a 1 kHZ tone was the normal means of communication. A good operator then was reputedly capable of reading Morse in adverse s/n ratios of 10–20 dB down, which is still better than an average data link, and is thought to be probably due to the effective tone selectivity and pattern recognition of the human brain.

For a voice system, modern analysis uses the articulation index, based on the hearing characteristic curves of the ear. In this the voice spectrum from 250 to 6000 Hz is divided into 20 unequal frequency sub-bands each contributing 5 per cent of the information to the Index, assuming a voice amplitude range of 30 dB between voice minimum and peaks. If noise is present, the contribution is reduced to $1/6 \times$ [(dB voice peaks) − (dB noise average)] thus if noise is 30 dB below the voice peaks, the contribution of the sub-band is 5 per cent. But if the noise is 10 dB below voice peaks, this reduces to $1/6 \times 10$ or 1.67 per cent. The sum of calculations for all 20 sub-bands then gives the total articulation index. If the noise has a flat spectrum (white noise), owing to the slight falling overall characteristic of the hearing response used, some improvement in articulation index is obtained with about 6 dB/octave pre-emphasis of the higher frequencies of the voice channel.

An American Air Force study of listening in high acoustic noise levels in combat aircraft established that a peaky headset response gave signals that when added to the ambient noise could drive the ear into the region of pain when the voice components came in the response peaks. Best results were obtained with a fully flat frequency response.

Several techniques are used to improve s/n ratio in voice systems. One is the so-called noise cancelling microphone, allowing sound to both sides of the microphone membrane. Distant sounds striking the membrane tend to cancel, whereas the near sound from the speaker tends to affect only the near side of the membrane. This effect is a function of the wavelength of the distant sound, giving an improvement of about 10 dB at 3 kHz falling to very little at low frequencies. The polar response is directional as a cardioid.

Normally the operator initiates voice transmission by the send/receive or 'Press-to-talk' switch. Later systems may use a voice operated switch (VOS) to reduce operator work load. This utilises a trigger circuit set to operate at a chosen microphone level. To prevent noticeable clipping at the start of a syllable, operation is less than 1–2 msec. This device normally works with an audio AGC system compressing speech amplitude variation by up to 10:1 in dB. Apart from 'attack time' similar to the VOS, a 'hangover time' of 200–500 msec is necessary to hold the level between syllables to obviate chopping. By using a separate noise amplifier, the VOS level may be set dynamically to suit varying noise conditions.

Voice clipping of the peaks of voice amplitude has been made use of in some AM transmitters to allow a greater depth of modulation of the voice waveform than normal. Voice waveform is very spiky with a high peak/average ratio, with consonants having the highest ratio and vowels tending towards sine waves. Clipping the peak level therefore affects mostly the consonants and squares-off the vowels to produce a 'gritty' sounding distortion but with articulation almost unaltered. Some filtering of the harmonics and intermodulation is usual if clipping up to 10 dB is used.

In modern aircraft with a variety of HF and VHF transmitter/receivers in use and particularly with military aircraft having a large operational crew, it is necessary to centralise these services and the intercom system so each crew member can operate with one microphone and headset and be able to select those services needed and communicate through them and/or with other members of the crew. Such systems are known as CCSs or communication control systems. The audio outputs of the aircraft receivers and intercom together with the microphone and key lines of the transmitters, are taken into a central junction box forming bus-lines.

A station control box connected to the junction box across the bus-lines is provided for each crew member, with press-button or rotary selector switch and gain controls enabling the outputs of the receivers to be selected and mixed at desired levels together with crew intercom. Control of those transmitters selected is by the press-to-talk key and microphone of the crew members headset.

An early system of 35 years ago provided these facilities by mixing the AF sources with hybrid transformers and double triode valves, in a central amplifier unit isolating the sources to prevent crosstalk. This gave intercom and a choice of three receivers selected from ten, and one transmitter/receiver combination selected from four at four crew stations, and with intercom only at six additional stations.

Voice systems

Later in a current system, with the advent of IC operational amplifiers, mixing and isolation are done through attenuator resistance pads working into virtual earth amplifiers, again in a central amplifier unit. Remote switching and gain are by diode attenuators on thick film ICs. This gives access to 15 receivers and 4 transmitters in a typical four crew station box installation. Other combinations are possible.

Recently a more sophisticated system switches the AF lines by multiplexer logic ICs used as analogue switches controlled by a microprocessor. AF gains are remotely controlled manually by variable gain op-amps in the central unit. This system enables five independent channels of intercom and up to 20 receivers and six transmitters to be operated from six crew stations. In addition a secure speech overlay system provides the extremely low crosstalk or 'Tempest' requirement. 'Tempest' is the technique for the prevention of unwanted radiation of un-encrypted secure speech or data. This AMRICS automatic management receiver and intercom system also enables the transmitter/receivers to be remotely set to a desired frequency by keying-in a channel number at the crew station panel. This is compared with the allocated frequency in a 'look-up' table in the microprocessor, and the appropriate signals are sent on the standard digital tuning lines to the transmitter/receiver, which is thereby tuned (Fig. 8.7).

Fig. 8.7 Computer auto-managed receiver and intercom system (receiver tuning not shown)

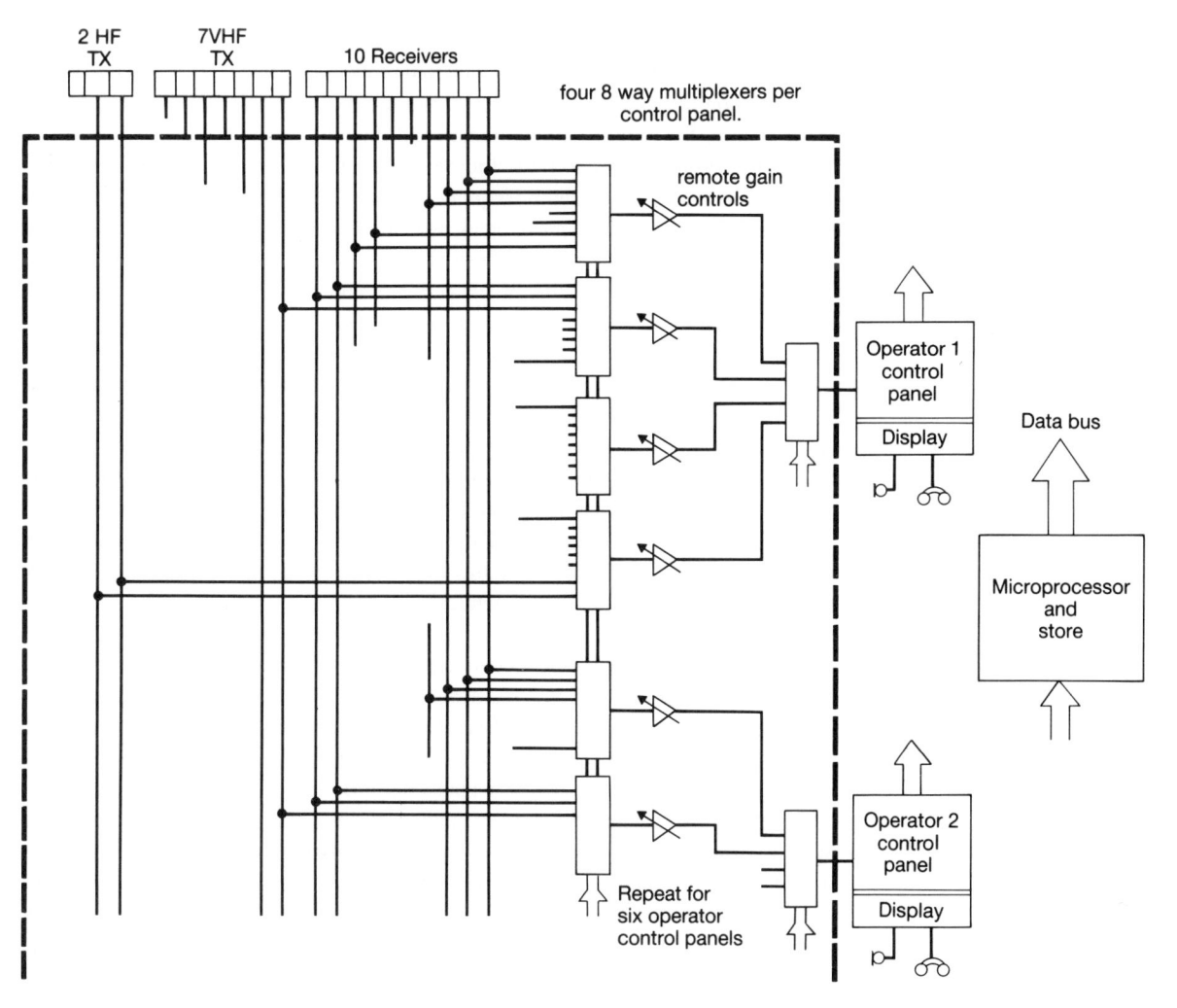

It is to be expected that future voice systems will digitise the analogue voice signals for transmission on a common bus-line, and probably utilise fibre-optic linking between units throughout the aircraft to reduce cabling weight, minimise pickup and simplify 'Tempest' problems.

TRANSMITTER MODULATION

Airborne communication equipment uses two systems, speech and data. Speech modulation of the transmitter is double sideband (DSB) or A3, single sideband (SSB) reduced carrier or A3A, SSB suppressed carrier or A3J and frequency modulation (FM) or F3.

DSB was the most common form of modulation, often low level, in which the audio frequency AF waveform is applied as bias to the grid of the RF power amplifier output stage or PA. This requires minimum modulation power, but the RF power is reduced to a quarter since the PA must be biased to half output voltage to enable equal upward and downward carrier swings with 100 per cent modulation.

High level modulation requires a modulator of the same power as the PA. The modulation is applied to the anode of the PA to drive the supply voltage up to double the normal value for positive modulation peaks Some semiconductor transmitters using broad-band tuning apply modulation to the power supply to all stages. Others used a 'floating carrier' technique that reduces the carrier level in the absence of modulation. DSB airborne transmitters are typically in the 20/100 watt output power region, the higher powers being used on HF.

The increased congestion on the radio spectrum has led to restriction of channel bandwidth and tighter frequency control. The power budget of a DSB waveform is carrier at half power and two sidebands one quarter each. Since all the information is contained in each sideband, by only transmitting one sideband, an improvement of about 12 dB in s/n at the receiver is obtained made up by 3 dB bandwidth reduction, 6 dB power improvement and 3 dB synchronous detection. Single-sideband modulation is obtained using a double-balanced modulator that produces sidebands with the carrier balanced out by about 40 dB; by passing the AF input through a Hilbert transform filter (having a constant 90 degree phase shift throughout the AF band) it is possible with another balanced modulator to cancel one of the sidebands to the same degree. In practice, this amount may not be sufficient to meet current tighter frequency restrictions on off-channel radiation, and will thus require costly RF bandpass filters with very sharp cutoffs in addition. The use of digital filters has been proposed in this application, as very sharp cutoff can be obtained economically.

The reception of an SSB signal requires replacement of the missing carrier before detection, and frequency error has to be less than about 10 Hz to give AF output of satisfactory quality. One aspect of this is that it is no longer possible to use an AF tone selective calling device, SELCAL, to call an unattended receiver, as the Selcal depends on a bank of AF tone filters or tuned reed relays with bandwidths of the order of 1 Hz. This problem can be partly obviated by calling from the transmitter on DSB, although the s/n ratio will now be reduced by 12 dB. Another aspect is that it is no longer possible to rely on the frequency stability of non-temperature compensated crystals to be adequate either between transmitter and receiver or absolutely in its channel allocation. A frequency synthesiser having a reference oscillator of better than 0.1 p.p.m. frequency accuracy is therefore essential.

On VHF, DSB is the norm, since s/n is not a problem, propagation being limited to line-of-sight (LOS), normally about 200 nm. Transmitter power greater than 20 watt is not necessary. With military equipment on low flying helicopters, due to man-made static, etc., it is thought that narrow-band FM gives better noise performance and has other advantages, particularly in the simplicity and physical dimensions of the modulator.

Data transmission modulation can be either keyed continuous-wave (CW) or frequency shift keying (FSK) or phase shift keying (PSK), or combinations of all these as in JTIDS waveforms. The present tendency in all communication links is to digitise analogue voice signals. There are several methods of data modulation, such as pulse coded modulation (PCM), delta, delta-sigma, pulse width and pulse amplitude modulation. PCM is the most common with delta and delta sigma providing a simple implementation for military mobile links. PCM normally provides a 16 kB/s bit stream in which each byte is a word of 4 bits defining the instantaneous amplitude of the speech waveform. All these systems use an A-to-D, analogue to digital converter to digitise, and a D-to-A converter to restore the voice signal. Once digitised, speech can be processed in several ways. One is encrypted, in which the bit-stream has a pseudo-random or 'key' bit-stream added by modulo-2 addition, making the transmission unintelligible to an unauthorised listener. This cyphered signal is then transmitted safely in any normal way, and when received at the authorised receiver end, it is decrypted by a second modulo-2 addition of the key bitstream (the code of which is held at each end) which then restores the original signal.

With both encrypted and plain digitised voice, the voice spectrum has excessive bandwidth, and this is compressed by processing the voice signal by a Voice Coder or Vocoder. This effectively analyses and then synthesises the voice signal with a bandwidth of at most one third of the original. The main systems are either Channel or LPC, linear predictive coding vocoders. One standard is 2.4 kB/s. These systems are complex; development commenced in the 1930s and is still very active. This and machine voice recognition have both made considerable strides in the last decade or so owing to the availability of microprocessing technology and recent developments in coding mathematics.

INFORMATION THEORY APPLICATIONS

The last 20 years have seen the impact of information theory on avionics communications with the rapid growth in subjects such as error-correcting coding, adaptive filtering, real-time channel evaluation, packet switching and various methods of digital modulation. Most of this stems from the introduction of digital ICs and microprocessors making the implementation of complex digital signal processing practicable (see Chapter Three).

The concept of signal processing has undergone considerable modification since the classic studies of Shannon, with extensive mathematical developments in the error-coding field in particular leading to powerful codes such as the Reed–Soloman family being used as normal adjuncts to communication channels.

Errors in data transmission can occur in several ways, straightforward loss of bits in Gaussian (white) noise, in bursts or peaks of noise or interference and by distortion due to differential propagation delays of harmonic components of the bit-waveform which cause a smearing or overrunning of components of bits into succeeding bits (intersymbolic interference).

Error coding is used for the first two noise cases and changes the normal plotted characteristic of bit error rate (BER) versus s/n of an uncorrected system to one having a much steeper slope. This is an improvement, but now a very small change in s/n gives a large change in BER. Typically in an uncoded channel, change of s/n from 4 to 8 dB might improve the BER from 10^{-2} to 10^{-4}. With coding the s/n would be 2 dB to 4 dB for the same 10^{-2} to 10^{-4}. However there are many kinds of codes, two main groups being classed as convolutional and block. The convolutional example above is suitable for Gaussian noise and BERs in the 10^{-3} range, this BER being acceptable for digitised voice (Fig. 8.8). A Reed–Solomon code, a block code, has a very much steeper curve, a change of 0.2 dB at 3 dB s/n would change the BER from 10^{-2} to 10^{-8} and this type is mostly used for noise burst interference where machine data requires BERs in the order of 10^{-10} accuracy.

Fig. 8.8 Effect of convolutional coding on data message bit error rate

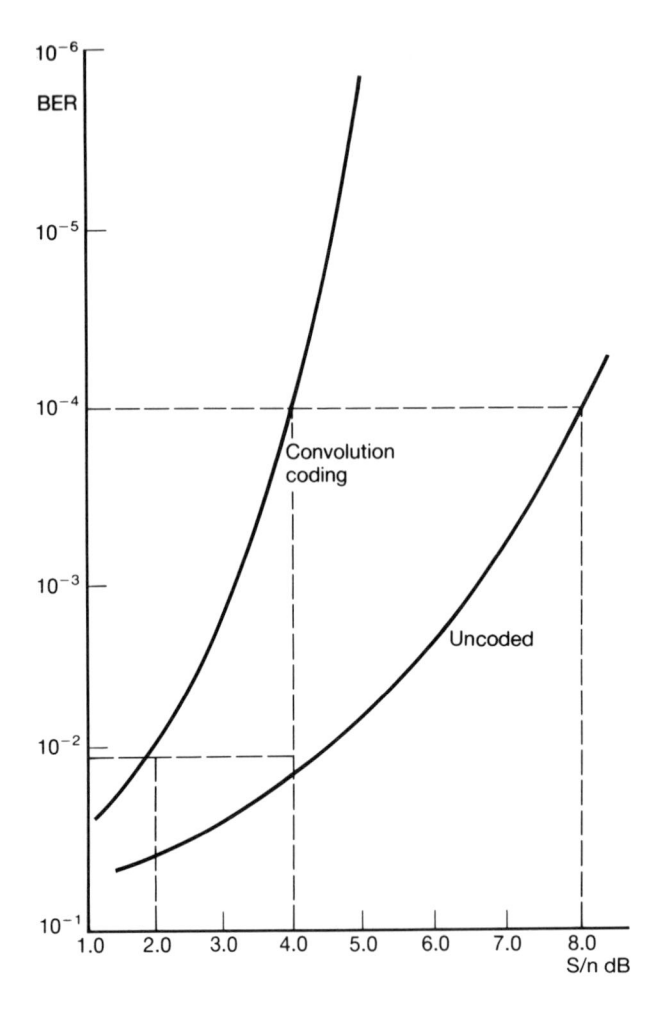

Other aspects of information theory coming to the fore are adaptive filters. These form an area of signal processing in which the system is modelled algorithmically and comparison is made with the actual input using an error output to adjust the system. These subjects are rather specialised and recondite and can be followed in the IEE and IEEE Proceedings.

DATA LINKS

Data link communication differs from analogue communication in that it tends to be communicating directly with a microprocessor system that conveys the information to the recipient indirectly through a display or other means, although it may be used more directly with digitised voice which may be processed or not by encryption or vocoding. The effects of fading and interference will differ from that on the analogue channel.

Speech and manual continuous-wave links operate successfully in poor s/n conditions owing to the ability of operators to recognise words and sentence patterns and interpolate missing elements by familiarity of words to make sense.

Aircraft speech messages often appear completely unintelligible to an ordinary listener, but are quite clear to an experienced operator who is anticipating only a limited number of familiar procedure words in a given operational situation. Because speech incorporates a high degree of redundancy the human brain assimilates this and integrates the information into a recognisable message.

Data messages however consisting of a serial bit-stream are peculiarly susceptible to instantaneous loss of signal due to fading or corruption by noise as these signals do not have the pattern-forming property of speech and loss of only occasional bits can wreck the information content.

The criterion of a link is its minimum bit-error rate, rather than the corresponding minimum received s/n ratio of an analogue channel, and most of the complexity of data-link systems is involved in the attempt to overcome this limitation. BERs of 10^{-3} and 10^{-10} are typical of voice and data channels respectively.

Propagation conditions are particularly bad in this respect on HF, and rapid fading causes instantaneous loss of data whose bit-time is small compared with the fade-out time. Several systems are used to compensate this, most incorporate forward error correction (FEC), in which additional parity bits are added to each digital word, in such a way that failure to meet a pattern criterion such as in the simplest case, an odd or even number of bits per word rejects the word. This is incorporated in a duplex system as Automatic Repeat Request (ARQ) where, on receipt of an error signal from the receiver, the corrupted word is re-transmitted.

Error coding is a highly specialised subject in current development, much of which has become practicable only recently with the availability of complex digital VHLSI packets. Coding uses either a block or a convolutional type system, and here again the attempt is based on the provision of redundancy and pattern analysis in the message structure.

Another way of reducing error is the use of special modulating systems such as kineplex and data link 11. Consider the case of 75 b/s data; each bit duration is 13 msec, a rapid fade will lose several bits. Suppose a transmitter were modulated with 16 AF tones keyed by the data-stream in bit sequence, each tone keyed for 16 times 13 msec, the effective bit duration is thus 208 msec, although the information rate is the same. This message is reassembled at the receiver to give the original data-stream. The arrangement is analogous to a serial compared with a parallel data bus. This has the advantage that not only has the chance of a bit being lost in an instantaneous fade been reduced but also there has been the advantage of a form of frequency–diversity transmission. There are however disadvantages, one being the reduction of carrier power for each tone. The system is effective for HF conditions, but in the case of VHF where this type of fading does not exist, better results would be obtained with simple FSK or PSK modulation.

In the case of military systems in particular, data transmission is used with a message protocol. The old Morse operator's 'Q' code (going back to 1912) is the progenitor of this. A digital message may consist of a series of words of say 224 bits, such a word as word (1) 'Navigation', could be followed by word (2), 'Engine state'; word (3), 'Weather'; etc., each word (n) being divided into six 'fields' of 32 bits plus 32 bits of error-correction. Each field contains data, e.g. field 1 latitude; field 2 longitude; field 3 altitude; field 4 time of arrival (TOA), etc. Operational organisations, on this basis, have libraries of messages (protocol) designed to meet all these requirements.

The power of this system is that being digitally coded, it may be machine interfaced with both aircraft and ground computer command and control systems to read out selected sensors automatically.

It has been proposed for civil transatlantic flights that automatic readout of all aircraft inertial and other navigational systems should be transmitted automatically continuously to a geostationary satellite by data-link and thence back to Air Traffic Control Centre. This system automatic dependence surveillance, (ADS) should obviate the occasional large position errors caused by delays in manual insertion during aircraft movement and allow a reduction in safe separation distance between aircraft.

A civil aircraft data link system is the ARINC communications addressing and reporting system (ACARS) used with existing airborne radio equipment to improve operational effectiveness and inter-company communications. Initially, in the US, ACARS utilised VHF line-of-sight on a dedicated frequency. Some airlines were interested in operating on HF also; this subsequently provided long range aircraft with enhanced capability.

The basic components of an ACARS are a management unit and a control unit. The management unit receives and transmits ground/air digital messages and gathers data for transmission from on-board event sensors. The control unit is the pilot interface, providing facilities to enter message text.

The major use of ACARS at present is for the receipt and transmission of operational data such as arrival and departure information. However, engine parameters – temperature, pressure, fuel flow, fan speeds, vibration levels – are sent routinely as elements in an airlines maintenance plan, early information of system malfunction allowing corrective action before incidents.

Future developments and a more complex use of the concept probably will be linked to the extension of satellite communication availability. Other aspects of data links are considered in the sections JTIDS, Information Theory and Transmitter Modulation.

JTIDS

JTIDS (Joint tactical information distribution system) is a military joint-service, jam resistant VHF communications system of advanced characteristics for interconnecting large numbers of operational aircraft with ground/sea battle commanders and units to exchange tactical information in a fast-moving situation. The present positions, weapon states, etc., of all participating aircraft, together with reports of the disposition of hostile aircraft or missile battery locations are automatically fed into an effective common pool from which they are drawn for display either on a digital readout or radar/navigation screen. Relative navigation, identification and digitised secure speech may also be implemented. Development of this system in the US has been carried out over the last 10–15 years and production models have been delivered since 1984.

The equipment is formalised for NATO military purposes as a multi-function information distribution system (MIDS). Communication, navigation and identification (CNI) functions are all carried out on a common digital data link operating a protocol, Link 16, using words of 75 bits of error-coded data. (See Data links and Chapter Three, Digital Technology.) As a radio system, the equipment is a digital frequency-hopping, direct-sequence transmitter/receiver system of 1536 TDMA time-slot channels operating as spread-spectrum across the 969–1215 MHz frequency band (Fig. 8.9). A TACAN navigation unit is also included in the JTIDS box as a convenient extra for use outside a JTIDS net.

Fig. 8.9 JTIDS installation units – transmitter/receiver (top left), data processor group (top right), data link control panel (centre) *GEC Sensors. Singer Electronic Systems Divisions*

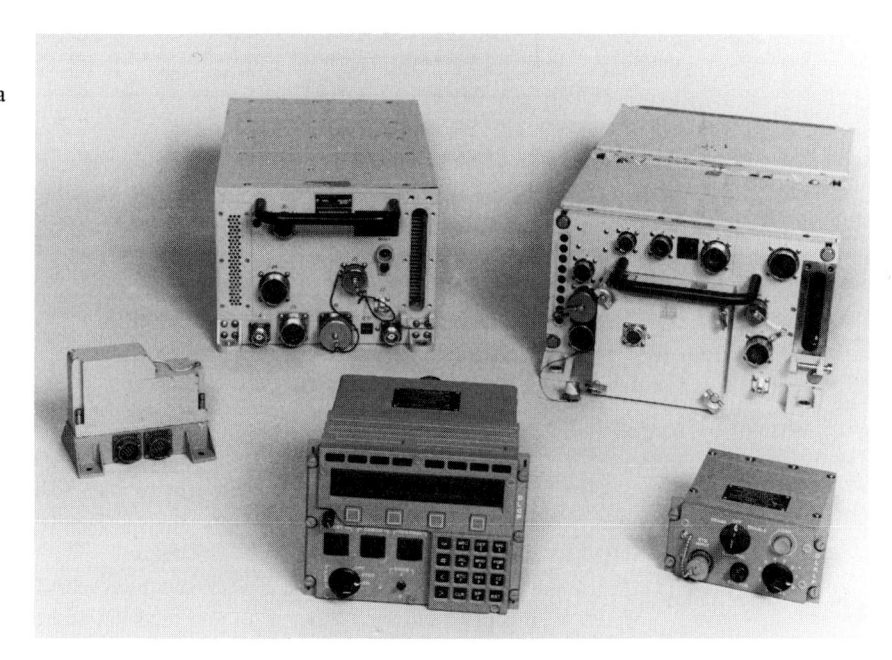

In operational use, a group of aircraft within line of sight participate in an intercommunicating JTIDS net, each aircraft being allocated its own period of transmission or time-slots from a continuous cyclic sequence of 1536 time-slots of 7.8125 msec duration. Each aircraft transmits only in its own time-slots, but all aircraft may receive in any time-slot. There is thus a number of transmissions from aircraft following continuously in sequence forming a time-division multiple-access (TDMA) system. This is analogous to each aircraft being allocated its own frequency channel in a conventional system.

At a rate of 128 time-slot/sec, 1536 time-slots constitute a 'frame' of 12 seconds, 64 frames form an 'epoch' of 12.8 minutes. Each time-slot can contain up to 12 Link 16 words of 75 bits of error-coded data which are updated at every frame, while the epoch is the basic repetitive net cycle (Fig. 8.10).

Transmission waveforms are pulses of 6.4/6.6 microsecond mark/space ratio, each pulse transmitted in sequence at a frequency chosen by pseudo-random code selection from a list of 51 frequencies spaced apart by 3 MHz to implement the 'frequency-hopping' and coding characteristics. In addition, each pulse is 'chipped' by 32 chips of phase modulation, again selected on a pseudo-random code basis. This spreads the spectrum between the 3 MHz spaced carriers and provides additional coding information for the time-slot.

Fig. 8.10 JTIDS net time-slot structure

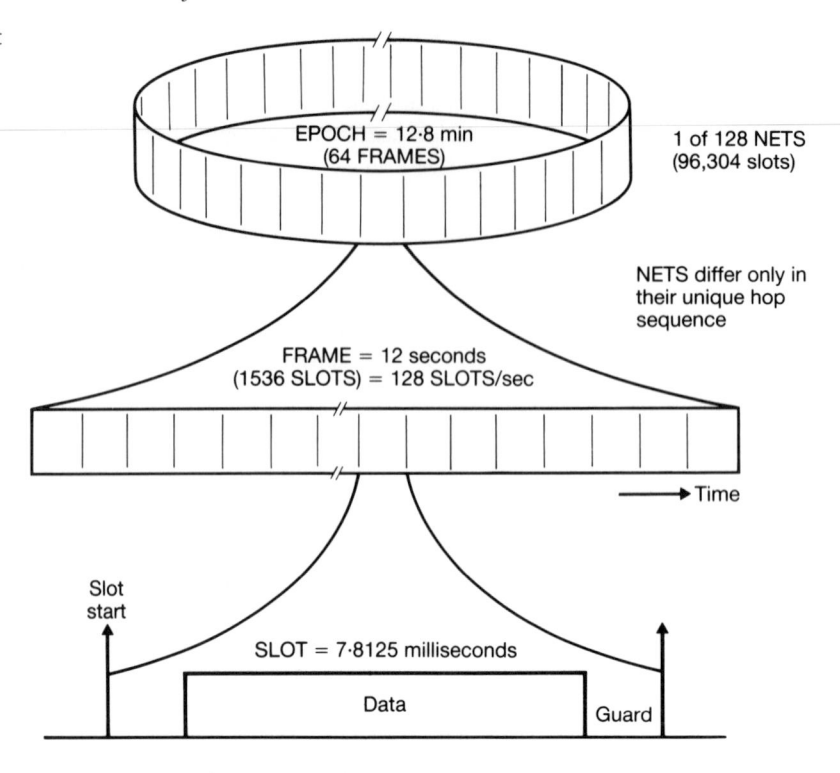

EPOCH = 12·8 min (64 FRAMES)

1 of 128 NETS (96,304 slots)

NETS differ only in their unique hop sequence

FRAME = 12 seconds (1536 SLOTS) = 128 SLOTS/sec

Time

Slot start

SLOT = 7·8125 milliseconds

Data | Guard

The result of this is a spread of noise-like signal across over 153 MHz, which is extremely difficult to detect, jam or d.f. Also, not knowing the pseudo-noise PN codes used, it is unlikely that a hostile receiver could lock on to or decode the transmission.

To ensure operation it is necessary to synchronise the start of transmission time of signals between transmitter and receiver, and this is done by a receiver awaiting the transmission having been set up previously to approximate 'net time'; when the transmission arrives it may be synchronised (in a similar manner to a digital watch being preset and keyed-on when the time-pip arrives). This synchronisation is completed in phases of coarse and fine and it is necessary to obtain coincidence between digital words and digits by means of transverse filters using surface acoustic wave filter devices (SAW) or shift registers.

The achievement of 'system' or 'net' time between members of a JTIDS net is a fundamental requirement. A member may lose radio contact with the net by going out of range for a number of hours, but can maintain system time with a local reference oscillator to be in coarse synchronisation, sufficient to pull into fine synchronisation in a few seconds after returning to the net.

When in full synchronisation, relative geographical position with respect to the net is established by round-trip timing (RTT) and Present Position Interrogation (PPI) which interrogate other members by individual address and then calculate the distance by the time delay between the known interrogation time and measured reply time and from the instantaneous present position declared by the interrogated aircraft. If the geographical coordinates of a ground station and/or the instantaneous position of at least one member are known, the system approaches an absolute rather than a relative navigation system (Fig. 8.11).

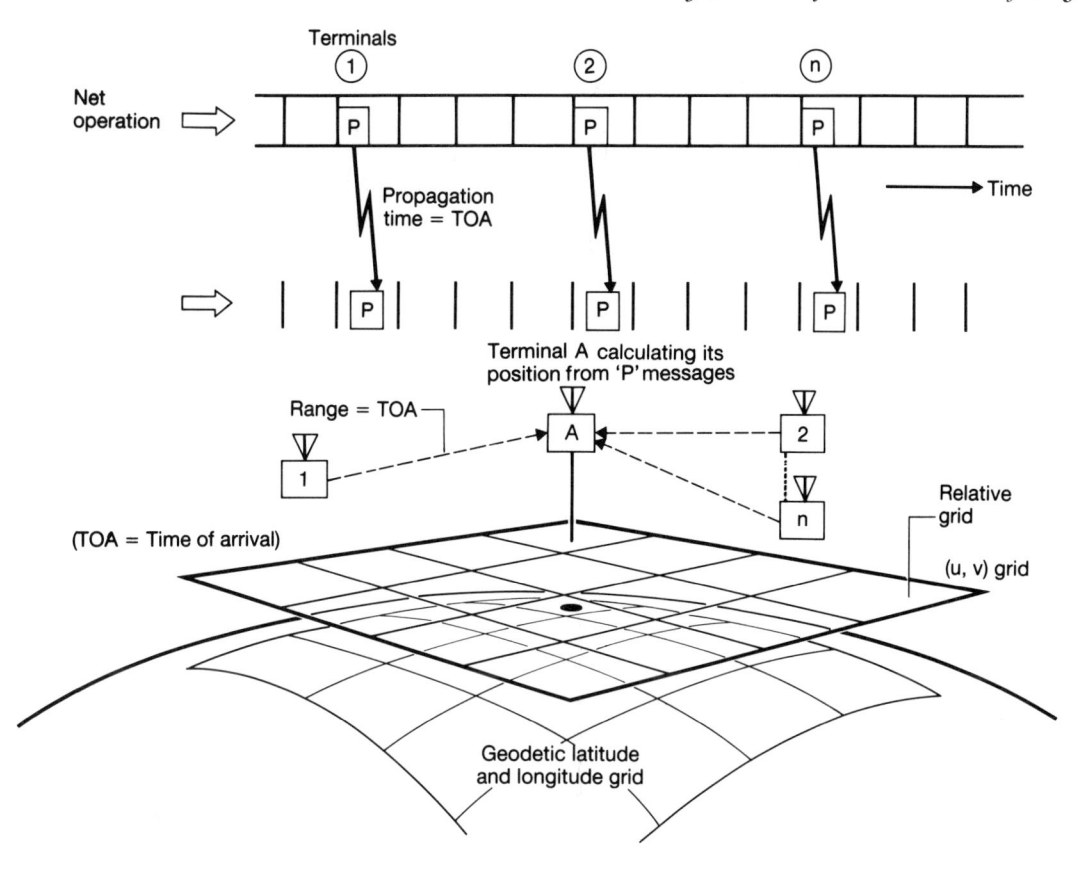

Fig. 8.11 Position fixing by time-of-arrival technique for JTIDS terminal

Since this very complex system includes individual members' addresses and hence identity, it forms a very efficiently coded identification or IFF system, although of course, limited to the network.

Owing to the high peak power of the pulses and the spread-spectrum waveform, an enemy will have difficulty in jamming the system, assuming a conventional single frequency jammer, the AJ ratio being very much greater than existing conventional equipment. Additional protection against noise peaks and the effects of jamming on data transmission accuracy are provided by a Reed–Solomon error-correcting code and other anti-jamming techniques.

With 1536 TDMA time-division multiple access channels available, there is a maximum effective overall data rate of 238 kB/s. This is reduced with error correction and AJ techniques in some modes, but still allows a very large message protocol to be implemented, and also permits operation of free text messages including digital voice by means of a vocoder at 16 kB/s or 2.4 kB/s per channel depending on the type of vocoder.

A net will normally support up to 32 transmitters, (more than one time-slot is usually allocated to an aircraft). However, owing to the pseudo-random distribution of frequencies, it is possible to operate more than one net using different codes in the same region without mutual interference. Such an arrangement termed 'stacked nets' is possible because the pulses from one net can interleave with the other with statistically negligible chance of two long codes coinciding and being decoded in the wrong net. Numerous nets (up to 128) may be stacked in this way.

SATELLITE COMMUNICATIONS

Military experiments are under way to evaluate the practicability of using satellite communications. Using L band frequencies about 1.6 GHz for the air/satellite link and C band frequencies about 4–6 GHz for the ground/satellite link, flight trials for civil aviation are taking place using the Marisat marine satellites with airborne installations fitted to a Boeing 747 to implement some of the facilities described in the Public Correspondence section. Additionally, it is hoped that improved navigational reporting accuracy will be obtained by automatic transmission of the inertial navigation outputs. Some of the difficulties with the air link are mentioned in the section on Aircraft Antennas.

AERONAUTICAL PUBLIC CORRESPONDENCE (APC)

Since the 1950s airlines have considered providing telephone facilities for passengers, especially the growing business-traveller community, used to being in constant telephone touch. One difficulty has been the lack of allocation of frequencies for public correspondence from aircraft. However with the decline of ship passenger traffic, aircraft could use the under-utilised ship/shore channels on the HF channels. With PTT cooperation, flight trials were carried out by a number of international carriers, using DSB amplitude modulation, with a calling channel to allocate a frequency for the telephone call. Setting up a call by the Flight Radio Officer was difficult and time-consuming, as by the time the frequency was established, propagation conditions had changed, the aircraft entering a noisy weather situation or air traffic control necessitating a retune to ATC channels.

Of the half-dozen or so airlines who tried this, only a few actually offered the service to passengers. By 1960 most airlines had dispensed with specialist radio officers; all radio now being pilot operated. With a reduced crew, the handling of passenger telephone calls via HF/DSB ship/shore channels was no longer feasible.

The next phase came in the early 1970s with the use of HF/SSB for the aeromobile service, which with frequency synthesisers gave access to a greater number of frequencies, allowing HF/SSB channels to be allocated for airline company operational control with licences for ground stations. PTTs were upgrading international and internal telephone trunk-dialling facilities to allow setting up calls automatically.

Many executive jets use these facilities either via some airline HF/SSB ground stations or those operated by some PPTs. On scheduled services many problems remain, as the technical crew still must set up a call using equipment for ATC channels. At the time of writing no major airline offers a telephone service to passengers although there are facilities on the flight deck for telephone calls if required by the crew.

In the USA, VHF/UHF ground mobile telephone services have been used to support an aeromobile telephone service. Airphone on the 900 MHz band is now installed in a number of aircraft operating internal services on heavily used routes, although full area coverage of the USA is not yet provided.

Satellite telecommunication is bringing a new development into the public correspondence field. With geostationary satellites (GEOS) spaced at 120 degrees longitude apart above the Equator, almost global coverage is available. Satcom can exchange voice and routine data message traffic between aircraft and air traffic control. It can also provide data communication between airline company headquarters and flight crew on topics from flight planning to bar uplift at the next station. An essentially new aspect will be provision of a passenger telephone service encompassing automatic charging and billing procedures through credit card readers and direct international dialling over digitised speech between aircraft and ground via the satellite. Plans are being evaluated for a small electronic PBX with access from the passenger cabins.

BIBLIOGRAPHY

Bennet, W. R, Davy, J. R. *Data Transmission*, McGraw-Hill.
Berlekamp E. R. *et al.*, 'Application of error control to Communications', *IEEE Com. Mag.*, vol. 28, April 87.
Betts, J. A. *Signal Processing, Modulation and Noise*, Hodder and Stoughton.
Brown, T. T., 'Transistor Multichannel VHF Com/Nav Receiver for Aircraft', *Proc. IEE*, No 7. July 63.
Burrows, M. L. *ELF Communication Antennas*, Peter Peregrinus, 1978.
Cowan, C. F. N., Grant, P. M., *Adaptive Filters*, Prentice-Hall.
Darnell, M. 'Channel Evaluation Tech. for Dispersive Comm. Paths', in *Communication Systems and Random Process Theory*, ed. J. K. Skirwzynski, Sigthof and Nordhoffs, 1978.
Evans, B. G. *Satellite Communication Systems*, Peter Peregrinus.
Lucky, R. W. *et al.*, *Principles of Data Communication*, McGraw-Hill.
Markus, J., Zeluff, V. (eds), *Electronics for Communication Engineers*, McGraw-Hill.
Reed, I. S., Soloman G. 'Polynomial Codes for Certain Finite Fields'. *Journ. S.I. Appl. Math.*, June 1960.
Rudge A. W *et al.*, *Handbook of Antenna Design*, Peter Peregrinus.
Sams, H. W. I.T.T. Reference Data for Radio Engineers.
Shannon, C. E. 'Mathematical Theory of Communication', *Bell System Tech. Journal*, vol. 38.
Terman, F. E., Pettit, J. M. *Electronic Measurements*, McGraw-Hill.
Weichbrod, J. 'Problems of High Altitude Communication', *Journal Acoustical Society of America*, July, 46.
White, 'Air–Ground Communications – History and Expectations', *IEEE Transactions on Comms*. May 1973.

CHAPTER 9 AIRBORNE RADAR

JOHN A. C. KINNEAR MA; FIEE; FIERE,
Marketing Development Manager; GEC Avionics Ltd

Mr Kinnear joined the research laboratories of Elliott Brothers in 1951, to work on microwave antennas in the centimetric and millimetric bands. He carried out the reduction into practice of the newly invented Elliott twist-reflecting Cassegrainian antenna. This was followed by the development of new microwave components and then the design of high grade instruments including automatic equipments for microwave component and system testing and for antenna near-field measurement. This led to the research and development of airborne radars for airborne interception and high definition mapping and terrain following with involvement in integrated concepts of overall avionics systems.

Since the first applications of radar in aircraft (see Chapter One), which were directed to giving night fighters the ability to locate their quarry in non-visual conditions, the versatility of radar has brought about its progressive involvement in most of the roles of aviation, so that, today, radar makes a major contribution in both civil and military operations.

In the civil aviation field, as a consequence of economic and logistic considerations, the philosophy is for the major facilities to be ground-based, concentrated along airways and terminal areas where they can provide services to large numbers of aircraft in the high movement density areas. Commercial aircraft, in general, carry only the specialised equipments which must necessarily operate where the aircraft is at any moment, such as altimeters and the airborne elements of ground-based systems such as secondary surveillance radar (SSR) transponders and distance-measuring equipment (DME) interrogators.

Military aircraft, on the other hand, cannot expect assistance from ground-based aids when operating in times of hostilities, and each aircraft must carry a comprehensive capability to ensure that it can perform its mission. Thus the modern attack aircraft carries navigation attack and terrain-following radars which assist the aircrew to follow a planned route to the target area and locate the target, while flying fast and low, even in darkness and low visibility, to minimise their exposure to air defences.

Airborne interception (AI) radar provides the means for the fighter aircraft to locate intruding aircraft at many tens of miles enabling it to compute and fly an intercept course and engage its quarry with radar-guided missiles from well beyond visual range.

Specialist radars enable surveillance aircraft to search immense areas, or volumes of airspace, providing a means to detect, and, in many cases, identify aircraft, shipping, and land forces. The airborne early warning (AEW) aircraft, for example can maintain continuous watch over well over a million cubic miles of airspace, detecting and tracking all aircraft movements within its coverage.

At the other end of the scale, radar enables reconnaissance aircraft or pilotless drones to scrutinise in close detail zones of particular tactical interest, relaying back observations in real time.

Apart from its technical sophistication and complexity, the modern airborne radar stands out for the number of specialist engineering disciplines reflected in its design. Power engineering is associated with converting the aircraft primary power supplies so as to satisfy the various internal voltage and power requirements of the radar, particularly those of the transmitter which may require inputs of many kVA. High voltage technology drives and supplies the high power transmitter tube; r.f. frequency synthesis generates the microwave waveforms which will be amplified in the r.f. power amplifier, and the local oscillator frequencies required in the superheterodyne receivers. Also needed are: microwave technology for the antenna, duplexing, and the receiver front-end; mechanical and servo-system engineering for antenna stabilisation and scanning; digital data processing for analysis of the signals and data handling as well as man/machine considerations in relation to presentation and control. All of these must be blended with an understanding of the properies of the environment in which the radar must perform and an awareness of its operational tasks and its users' operating procedures.

To understand how the radar is designed to play its part in the succesful performance of the aircraft's mission, the factors which affect the radar's performance are examined first, both those which the designer can control and those over which he has no control and must live with. Modern radars are extremely complex, so the objective throughout is to outline the most significant aspects, without involving the intricacies (Fig. 9.1).

Fig. 9.1   Principles of radar

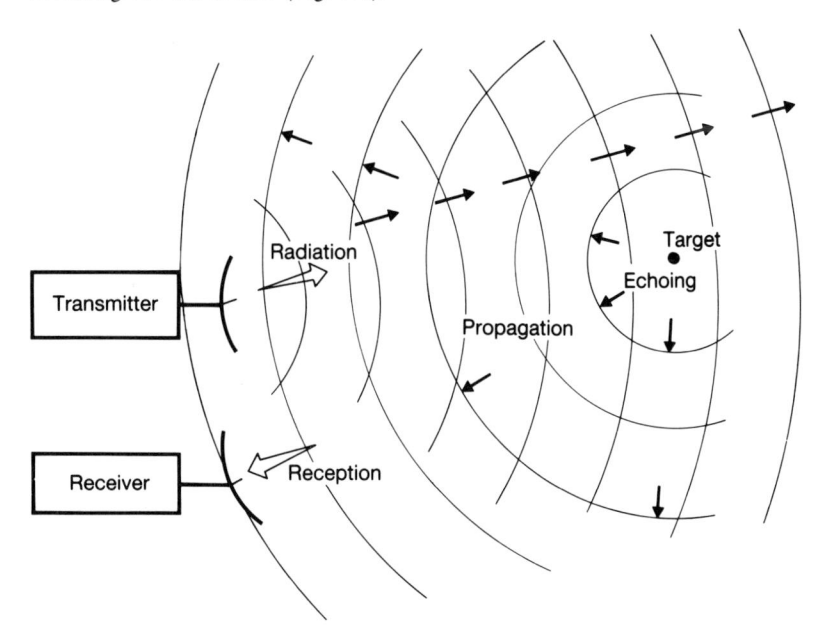

PROPAGATION

Once launched from the radar antenna, the radio energy radiates through the atmosphere, concentrated within the radar beam. Following reflection by a radar target a proportion of the energy re-radiates back to be captured by the receiving antenna (usually the same antenna). During this propagation phase five significant phenomena affect the radar performance.

The first, 'propagation loss' is the reduction in power density of the radar transmission as it radiates from the radar antenna towards a target, together with that which it incurs, after reflection by the target, as it returns towards the radar's receiving antenna—assuming the intervening medium to be homogeneous and lossless.

The second is 'atmospheric attenuation'; the absorption and scattering of energy from the beam by constituents of the atmosphere through which it is propagating.

The third, 'radar reflectivity' relates to the effectiveness of the target in reflecting energy back towards the radar's receiving antenna, a function of the size, shape and reflective efficiency of the target.

The fourth, 'clutter' results from undesired scattering back to the radar's receiving antenna from other objects within the radar beam, such as raindrops or, in the case of low flying targets, terrain simultaneously within the radar beam.

The fifth factor, 'shadowing' results from the straight line propagation properties of high radio frequencies, as a result of which objects below the horizon, or behind high ground, buildings or other substantial obstacles, will not be detected.

Propagation loss

In radiating towards a target the power density of the radar transmission falls off according to the usual 'inverse square of distance' law of radiation. The power density of the wave incident on the target therefore is $1/(\text{target range})^2$.

Some of the energy incident on the target is scattered back towards the radar antenna; the power density of this re-radiated signal, in turn, falls off in proportion to the inverse square of the distance from target to radar. Accordingly the energy density at the radar antenna, after radiation to a target and back, falls off as $1/(\text{target range})^4$.

Atmospheric attenuation

The radars being discussed all operate within the earth's atmosphere, which is constituted of varying amounts of gases and vapours. These absorb radio waves and thus attenuate the radar beams in proportion to the distance they travel. The absorption introduced by these mechanisms increases progressively towards higher frequencies but with the superimposition of peaks resulting from molecular resonances. This effect becomes significant at frequencies above about 1 GHz and may dictate a choice of operating frequency which is a compromise between the interests of long detection range and target resolution.

Radar reflectivity

According to its role an airborne radar may direct its beam toward a 'point' target which subtends only a small part of the radar beamwidth, or a 'distributed' target, such as conurbations, woods or even the earth's surface itself, which may fill a substantial part of the beam.

An aircraft target is in the former class, and its effective reflectivity in a particular direction is conventionally given in terms of the cross-sectional area of a uniform, perfectly reflecting, sphere which would return an echo of the same magnitude back towards the radar receiving antenna. As an indication of the magnitude of the reflection, a fighter aircraft might have an echoing area varying from a few square metres in the head-on or tail aspects to many square metres broadside on.

This same yardstick is used as the measure for the overall reflection from distributed targets, such as those on the earth's surface and concentrations of raindrops.

The nature of radar returns

While an aircraft can be considered a point target from the viewpoint of its dimensions in relation to the radar coverage and pulselength, it is large in relation to the wavelength of the radiated waves. The overall echo from an aircraft results from reflections from parts of the structure separated by many wavelengths, such as the fuselage, engine intakes, wings, fins, etc. The magnitude of each of these individual echoes will vary with the aspect of view, and because of their separation, the relative phases of their several contributions will change rapidly with aspect angle, so that their sum will fluctuate. As a consequence, as a target moves relative to the radar, the radar echo will be subject to rapid fluctuations in amplitude, known as 'scintillation' or 'fading', and the echo will appear to come from varying positions around the aircraft, a phenomenon known as 'glint'.

The fading and glint effects are common to most real-life radar targets, since they are usually large compared with the wavelength and complex in shape, be they aircraft, ships, man-made structures, or even terrain features or the sea surface.

Clutter

Clutter is the term applied to returns from objects which reflect energy back to the radar which may mask the returns from desired targets.

For example the radar return from an individual raindrop is minute and not separately detectable by a radar, but the aggregated echoes from the myriad of drops in a raincloud or area of precipitation are detectable and can be of such a magnitude as to swamp other returns. These returns constitute 'rain-clutter'. Similarly, undesired echoes arising from the earth's surface and objects on it, and from the sea surface are termed 'land-clutter' and 'sea-clutter'.

Atmospheric refraction

In a uniform medium, radio waves at microwave frequencies propagate in straight lines. However airborne radars operate in the atmosphere which varies in density with altitude. Furthermore the earth's surface over which the radars operate is curved. As a consequence a radar beam directed downwards from an aircraft towards the horizon experiences a progressively increasing refractive index which causes it to be bent downwards. This has the effect of extending the onset of horizon shadowing to beyond the geometric value. This effect is usually allowed for in range calculations by assuming a scaled-up value for the earth's radius using a factor of between 5/4 and 4/3, as is the practice in surveying.

Another effect of atmospheric refraction occurs in the meteorological condition known as an inversion. This occurs when warm air over the sea or land is overlaid by colder, more dense air. This causes a radar phenomenon known as 'ducting' in which radar waves from a radar near the ground, entering the area of the inversion at a grazing angle are bent down from the denser layer and may be trapped or 'ducted' by being reflected successively at the earth's surface and the inversion, resulting, sometimes, in detection of targets beyond the horizon. The radar from an aircraft incurs this phenomenon from the other side: a radar beam approaching the earth at a shallow grazing angle is bent upwards by the density gradient leaving areas where the beam may not penetrate so targets may be undetectable.

FUNCTIONAL ELEMENTS OF A RADAR

Having reviewed some of the factors affecting the capability of a radar, attention can be turned to its various major constituents, their contribution to the capabilities of the radar and the limitations and opportunities they offer in the overall functional design of the radar.

Fig. 9.2 Basic elements of an airborne radar

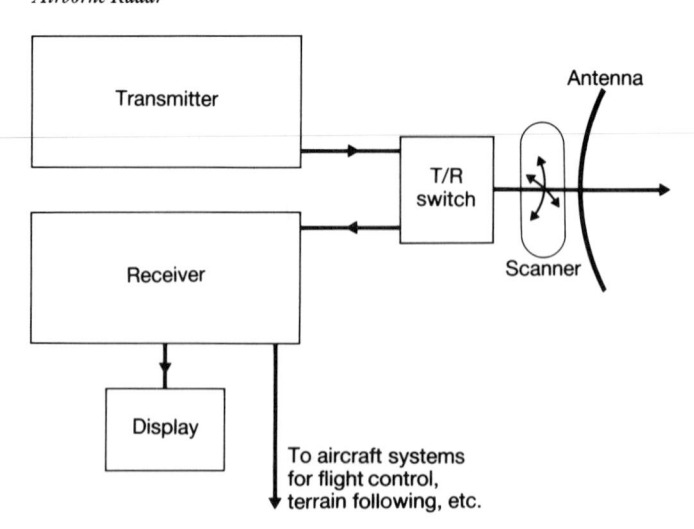

The basic elements of a radar are shown in Fig. 9.2. Their functions are closely inter-related in a way that is highly dependent on the overall purpose of the radar.

Each element has practical limitations to its capabilities, and these limitations, in combination, dictate or restrict the optimisation of the design, and may ultimately be reflected in limitations to the performance that can be achieved by the whole system.

The transmitter is the source of the radio frequency energy by which the radar performs its function. The mean power it generates may range from a few watts in the case of an altimeter, to tens of kilowatts in an airborne early warning radar.

The transmitter power tubes usually require operating voltages of tens of kV. The transmitter power supplies are bulky and heavy, and a considerable amount of waste heat may need to be extracted. Thus the transmitter is usually a dominant contributor to the overall size and weight of a radar.

Energy from the transmitter is fed through the transmitting antenna which radiates it into space. Depending on the tasks for which the radar is designed the antenna may concentrate the radiated energy into a cone ('beam') so that the energy density radiated towards the target will be many times higher than if the antenna radiated in all directions. By confining the radar energy into a beam the antenna also defines the direction of any target that is illuminated by the beam.

In order that the radar beam may be moved so as to point in a desired direction or to sweep through a required coverage sector, the antenna may be carried on a movable mounting, known as a 'scanner'. This is usually equipped with controllable motors to direct the beam and electrical position sensors to indicate the direction in which it is pointing. This platform may be controlled by position signals from the aircraft's attitude and position reference system to hold the beam in a particular direction irrespective of aircraft manoeuvres ('stabilisation').

The radar energy, as it radiates to and from the target suffers a reduction in strength as a result of interaction with the constituents of the atmosphere. Only a small proportion of the energy in the beam is incident on an aircraft target since its dimensions are small compared with the spread of the beam at operating ranges. According to its size, shape and orientation the target scatters some small proportion back towards the radar. Depending on the area of the receiving antenna some part of this energy is collected and fed to the receiver where it is amplified and 'detected' to give a low frequency signal indicative of the presence of the target.

Some processing may be carried out on the signal to derive information about the target, such as its range, range-rate and height.

The signal outputs may be passed on as electric signals to enable measurement or control for such purposes as automatic landing, navigation, terrain following or weapon priming and release. Alternatively they may provide information for presentation to the crew on an instrument or display or may be transmitted over a data link to another aircraft or the ground.

In looking at the individual parts of a radar in closer detail it is convenient to start by considering the receiver, because it is here that the fundamental limitation to the radar's ability to detect the smallest echo signals resides. This limit is set by the 'noise' power inherent in the early stages of the receiver, caused by thermal agitation.

Receiver noise level The level of the noise power in the receiver is determined by the product of four terms.

The first, the 'Boltzmann constant' is a fundamental physical constant with a value of 1.38×10^{-23} °K.

The second is the 'noise temperature', the equivalent temperature of the receiver, arising from the effective temperature of the various receiver components contributing to the overall noise level.

The third 'noise bandwidth' is the receiver frequency bandwidth, involved because thermal noise power is spread uniformly throughout the whole frequency spectrum and therefore the wider the receiver bandwidth the more noise power is accepted.

The final term is the receiver noise factor' a multiplier which is a measure of how closely the receiver circuits approach the theoretical level of noise.

Of these factors, the Boltzmann constant is, *ipso facto*, fixed as far as the radar designer is concerned. The minimum practical receiver bandwidth is imposed on him implicitly by system requirements. Noise temperature is determined by the working temperature of the components contributing to the system noise. The designer can select the configuration and components of his receiver to achieve a good noise factor for example, by including a low noise amplifier at the front-end, which by amplifying the received signals, makes the contributions of the succeeding stages proportionately less significant.

If circumstances permit, the designer may elect to provide cooling to reduce the temperature of the most critical contributors, such as the microwave amplifier, but frequently this is not possible or economic.

Before the parametric amplifier came along the major contributor to receiver noise was the microwave superheterodyne mixer. Mixer noise factors have fallen progressively from, for example 100:1 (20 decibels) in the 10 GHz pulse radars of the 1940s and 1950s to around 5:1 (6 to 7 decibels) in present day systems. By preceding the mixer by a parametric amplifier these figures can be approximately halved.

In the receiver of a pulsed radar the best signal-to-noise ratio is obtained when its bandwidth is matched to the pulselength. A bandwidth slightly greater than the reciprocal of the pulselength has been found to be optimum; for example, for a pulselength of 1 microsecond a bandwidth of about 1.2 GHz, for 0.2 microsecond a bandwidth of 6 GHz. In a pulse Doppler radar where the Doppler signals of individual targets are separated out in a narrow band frequency analyser, noise bandwiths of a few hundred Hertz will be obtained.

Signal to noise ratio The thermal noise level, after detection in the receiver, has a randomly fluctuating value varying from instant to instant. As has already been mentioned the signal from a target is also subject to fluctuation as a result of fading. The ratio of the signal to the noise power at any moment is therefore varying in a complicated manner, so that the radar's ability to distinguish the target return can only be expressed in terms of a probability that any pulse received from a target will exceed the instantaneous noise by any chosen margin. This margin has to be carefully chosen. A threshold level is set within this margin. Since the noise signal is random there is always some probability, however small, of the noise exceeding the threshold, and being detected as a 'false alarm'. If the threshold is set too low this probability may be uncomfortably high; if it is too high there may be too great a probability that the target echo may not exceed the threshold and the target will go undetected. The objective of the radar designer is to set a threshold at which the probability of a false alarm is negligible and at the same time to adopt values for the significant radar parameters, for example antenna size, transmitter power, etc., to achieve an adequate signal from the stipulated target at the maximum design range. Then, because of the $1/\text{range}^4$ law the signal from a target will rise rapidly as it approaches from maximum range and the probability of detection will rise also.

In the case of a simple pulse radar where the radar output is shown as bright-ups directly on a cathode ray tube display the threshold is applied by adjusting the video gain until the noise peaks only just paint on the screen with an acceptable probability; the signals, which are stronger will then paint with a greater probability. In more sophisticated radars the signal is gated electronically in the signal processing.

Because the received signal from a target increases rapidly with the closeness of the target because of the fourth power relationship with range, the dynamic range required of the receiver is immense. To avoid overloading, it is common in low p.r.f. radars to introduce a synchronised or 'swept' gain control in the early stages of the receiver amplifier to introduce a high attenuation immediately after each pulse is transmitted, reducing rapidly at such a rate as to compensate for the $1/\text{range}^4$ fall off. This is obviously not possible in a radar with range ambiguity since returns from different ranges resulting from different transmitted pulses will be intermingled. The receivers for medium and high p.r.f. radars, therefore, need to be specially designed to handle the very large dynamic ranges encountered.

Signal Integration The effective signal-to-noise level can be improved if successive pulses from a target can be summed (integrated) before thresholding. The number of pulses which can be integrated effectively is limited by the rate at which the target echo is fading and the length of time that the radar beam illuminates the target. Under these circumstances the signal at the target is substantially invariant while the noise signal is varying randomly.

If the integration is performed before detection, while both the phase and amplitude of the signals can be taken into account 'coherent integration' is possible, by which the signal-to-noise ratio is increased directly as the number of pulses integrated. If the integration is performed post-detection and/or takes no account of signal phase, the integration is 'non-coherent', and the signal-to-noise is improved only in proportion to the square root of the number of pulses.

The signal-to-noise required to give the desired probability of detection with an acceptable false alarm rate, taking into account the benefits of any integration, determines the minimum signal levels that must reach the receiver from the radar

Range resolution

The smallest increment in range at which the echoes from two point targets can be separately resolved in a pulse radar is when they are separated by one half of a spatial pulselength – the distance that the radar waves travel in the duration of one pulse – for this is the separation at which a pulse received from the nearer target will have just finished before the leading edge of the pulse from the further target arrives.

From this is derived the concept of dividing the distance out to the maximum range of the radar into 'range cells', each of a length equivalent to one half the spatial pulselength. In simple radars these merely represent the resolvable range elements, but in the more sophisticated radars the returns in each cell are gated and stored for processing or analysis and display.

Amplification and detection

To raise the minute incoming signals to a level at which detection can be performed efficiently they need first to be amplified. Microwave amplifiers have only been available for the present generation of radars; even now they are a relatively bulky and expensive way of obtaining amplification and are, therefore, only used as pre-amplifiers to achieve low noise factors. The incoming microwave signals are therefore converted to a lower radio frequency convenient for efficient amplification (known as an Intermediate Frequency or IF) by beating them with a locally generated microwave signal (local oscillator signal) having a frequency differing from the transmitted frequency by the IF. The difference product which is separated and amplified prior to detection, carries the pulse envelope and the relative phase of the received signal.

The majority of low p.r.f. radars have used Magnetron transmitter tubes, the output of which is not highly stable in frequency. The local oscillator signal has usually been provided by a low power Reflex Klystron oscillator, whose frequency has to be controlled by a frequency discriminator/automatic frequency control loop-to ensure that its output frequency remains at the correct difference from the transmission. Now, most modern Pulse Doppler radars are based on power amplifier tube transmitters driven by highly stable crystal controlled frequency synthesisers which also provide local oscillator outputs precisely related to the transmission in both frequency and phase.

THE ANTENNA

Virtually all airborne radars employ the same antenna for both transmission and reception, because the size and aerodynamic shape of an aircraft often make it difficult to find even one location sufficiently large and with a clear view for the radar to cover the required aspects.

The purpose of the radar antenna is to launch the radar transmission efficiently into space within the required coverage zone, and to 'collect' the energy incident on it by reflection from any targets within the coverage. At the same time, the antenna is usually designed to give a degree of directivity to the transmitted energy, confining it as a beam of radiation. This serves two valuable purposes; by concentrating the radiated energy within a narrow cone, the density of energy incident on targets is increased, the degree of concentration being termed the antenna 'gain'. At the same time the narrowness of the beam enables the direction of a target returning energy to the radar to be more precisely determined.

The formation of the beam is a wave-diffraction effect, and the degree to which an antenna of a particular size can concentrate the energy, and therefore the beamwidth, is a function of the dimensions of the antenna aperture measured in wavelengths – the greater the aperture and/or the shorter the wavelength, the narrower the beam. Thus, for example, an antenna 0.9 metres in diameter operating at 10 GHz (3 cm wavelength) having an aperture width of 30 wavelengths, would have the same beamwidth as a 3 metre antenna operating at 3 GHz (10 cm wavelength).

It is commonly found convenient to define an antenna by its dimensions in two orthogonal planes, one at least of which runs through an axis of symmetry, referred to as the azimuth and elevation planes.

At any chosen operating frequency, the concentration of the beam in the azimuth plane is inversely proportional to the beamwidth and hence directly to the azimuth plane dimension of the antenna; the concentration in the elevation plane varies directly as the elevation dimension. Consequently the overall concentration of energy is proportional to the product of azimuth and elevation dimensions. In the most common case, that of a circular antenna, the antenna gain is thus a direct function of the square of the diameter and, therefore, proportional to the aperture area of the antenna. The power incident on a target in the beam and reflected back to the radar antenna is proportional to the antenna gain and the proportion of this which is collected by the radar is proportional to the aperture area of the antenna. Consequently the received power, and hence the ability of the radar to detect the target, is directly proportional to the fourth power of the antenna diameter. For this reason antenna size is a very significant parameter in the design of a radar, and it is usually necessary to achieve the largest possible aerial in an aircraft.

For a chosen dimension of antenna, the beamwidth in each plane varies directly with wavelength and the antenna gain varies inversely as the square of the wavelength. The wavelength is therefore a strong factor influencing the detection capability of a radar and the highest practical frequency is usually adopted. Because propagation effects cause the path loss to increase with frequency a compromise has to be struck in relation to the particular operational requirements for the radar.

At the time of the first developments of radar it was already possible to draw on antenna technology which had been established for radio communications and broadcasting.

The design concepts for both omnidirectional antennas and directional arrays existed at the longer wavelengths used for the first surveillance radars. As the drive towards equipments operating at shorter wavelengths yielded oscillators and components in the decimetric and centimetric wavebands there were techniques which could be drawn on and extended. End-fire and broadside arrays of radiating elements were carried down into the decimetric waveband and used in the first airborne radars. When operation in the centimetric wavebands became possible, and the number of elemental radiators in an antenna of practical dimensions became inconveniently high, the concept was adopted of the parabolic reflector fed by a small primary feed at its focus. This practice was already established for sound ranging, in which the acoustic wavelengths involved also included centimetric wavelengths. Except for a few systems operating in the lower frequency bands arrays of dipoles are little used today.

The parabolic reflector with focal plane feeds has been developed in a number of forms and has been extensively used in airborne radars.

In recent years planar slotted waveguide arrays have come into wide use because of their higher efficiency and the better control they allow the designer in the characteristics of the radiated beam pattern.

Current developments are aimed at exploiting progress in microwave semiconductors and integrated circuits, with the aid of fast data processing, to make planar arrays where each radiating element is integrated with its own individual miniature transmitter, duplexer, phase-shifter and receiver, the necessary phase control to point and scan the beam or beams being controlled by computer. The multiplicity of elements allows a high mean power to be achieved despite the low r.f. power limitations of individual semiconductor devices. The particular benefit of these techniques will be the possibility of forming more than one beam simultaneously, the speed at which the beam can be moved, and fault tolerance because failure of a few random elements will have little effect on the overall performance.

The 'pipe dream' for the future is conformal arrays in which the individual elements of the array are not planar, but are contained in the natural surface contours of the aircraft. The benefits of this to the aircraft designer appear considerable, but the technical problems of implementation are formidable, except in the simplest of situations.

Antenna characteristics

The simplest antenna is the 'dipole' which provides an omnidirectional radiation pattern in its equatorial plane and a figure-of eight in the polar plane. A number of dipoles can be positioned in line which, when fed with r.f. energy in appropriate phase, form a narrower beam along the line of the elements. Much greater directivity can be achieved by arrays placed broadside to the direction of the required beam and fed in the same phase.

Dipole arrays became mechanically much more difficult to implement when operating wavelengths moved into the centimetric and millimetric wavebands. Much simpler was the expedient of locating a single dipole at the focus of a parabolic reflector. By this means antennas of any desired aperture size could be achieved without greater complexity.

Because waveguides became the preferred method of transmission of microwave energy and signals, the dipole fed by coaxial cable gave way first to a dipole energised from the waveguide, then to direct radiation from the waveguide as the method of illuminating the parabolic reflector. By flaring or narrowing the extremity of the waveguide in one or both dimensions the angular spread of the radiated energy could be varied and the simplest and most used form of primary feeder, the 'waveguide horn' was devised. The use of a horn at the focus of the paraboloid allows the spread of energy across the aperture of the paraboloid to be adjusted to match its size.

The objective of the radar designer is to feed the focusing reflector in such a manner that the reflected radiation leaving the aperture of the antenna is at a uniform phase. This aperture distribution is transformed by the diffraction process operating in the transition region known as the 'near field' into a beam pattern in the 'far field'. The diffraction process, however, produces not just a single beam but also a pattern of lesser beams surrounding the desired 'main beam', and known as 'sidelobes'. As well as taking some of the energy that should go into the main beam, these sidelobes can be detrimental to the operation of the radar by receiving spurious reflection from targets and clutter echoes from the ground (Fig. 9.3).

The sidelobes fall off in magnitude, away from the main beam and become negligible in their effect, but in practice there is usually a random pattern of low level lobes at all angles caused by tolerances and imperfections in the antenna.

The main beam is at its narrowest and the antenna gain at its highest when the

Fig. 9.3 Antennas – centre-fed and off-set

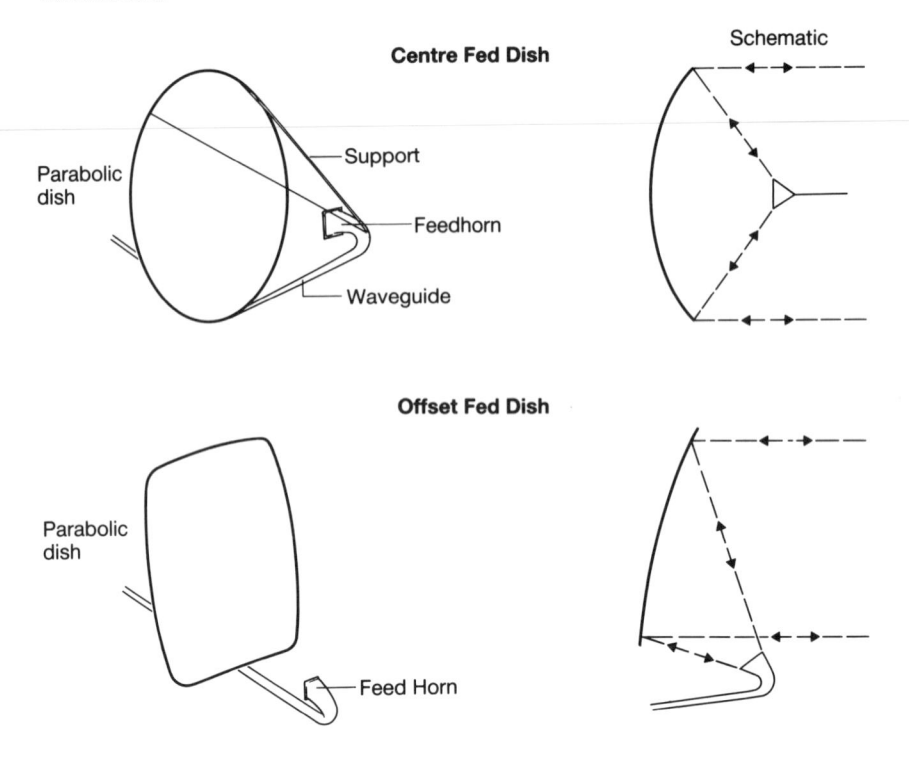

illumination provides a uniform power density over all the antenna aperture. However this condition also generates the highest level of sidelobes, and, anyway, is impossible to implement from a feed-horn at the focus. A practical arrangement usually adopted is to size the feed-horn so that the illumination falls by about 10 to 1 between the centre of the parabola and its edges. This gives a satisfactory compromise between reduction in gain and broadening of the beamwidth, the sidelobe levels, and the proportion of the energy from the feed-horn lost by falling outside the edges of the reflector – so called 'spillover'.

A shortcoming of the parabolic dish and feed type of antenna is that the feed at the focus and the waveguide feeding it, together with any supports, block a part of the aperture area by intruding into the outgoing beam. This causes a deterioration in the performance and an increase in the level of the sidelobes.

A number of configurations have been devised to counter this difficulty. These include the microwave Cassegrain antenna and offset feeding of the antenna (Fig. 9.4).

Many of the latest generation radars employ a planar slotted array antenna (Fig. 9.4) comprising a flat conducting plate through the front of which lines of resonant slots have been cut, each coupling with waveguides on the back of the plate. The manner of feeding and coupling energy through each individual slot enables the designer to arrange the distribution of radiation over the face of the plate, which constitutes the aperture of the antenna. He thereby controls the antenna gain and beamwidth, and the levels of the sidelobes to suit the purposes of any application. An additional benefit of the planar array is that no energy is lost by spillover and a higher efficiency and antenna gain can be achieved.

Radar coverage There is a range of ways in which a radar can provide coverage of a desired zone of sky. At its simplest a radar beamwidth can be chosen which can be directed

Fig. 9.4 Antennas – Cassegrain and planar slotted array

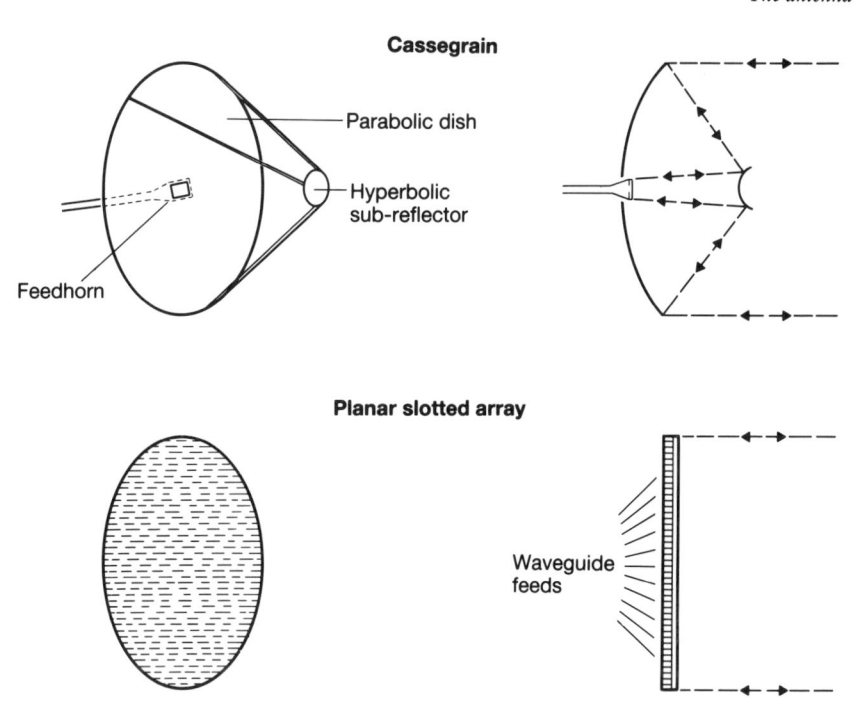

Cassegrain
- Parabolic dish
- Hyperbolic sub-reflector
- Feedhorn

Planar slotted array
- Waveguide feeds

to cover the whole of the zone required. This is the only practical approach in some types of radar, including sideways looking radars and altimeters. For search and surveillance radars however this has two disadvantages – that the gain of the antenna is low, and that the direction of a target within the beam cannot be accurately measured. By using an antenna with a narrow beam the gain and directivity are increased but in order to maintain coverage the antenna beam must be 'scanned' to cover the requisite zone. This results in a target only being illuminated for a part of the time, while the beam is passing over it.

If the transmission is pulsed there are two restrictions on the rate at which the radar beam can scan the coverage zone. First, if the pulse is long the beam must not sweep through the target in less than the pulse duration or else the reflected power will be reduced; second, the beam must remain on a target at maximum range for at least long enough for the desired number of pulses to transit to the target and return, or else the antenna will not be 'looking at' the target when the echo arrives.

Another consideration is that the pattern of scanning of the coverage zone must be repeated at such a rate that a target does not intrude too far into the coverage zone before it is detected. Thereafter it must continue to be 'plotted' at frequent enough intervals for its course to be tracked without gaps and confusion with other nearby detections.

Radars employing multiple p.r.f.'s such as are used to eliminate range ambiguities require the dwell to be sufficient to cover fully the transmission and reception of a complete sequence of p.r.f. bursts.

Monopulse In the early development of radar the accuracy of locating a target was determined by the width of the radar beam, but a greater accuracy in angular measurement was clearly desirable. A technique, known as 'lobing' was devised to measure the angular position more accurately by a differential method. The radar beam was

Fig. 9.5 Monopulse ground ranging

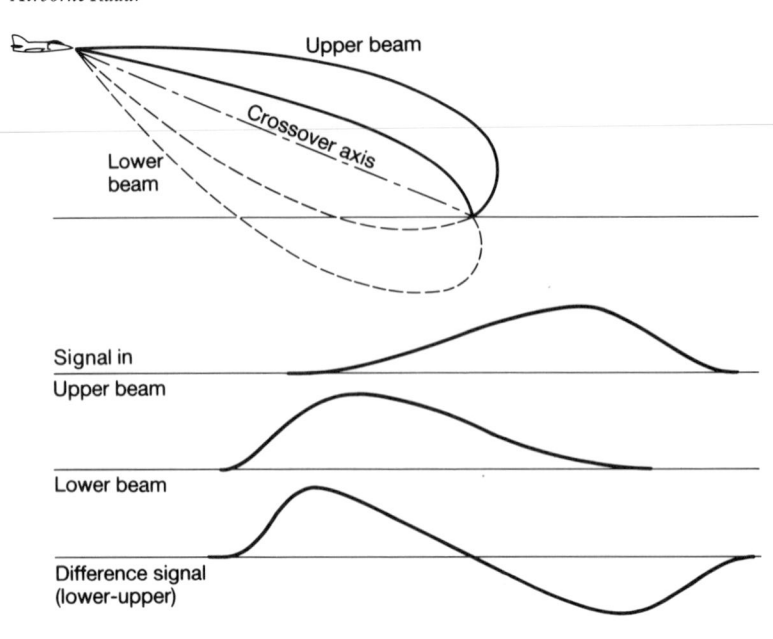

[Figure: antenna pattern showing upper beam, lower beam, crossover axis, and signal traces for upper beam, lower beam, and difference signal (lower-upper)]

stepped each side of the coarsely determined bearing so that the signal was received alternately on the opposite slopes of the antenna beam. By comparing the strength of signal received on each flank of the beam, position could be measured more accurately, as long as the signal was sufficiently strong.

In principle this offered an improvement in accuracy, but in practice if the signal was subject to any fading between the two positions the measurement was spurious.

The monopulse technique was developed to surmount this limitation. This employs an antenna generating two overlapping beams with an angular separation. The signals received by each beam on each pulse are combined in a microwave receiver giving 'sum' and 'difference' outputs. The sum channel is used for transmission and, on reception, for detection and ranging. The difference channel signal is at a minimum on the radar axis and rises each side as the signal from one of the two beams predominates. By comparing the phase and amplitude of the difference channel signal with those of the sum channel signal the difference channel signal is normalised and gives a measure of the deviation and its direction. The normalised difference channel signal is linear over most of the beam overlap and can be used to measure the off-boresight angle of a target or can provide an error signal to servo the antenna direction to the target.

According to the role of a radar the monopulse technique may be applied in either the azimuth or elevation plane of the antenna, or by use of a four beam cluster, in both planes.

Monopulse ground ranging

So far the use of the monopulse technique to determine more accurately the position of an airborne target has been considered. The same technique can be used to determine the direction of a point target on land or on the sea, so long as the echoes from the surrounding terrain (clutter) are very much smaller than those from the target.

In the ground ranging mode the method is applied in an inverse manner. The radar boresight is directed at a point on the ground whose range is to be determined. Because of the low incidence a simple pencil beam would illuminate a very long strip of ground and echoes from within this would be received from many pulselengths so range measurement will be inaccurate.

Accurate ranging is possible using elevation plane monopulse, in the manner indicated in Fig. 9.5. The method utilises the fact that the signals received from the upper and lower beams have opposite phases when compared with the sum channel signals from the same range elements.

After the transmission of each pulse, echoes are received initially more strongly in the lower beam, according to the range at which they are reflected. Then the response in the upper beam grows steadily until, at the boresight range it is equal to that in the lower beam, at which time the difference channel signal reduces to zero then starts to grow with opposite sign. Various circuit techniques have been devised to sense the moment of crossover and thence to determine the range.

THE TRANSMITTER

Early radars used thermionic amplifier tubes for the generation of their r.f. power. These have an upper frequency limit to their operation, and the drive to achieve higher frequencies called for new devices. Two of these, in particular the Klystron and the Magnetron, enabled the centrimetric wavebands to be opened up.

The klystron amplifier consisted of two or more cavities, resonant at, or tuned to, the operating frequency, spaced along the axis of the tube.

An electron gun, like that of a cathode ray tube, generated a beam of electrons, focused and accelerated by high voltages applied to the electrodes of the gun and directed along the axis and through the successive cavities. A small r.f. signal, at the resonant frequency, coupled into the first cavity causes bunching of the electrons passing through the cavity. This bunched beam then interacts with the next cavity giving up some of its beam energy resulting in an amplified signal in the cavity. Power is coupled out of the final cavity. Klystrons have been designed to deliver up to tens of kilowatts.

A disadvantage of the klystron is the very high voltages required for the generation, focusing and acceleration of the electron beam.

A low power oscillator variant, the reflex klystron has found extensive use in radar receivers to generate the local oscillator signals required for superheterodyne operation.

The cavity magnetron, when it came along in 1940, was a much simpler device. It comprised a metal block into which were cut a number of resonant cavities arranged in a ring around a central cylindrical cavity, to which each was coupled by a slot in the common wall. A thermionic cathode was located along the axis of the cylinder, and the block was located between the poles of a powerful magnet, so that the field was directed along the axis. The block constituted the anode and for operation was raised to a high voltage with respect to the cathode.

Electrons emitted from the cathode were accelerated towards the anode by its high potential. However they were deflected by the magnetic field into a curved path. With the correct potential and magnetic field the curved path of the electrons past slots in the cavities coupled energy from the moving electron beam into the cavities and oscillations were built up. A coupling out of the cavities provided the microwave power.

The only voltage needed for this type of oscillator is that applied to the anode block, so that it is much simpler to use than the klystron. In pulse radars this voltage is applied as a short high-voltage pulse delivered to the tube by a modulator unit, the triggering of which is controlled, at the required pulse repetition frequency, by a synchroniser unit.

This is a central master timing unit which also provides correctly related timing pulses for other purposes, such as the ranging circuits and display timebases. The modulator is a very high power, high voltage, fast switching device and modulator design has always been a challenging and difficult task.

A magnetron only bursts into oscillation when the voltage pulse applied to the anode block has brought its potential above a critical level. As a consequence the phase of the r.f. output is completely random from pulse to pulse, and the exact frequency is affected by the block temperature and the effects of reflections in the waveguides that couple it to the duplexer and antenna.

It is, therefore, not suited as the power source for coherent radars where it is necessary to compare the phase and frequency of returning pulses with a reference signal consistent with the original transmission.

For coherent radars the synchroniser and modulator of the pulse radar give place to a master waveform generator, which has already been described in connection with the receiver, and a power amplifier.

The travelling wave tube comprises an electron beam gun which produces and accelerates an electron beam along the axis of the device which is encompassed by a concentric solenoid or permanent magnet system producing a field along the axis. This constrains the electron beam, while it passes along the centre of a 'slow wave' structure. This is the equivalent of a loaded transmission line and is used to propagate the small input signal along the structure. The electron beam, as, it passes along the tube, interacts with the field due to the input wave causing progressive bunching of the electrons in the beam and the abstracting of energy from the beam to the wave as it passes through the structure. The wave energy is coupled out of the tube at the end of the structure to provide the output power.

TYPES OF RADAR

Primary & secondary

In the normal applications of radar, the system transmits a beam of r.f. energy and subsequently receives the minute proportion of this energy which has been echoed back to it by the target. This is referred to as primary radar.

In the early days of radar, when the need was seen to distinguish 'friendly' aircraft from 'hostiles', the idea was conceived of carrying a small receiver/transmitter (transponder) on all friendly aircraft, which, on detecting illumination by radiation from a radar beam, would transmit a signal back to the radar. Since this would be much larger than the reflected signal the target would be distinguished as friendly.

This concept of a responding transmitter on the target aircraft has evolved into one in which the radar transmission comprises a characteristic group of pulses recognisable to the transponder in the target aircraft which then responds after a predetermined precise interval with a coded train of pulses, which identifies and/or provides information about the aircraft. This is known as secondary radar.

Monostatic, bistatic & multistatic

In most radars the same antenna and feed system is used for both transmission and reception, a form known as 'monostatic'.

In the military context this has the weakness that an enemy with the ability to direction-find on the transmissions can direct counter-measures against the radar, either by physical attack or by electronic jamming. This can be obviated if the transmitter is in a separate aircraft (or location) from the receiver(s) which can be totally passive so that the enemy is unable to direct his counter-measures at them.

This is termed bistatic in the case of one receiver vehicle or site and multistatic if there are more (see also Chapter Ten).

The semi-active radar guided weapon is a simple example of bistatic operation as it carries only a radar receiver and homes on radiation scattered by the target as a result of illumination by the radar in its launch aircraft.

There are formidable technical problems associated with ensuring simultaneous coverage of the target by both the transmitting and receiving beams from separate surveillance or fighter aircraft, and with determining target range passively at the receiving aircraft, but modern fast data processing offers the prospect of achieving this.

It is interesting that the earliest British airborne radar attempt, carried out in December 1936, was bistatic, with the transmitter on the ground and the receiver in a Heyford aircraft. (see Fig. 1.4).

The term bistatic is usually reserved for the case when the receiving system is at a remote position from the transmitter or is separately mobile, as distinct from two antenna radars which have separate antennas for transmission and reception.

Continuous wave (CW) and pulsed radar

The earliest indications of the potential of radio transmissions to detect aircraft were the observations of fluctuating signals of broadcast radio waves received in sensitive receivers when aircraft passed overhead, an occurence commonplace today as the 'jittering' of the TV picture seen as an aircraft flies nearby.

The receipt of a return from a target illuminated by a continuous transmission of this type shows the presence of a target, it can indicate direction if the transmission is confined to a narrow beam, but gives no measurement of distance unless several separately located receivers are positioned so as to make triangulation possible.

To measure range directly from a single location use is made of the fact that the radio energy travels at the speed of light, approximately 300,000 Km per second. The time taken for the radar beam to travel to the target and return is measured: since the wave must travel the path in two directions the range will correspond to 150,000 times the transit time or, conversely, the delay will be roughly 7 microseconds for each kilometer distance.

In order to measure the echo delay it is necessary to 'tag' the transmitted signal in some way so that the delay between the transmission of an identifiable epoch of radiation and its subsequent moment of reception can be determined. The most commonly used method of doing this has been to break the transmission into short pulses. The rate at which successive pulses are generated (the pulse repetition frequency or p.r.f) is usually arranged so that the next pulse is not radiated before the previous pulse has had time to return from the maximum distance to which the radar is effective. Otherwise, echoes from targets at longer ranges will overlap those from nearby and the range measurements will be 'ambiguous'. Radars which are unambiguous in range are termed 'low-p.r.f.' radars.

As an example a low-p.r.f. radar with a maximum range of 100 Km will receive signals for up to 667 microseconds after each transmitted pulse. The pulse repetition rate cannot, therefore, be greater than 1500 pulses per second.

When a target is in the beam of a pulsed radar each pulse will illuminate it for the duration of the pulselength and the echo will have this duration. In general the shorter the pulselength the more precisely the range of the target can be determined and the finer the resolution of the separation of the echoes of adjacent targets, will be. However the smaller the pulse duration as a proportion of the interpulse interval the lower is the duty cycle of the transmitter and the higher the peak power in relation to the mean power. As a consequence low-p.r.f. radars have largely depended on the use of magnetron transmitter tubes, since these are capable of generating short pulses of very high peak power. Ultimately the limit is set at which breakdown occurs within the waveguide circuits used to connect the transmitter valve to the antenna. Since the waveguide dimensions are related to the wavelength the available transmitter power falls off with increasing frequency. Typically, peak powers of, at most, one or two megawatts are available in radars operating at about 3 GHz and $\frac{1}{4}$ Megawatts in those operating at 10 GHz. In typical applications these high power levels are associated with p.r.f.'s of about 400 p.p.s. so that the mean power levels are only a few hundred watts.

The performance of a radar is a complicated consequence of the values assigned to a number of interacting parameters, which include such factors as how long the radar beam dwells on the target as it scans through its coverage zone, how many pulses are transmitted in this interval, etc. When values have been determined for these, the range at which a radar can detect a target is, broadly speaking, related to the mean power transmitted. Since the mean power of a low-p.r.f. radar is determined by the maximum peak power and by the p.r.f. (itself a consequence of the maximum effective range) a limit is set to the radar's detection performance.

One approach used to surmount these limitations has been to operate at a higher p.r.f. and to accept the added complexities of introducing a means of resolving the ensuing range ambiguities by logical processing of the received signals, usually achieved by using controlled sequential variation of the p.r.f. and the correlation of the timing of the received pulses arising from each pulse interval. Radars which are ambiguous in range are termed medium p.r.f. or high p.r.f. radars. The distinction is explained in a later section.

An alternative method is to employ frequency modulation of the transmitted radiation instead of pulse modulation. For this the frequency of the transmission is swept repetitively over a period longer than that of the maximum path delay. As a consequence, signals received back at the receiver are at a different frequency from that being transmitted at that instant, and, by comparing the frequencies in the receiver, the range of the target can be determined. If, however, the target is moving towards or away from the radar this motion will superimpose a Doppler frequency shift upon the echo. These shifts can be calculated out by employing upward and downward shifts.

Unfortunately one of the difficulties inherent in a CW radar is that of protecting the receiver from overloading by breakthrough of power from the transmitter. This is no problem in low-power radars such as can be used for radar altimeters and airborne Doppler navigators, while in ground radars dual separated antennas can be used for transmission and reception. However in an aircraft there is no possibility of duplicating the large aerials needed to achieve the performance of an interception or airborne early warning radar, nor has it proved possible to achieve satisfactory duplexing devices akin to the duplexing switches used in pulse radars. Consequently pure CW radars have not found succesful application in high power airborne radars.

PULSE DOPPLER

In certain circumstances, for example, when a radar is looking for low flying aircraft, the radar beam or its sidelobes may strike the ground around the same range as the target. Because of the very large area of the surface which is illuminated the returned clutter echoes may be many times larger than the target echo and obscure it. To overcome this disability, techniques are employed which distinguish the signals from a moving target from those from the stationary terrain around it, by virtue of the different Doppler frequency shifts the motion imposes on the radar signals. Radars using these techniques are known as 'pulse Doppler' radars.

These make use of the fact that when a target moving towards the radar is illuminated by the radar waves, the signal returned to the radar in each successive epoch originates from a slightly nearer point than its predecessor and therefore returns in a slightly shorter time. As a consequence, more cycles of the wave reach the radar in any second than would have been the case if the target had been stationary; this is perceived as an increase in the frequency of the wave. If the target was receding the frequency would be shifted downwards.

The frequency differences are quite small; for example 30 Hz for each kilometre per hour of closing speed at 10 GHz (10,000,000 Hz), so that to enable them to be measured the radar needs a transmitter with a very stable and pure output which can provide an accurate frequency reference against which the target returns can be related after the time delay of propagation to the target and back. Early coherent radars employed a method known as COHO/STALO (COHerent Oscillator/STAble Local Oscillator). For this method a sample of the transmitter output was used to reset accurately the frequency of the receiver local oscillator on each successive pulse, this having a sufficiently high stability to retain the frequency throughout the propagation interval.

With the development of semiconductors capable of operating into the microwave band, this method has been superseded by methods based on a master frequency synthesiser. Oscillations generated in crystal-controlled oscillators at convenient radio frequencies are stepped-up by a succession of harmonic generation and frequency mixing (summing or differencing) to generate stable microwave signals for amplification in a microwave power amplifier (occasionally a klystron amplifier, but more often a travelling wave tube), and correctly related local oscillator signals for the purposes of the receiver.

CIVIL AVIATION APPLICATIONS

Weather radar The *en route* phase of commercial aircraft operations is a regime of comparatively low risk. The major avoidable hazard in this phase of flight is the inadvertent intrusion too near to the core of a storm. Whereas in many applications the reflection of substantial echoes by areas of rainfall is detrimental to the performance of a pulse radar, this same phenomenon is turned to advantage in the weather radar. This is designed specifically to distinguish these echoes and by scanning a wide sector ahead of the aircraft, sufficient to indicate the extent of a storm, enabling a safe path to be selected round or through it.

The scattering of radio waves by raindrops becomes greater at higher frequencies, as the wavelength gets nearer the dimensions of the drops. However, the result of using too high a frequency can be massive reflections from the near face of the rainstorm, but with all the energy being scattered and absorbed rapidly, the regions behind are shadowed and the storm core may be undetected. A compromise between adequate echoes, resolution and position accuracy, on one hand, and penetration of rainfall zones, on the other, is the aim. The compromise results in the use of frequencies around 5–10 GHz, the latter invariably being used in the smaller equipments used in General Aviation aircraft.

The earlier generations of weather radar identified areas of rainfall to the aircrew as patches of bright-up on a plan position CRT display. In more modern radars signal processing is introduced to highlight zones where there is a sharp gradient of rainfall rate, to give a better indication of the most severe conditions in the storm. Alternatively the introduction of colour TV tubes allows the 'synthetic colour' presentation of bands of different rainfall intensity.

Radar altimeters

Barometric altimeters are extensively used in the flight operations of both civil and military aircraft. They have the disadvantage that they do not measure altitude directly, but deduce it from local atmospheric pressure, as a result of which their height readings may be inexact due to variability in atmospheric conditions and errors and inaccuracies in their datum settings.

While these deviations are not, in general, serious at the higher altitudes, they are of crucial importance at low altitudes where terrain clearance and landing are concerned. Radars, mounted in aircraft and looking vertically downwards, provide a means of height measurement directly referenced to the land or sea surface and capable of sufficient accuracy to control aircraft automatically to a safe touchdown. Typically they are designed for the lower altitude regime, usually up to a maximum of 2000 or, perhaps, 5000 feet.

Both pulse and CW methods have been applied to altimetry. Because of the difficulty of switching fast enough between transmit and receive the earlier altimeters adopted CW using frequency modulation for distance measurement. Because the land and sea surfaces have a very high reflectivity at near-normal incidence, altimeters can operate at low power levels, typically around 1 watt radiated.

Because of the short ranges involved the higher frequencies can be employed and frequencies in the region of 16 GHz have been allocated for this purpose.

Height must be measured in the presence of aircraft manoeuvre and to avoid the expense and complication of providing stabilised mountings the antennas are designed to have wide beamwidths, sufficient to encompass the necessary manoeuvre angles in pitch and roll.

Because the antennas are small it is possible to solve the problems of duplexing by employing separate antennas for transmission and reception, hard-mounted in the bottom of the fuselage and spaced wide enough apart for the leakage of transmitted energy between them not to damage or inhibit the receiver.

Typical separations of 1/3-1 metre and beamwidths of typically ±60° ensure that the antenna coverages overlap at operating altitudes, but the leakage path can set a lower limit to the measurable altitude, important in the case of automatic landing.

The advent of semiconductor technologies has enabled pulse radar techniques to be employed, through the ability to generate very short pulses, and to handle them in the receiver. The low power needed for the transmitter enables altimeters for some purposes to use semiconductors for the generation of the r.f. power.

Secondary surveillance radar (SSR)

Secondary surveillance radar, the system adopted by the International Civil Aviation Organisation (ICAO) to provide an identification capability within the world's air traffic control systems, was derived from the military 'identification friend and foe' (IFF) system with which it coexists and inter-operates. The system comprises interrogating radars on the ground and Transponder beacons carried in aircraft.

SSR offers particular benefits in areas where ground-based primary radars suffer from clutter, and where a high density of aircraft movements makes it impossible to associate notified aircraft movements with the appropriate primary radar plot.

Interrogation radars have usually been co-sited with the air traffic control primary radars, their antennas sometimes being attached pick-a-back on the primary radar antenna, but the value of the system has become such that separate sites at favourable locations are now sometimes adopted.

Interrogation is carried out by the transmission of a pair of pulses 0.8 microseconds in duration on a frequency of 1030 MHz, with a characteristic separation. The transponder carried by the aircraft receives and recognises the interrogation and responds by transmitting, after a precise pre-determined interval, a group of code pulses within two framing pulses.

There are four modes of interrogation, denoted A, B, C and D. The mode for which a response is required is called up by selection of the spacing of the interrogating pulse pair, the pulse intervals being 8, 17, 21 and 25 microseconds respectively.

An additional pulse may be transmitted between the interrogating pulses, 2 microseconds after the first pulse. This is transmitted using a broader lower gain antenna beam, and is to enable the transponder to recognise spurious interrogations being received through the sidelobes of the interrogator antenna, which are not responded to.

The response from the aircraft is at a frequency of 1090 GHz. The two framing pulses are separated by 20.3 microseconds, the interval between them allowing for codes of up to 12 pulses at 1.45 microsecond spacing, the pulse at 10.15 microsecond being disallowed. Interrogation modes A and B are used for aircraft identification each mode providing 4096 response codes. Mode C is used to report the altitude of the aircraft automatically, a barometric altimeter capsule independent of the aircrew's navigation system being provided for this purpose, preset to an arbitrary datum. The 4096 codes provide for altitude reporting by 100 foot intervals. Interrogations in modes A or B can be followed immediately by an interrogation on mode C. The fourth mode has not yet been committed to use but the possibility of using this for a two-way data link is envisaged.

Unlike primary radar, SSR can distinguish individual responding aircraft in a crowded environment such as the terminal area around an airport. To exploit its capability to supplement primary radar in this role some SSR interrogators are now being equipped with a monopulse capability which considerably improves the accuracy of tracking.

Airborne Doppler radars Aircraft operating over the oceans, and over the less developed areas of the world where they do not have the benefit of ground navigation beacons and Air Traffic Control monitoring, have to depend on on-board instrumentation and dead-reckoning for knowledge of their position. As a result substantial errors can build up on long flights since the system is not in any way ground-related, and landmarks which could guide them may be missed. The airborne Doppler radar was conceived initially against military needs for accurate autonomous navigation, as a means of providing continuous ground-referenced navigation information (see Chapter Seven, Doppler Navigation).

Distance measuring equipment (DME) (see Chapter Seven) This is a secondary radar system, usually used in conjunction with VOR directional beacons. In this case the interrogator is carried in the aircraft to interrogate ground beacons which respond after a defined interval, enabling the aircraft's distance from the beacon to be determined.

Air traffic control radars (see Chapter Seven) Although these are not airborne systems they are associated with flying, and for this reason they are frequently included in the classification of avionics.

Being located on the ground they do not have the restrictions on antenna size and available power and cooling to which airborne equipments are subject. Thus they employ very large antennas, and are able to operate at lower frequencies to minimise the effects of atmospheric attenuation and rain clutter.

MILITARY APPLICATIONS

Whilst the preponderance of civil aircraft have a common single role of transportation and employ radars for the purposes of facilitating this role, military aircraft are designed for a wide range of different roles. Some are specialised, such as for AEW, airborne interception, anti-submarine warfare, reconnaissance and deep interdiction, while some, such as the tactical fighter, are multi-role, giving air support to the army, ensuring air superiority, conducting tactical ground attack and providing airfield defence.

In turn the aircraft's radars may be specialised or multi-role. We will first take a look at radars for some of the specialised roles, and conclude with reference to multi-role systems.

Airborne interception Until the 1960s high power pulse radar provided an effective means for night and all-weather fighter aircraft to detect their quarry at sufficient range to be able to manoeuvre into a firing position for their weapons. However, the rethinking brought about by the introduction of highly effective surface-to-air guided weapons, led to the tactic of low level penetration at heights of a few hundred feet. As a result when the radar beam of an interceptor aircraft searched for a low flying intruder, its radar would illuminate and receive echoes simultaneously from the target and the ground close to it. The area of ground within the beamwidth and sidelobes, and simultaneously within the pulsewidth, could return echoes (ground clutter) large enough to obscure the target return. To overcome this disability the Pulse Doppler form of radar, described earlier, has been developed.

The range of Doppler frequency shifts encountered is a function of the maximum operating speed of the radar aircraft and the maximum target speed to be handled. For example if the maximum speeds of both fighter and target are say 1200 knots the closing speed could be 2400 knots, and the doppler shift, for a radar operating at 10 GHz, some 72,000 Hz.

To avoid ambiguities in range the p.r.f. of the radar must be over twice this figure, in excess of 144 KHz, if there are not to be ambiguities in the frequency spectrum as a result of p.r.f. sidebands falling within the Doppler frequency band or Doppler frequencies folding about the sidebands. A radar wherein the signal spectrum is unambiguous in velocity is known as a high p.r.f. radar.

It is possible to operate in a range between low p.r.f. and high p.r.f., known as medium p.r.f., by employing bursts of several different p.r.f.'s within each dwell-time and employing signal correlation processing to resolve both target range and Doppler velocity ambiguities. Fast Fourier Transform (FFT) processors are commonly used to create a pseudo-bank of Doppler filters for each range cell within the coverage, replacing the banks of hardware filters used in earlier systems.

The functional characteristics of high and medium p.r.f. systems have a number of important differences. Broadly speaking the high p.r.f. system offers clutter-free detection of on-coming aircraft but may incur clutter in the broadside on and tail chase aspects. Medium p.r.f. can offer clutter free operation at shorter ranges in any aspect, but there may be some speeds at which the radar is ineffective at the longer ranges. While the inclusion of both modes of operation provides a robust solution, weight, size and complexity limitations do not allow this in radars for smaller aircraft.

Airborne early warning Originally developed to enable more effective air defence of the US fleets in the Pacific Ocean in the later stages of the Second World War, AEW has, until the last decade been developed primarily for naval purposes. However the increasing speeds of modern aircraft and the adoption of low altitude contour-hugging tactics by deep-strike aircraft has created a vital role for AEW in overland air defence.

The objective of AEW is to provide much earlier detection of potentially hostile air movements than can be achieved from ground-based sensors so that the scale and tactics of a hostile force can be assessed, and the necessary defence resources committed and directed effectively into successful engagement with the enemy.

The detection range of radars of the size that can be incorporated into interceptor and fighter aircraft is only a few tens of miles so the AEW aircraft must provide the means of bringing the fighter within radar range and in a position where its speed and manoeuvre capabilities can ensure an engagement.

To achieve the search range required for effective early warning the AEW aircraft must carry as large an antenna and generate as much r.f. power as possible. For naval purposes the size limits are set by the needs of operation in aircraft carriers, but for the overland role large transport aircraft are employed.

Because of the long ranges, the higher radar frequencies are precluded by atmospheric and cloud and rainfall attenuation, and frequencies in the 3 GHz and VHF region are usually employed.

VHF operation has the benefits of virtual immunity to the effects of cloud and the rates of rainfall encountered in any area of the world, lower levels of clutter return from rough seas and land, but the disadvantages of wider antenna beamwidth leading to poorer resolution, greater vulnerability to hostile jamming, and somewhat lower maximum ranges.

The 3 GHz frequency band gives better resolution and longer ranges, and is unlikely to be significantly affected by cloud and rainfall in temperate climates.

Automatic tracking of targets by computer in the aircraft is employed and target information is usually provided for a team of fighter directors in the aircraft, but the target information may also be data-linked to the ground based air defence system.

Navigation/Attack There are several ways in which radar assists the crew of a ground attack aircraft to find their way to their assigned target with sufficient accuracy to be able to pick it out and attack.

In the *en route* phase the radar may scan and display a 'radar' map of the terrain for many miles ahead of the aircraft to assist in identifying features and landmarks. The radar may be directed against identified features on the ground and used to derive an accurate instantaneous range and bearing measurement which is used to update the on-board navigation system. This same method can be used if the target is hard to distinguish, but is at a known position relative to a nearby identifiable feature. An accurate positional fix is taken on this, from which the target position can be derived with precision by the on-board navigation system.

In the situation where the target can be visually distinguished by the pilot, the monopulse ground ranging mode can be used in conjunction with the aiming sight to determine the exact moment for bomb release, thus taking out uncertainties in knowledge of the aircraft's height above the target in rough terrain.

Ground mapping   In the early stages of the Second World War the operational emphasis of radar development was directed towards air defence, initially to ground fighter direction radar then to radars for airborne interception and ship and submarine detection and attack. It was nonetheless noted that airborne radars detected echoes from objects on the surface of the earth, and that these varied greatly in their magnitude and extent according to the nature of the objects causing them. Large conurbations were readily distinguishable from the surrounding countryside. Thus, when the emphasis turned to offence and the need arose for navigation aids which could operate at ranges beyond those attainable from ground-based beacons, the potential of airborne radar for ground mapping had been recognised. The H2S ground-mapping radar was conceived, operating first with some success at 9 centimetre wavelengths, then at 3 centimetres with an ability to map terrain out to some 50 miles around the aircraft (Fig. 1.6).

There are two major developments of the mapping radar in present day use. One is provided as a mode of operation of the forward looking nose radar in a ground attack aircraft, limited by the location to an antenna width of about $\frac{1}{2}$ to 1 metre giving a beamwidth of about 2–3 degrees. The other utilises comparatively large (up to 10-metre long) antennas, located along the length of an aircraft fuselage or carried in a pod beneath it, and looking broadside onto the direction of flight. These latter, 'sideways looking airborne radars' (SLAR), are relevant to the reconnaissance role rather than navigation/attack and are described separately later.

Mapping radars of the 1940–60 period were intended for aircraft flying typically at a height of 15,000 ft or above where the horizon limited range is in excess of 150 miles. Operation at 3 cm wavelength was preferred because it gave a good compromise between mapping resolution and the limitation of attainable range imposed by atmospheric attenuation.

From a height of 15,000 ft, the pencil beam from an antenna would illuminate the ground between some 30–50 and 150 miles. Because of the rapid fall-off in returned signal with range the returns from towards the nearer end of their coverage would be many times stronger than those from the far end, which introduced difficulties in handling their signal and displaying the radar map. The configuration of the antenna was therefore modified to extend the lower flank of the radar beam so that the signal strength would fall off at a rate which would result in approximately the same signal from equal target echoing areas through the beam coverage. Because of the law of variation of the antenna gain with the depression angle of the beam these became known as cosecant squared ($Cosec^2$) antennas.

The advent of the long-range ground-to-air guided weapon has denied the high and medium altitudes to attack aircraft. Many of the attack aircraft of today will fly at altitudes of a few hundred feet to penetrate below the beams of ground radars in the region where they are limited by the local horizon, or by ground clutter. From an altitude of 300 feet the horizon distance is only some 20 miles. These shorter ranges enable much shorter wavelengths to be used, which, with their finer definition, allow a better prospect of identifying features and target during the approximately two minutes in which the aircraft will travel the 15 or so miles that the mapping radar will display.

Ground collision avoidance The development of short pulse radar and monopulse ground ranging technologies has enabled modes to be provided in attack aircraft to safeguard the aircraft from collision with the ground when flying at levels of a few hundred feet. These can take a variety of forms of varying complexity. The titles used to distinguish these are not universally consistent; those used here are given primarily to distinguish the different forms.

Terrain clearance The simplest form of air-to-ground avoidance, this takes the form of an elevation plane monopulse radar operating in a sector scanning mode akin to ground mapping. The boresight is scanned in a stabilised plane looking forward and the penetration of terrain above the beam's crossover is sensed and used to paint on the pilot's display those areas which are above the aircraft flight vector.

In a more sophisticated implementation, the monopulse receiver measures the angular deviation of every echo relative to the crossover axis, and these values are used in conjunction with the associated range and the selected depression angle of the boresight to compute the distance of the reflecting features from the flight path plane. A clearance height is selected, and those areas less than this distance below the flight path painted on the pilot's display. This guides him in flying the aircraft to avoid those hazards either by lateral deviation or by climbing above them.

Terrain following In this function the radar is used in the ground ranging mode to determine the profile of the terrain ahead of the aircraft by measuring the range and associated depression angle of each point along the profile. The elevation angles of successive peaks ahead of the aircraft are computed to determine the climb or dive angle necessary to clear them. This information is presented to the pilot in flight director form or may be coupled to the aircraft's automatic flight control system to provide automatic control of the aircraft in the pitch plane – 'automatic terrain following'.

There are many factors influencing the detail design of practical terrain-following systems.

In a manned aircraft it is usual to limit the pitch plane acceleration demanded, to typically $\frac{1}{2}$ g negative, and a few g positive, in order to avoid giving the crew an unnecessarily uncomfortable ride which would detract from their ability to carry out their missions efficiently. This enables a template to be visualised which reflects the potential of the aircraft to climb over obstacles ahead of the aircraft without exceeding the selected maximum g loads. The elevation angles of peaks detected ahead of the aircraft by the radar are compared in the terrain-following computer with a stored version of the template and only pitch demands which penetrate the profile are fed through to the flight director display or the flight control system. This ensures that the aircraft does not climb earlier than is necessary to clear obstacles ahead, having regard to the g limits selected, and thereby minimises the exposure of the aircraft to ground-based air defence weapons.

In practice the terrain whose profile ahead of the aircraft must be measured to provide effective terrain-following within the usual limits lies between some 5–10 degrees above the flight path and 10–20 degrees below it; this range determines the vertical scan pattern of the radar. The gust excursion in aircraft heading in normal straight flight may be several degrees. If the radar azimuth beamwidth is less than this, a two bar vertical scan may be used to ensure that the terrain-following system does not miss any terrain ahead that the aircraft may pass over.

The climb rate limits, and the speed of the aircraft, determine the number of scans per second which need to be performed to ensure that clearance requirements are detected soon enough.

Terrain-following radars enable aircraft to fly at low altitudes in darkness or bad visibility conditions in which the pilot could not operate. A consequence of this is that a malfunction or failure of the equipment could leave the pilot in a situation from which he could not recover.

Much care is therefore taken in the system and equipment design of terrain-following systems to ensure that failure survival and fail-safe capabilities are incorporated in the design so that operational failures of any form are detected, and, for instance, an automatic climb demand is generated at the moment of failure.

IFF and SSR In the earliest days of air defence radar it was realised that there was a need to distinguish the radar plots of friendly aircraft from those of hostile ones. No fundamental differences could be deduced by which the radar echoes of one could be distinguished from the other, so it was decided to introduce a deliberate difference. This was effected by equipping friendly aircraft with transponders which would respond to an interrogating radar transmission by transmitting a recognisable return signal. From this concept have evolved the more advanced IFF systems used in current military aircraft and the secondary surveillance radar (SSR) system used for civil Air Traffic Control.

The essential features of IFF are the same as those described under SSR. IFF differs in its modes of interrogation and response. There are three modes, denoted Modes 1, 2, 3. Mode 3 is virtually identical with Civil Mode C. For Modes 1 and 2 the interrogation pulses have separation of 3 and 5 microseconds respectively. The responses to these modes are equivalent to the responses to civil modes A and B but are used for ephemeral 'password' codes rather than to give the identity of the interrogated aircraft.

Reconnaissance Information on the locations of enemy forces and facilities is of vital importance to military commanders. The aircraft has provided the means to take a look at enemy territory where ground patrols cannot reach.

The role is of sufficient importance for special versions of aircraft to have been developed specifically for the reconnaissance role. There is a variety of sensor types, each of which has a part to play in the reconnaissance task. These include the human eye, downward and oblique cameras, both in the visual region and in the infra-red, sideways looking infra-red linescan, and high definition radar. Radar has a special feature, in that all the other sensors see a progressive foreshortening of the view when looking towards the distance, whereas radar separates features uniformly according to distance, since it measures the time delay of the echoes from features and this delay is directly proportional to their range. The normal plan position presentation on a radar display therefore gives an unforeshortened view.

Obtaining the maximum of intelligence requires the radar to have the highest attainable resolution. Operation at very short pulselengths can be used to give excellent resolution in range but the nose location of the radar in most fighter aircraft limits the available aperture to about 1/2 to 1 metre. The specialist reconnaissance aircraft carries the antenna mounted along the length of the fuselage, giving radar beams broadside to the aircraft and enabling antennas of up to 10 metres in length to be carried. Instead of scanning the radar beam, the motion of the aircraft is used to sweep the narrow radar beam over the zone to be examined.

The radar presentation, as well as being displayed to the crew, is photographically recorded across a continuously moving film strip, so as to build up a continuous photographic map as the mission progresses.

Some aircraft are designed for high-altitude reconnaissance where range is not limited by the horizon, and for this purpose frequencies around 10 GHz give a compromise between high resolution and long range through the atmosphere. Other types of aircraft are designed to penetrate enemy territory at very low altitudes. For these the horizon limits the possible range to a few tens of miles and millimetre wavelengths may be used, giving excellent resolution.

The advances in coherent signal technology and in fast signal processing have made possible the use of 'synthetic aperture' techniques. Because the patch of ground illuminated at operating ranges by the radar beam is wide compared with the length of the antenna, the synthetic aperture radar is able to process the coherent radar returns from any feature while the aircraft is moving through several times the actual length of the antenna and thereby to simulate a correspondingly longer antenna.

Multimode radars We have already described a number of ways in which radar can assist the aircrew in performing in various missions. The tactical fighter is a versatile aircraft employed for many purposes, from ground attack to air superiority and air defence. The recent great reductions in the size of electronic circuitry have enabled various air-to-air and air-to-ground capabilities to be combined within a single 'multimode' radar small and light enough to be fitted in the restricted nose space of a fighter aircraft.

REFERENCE

Bowen, E. G. *Radar Days*, Adam Hilger, 1987, p. 36.

CHAPTER 10 ELECTRONIC WARFARE – AN OVERVIEW

BRYAN R. DRAKE. Product Group Manager, Electronic Warfare and Weapons Electronics, Plessey Avionics Ltd.

Bryan Drake's first encounter with electronic warfare was as an Air Electronics Officer flying on Vulcan Mk 2 bomber squadrons with the RAF in 1966. He was involved in EW and air defence exercises throughout the world, including USA, Australia, Cyprus, Malta, Singapore, Libya, Canada, Scandinavia and Germany.

He later became Station Electronics Warfare Officer at a Vulcan base, responsible for training and examining operational aircrew in EW. This also included participation in simulated combat exercises. In 1979 he was posted to the RAF Central Tactics and Trials Organisation where he managed airborne jammer trials in UK, California and against RN warships.

Taking up his present appointment in 1982 Mr Drake has responsibility for all aspects of EW including ECM, ESM, laser and missile warning, radar decoys, system programmers and interface equipments.

To the uninitiated, Electronic Warfare (EW) is a black art – a subject bedevilled with jargon, abbreviations, jiggery-pokery and catchwords. To the enthusiast, it is an exciting and challenging application of the principles of physics combining a blend of innovation, low cunning and deceit in a cat-and-mouse game of measure and countermeasure.

In this chapter the meaning and nature of EW is discussed, followed by a study of the extent and characteristics of the threat before examining in detail the various components of typical airborne EW systems. Finally, future traits and trends will be considered.

So what is EW? Probably the best all-round definition is: 'The means of determining, exploiting or disrupting the enemy's use of the electromagnetic spectrum while safeguarding its use for friendly forces.' To the strategist, this means minimising the enemy's potency by force reduction. To the Air Commander, EW is a force multiplier – enhancing the effectiveness of his combat aircraft. The accountant views EW with more jaundiced eyes; it is as an expensive addition to the weapons arsenal which lacks visible justification. For aircrew, it is the key to survivability – reducing the time available for hostile weapon systems to acquire and kill.

The EW spectrum shown in Fig. 10.1 spans from near DC to light. Wherever communications or radar search, detection and tracking weapon systems operate, EW equipment must be able to intercept, analyse, follow and counter. Thus, in addition to the radio bands, EW must respond wherever radar, laser, infra-red and other electro-optical (EO) sensors are found.

It is hard to pinpoint exactly when airborne EW came into fashion. Certainly, there are numerous examples of its use during the Second World War. Micro-

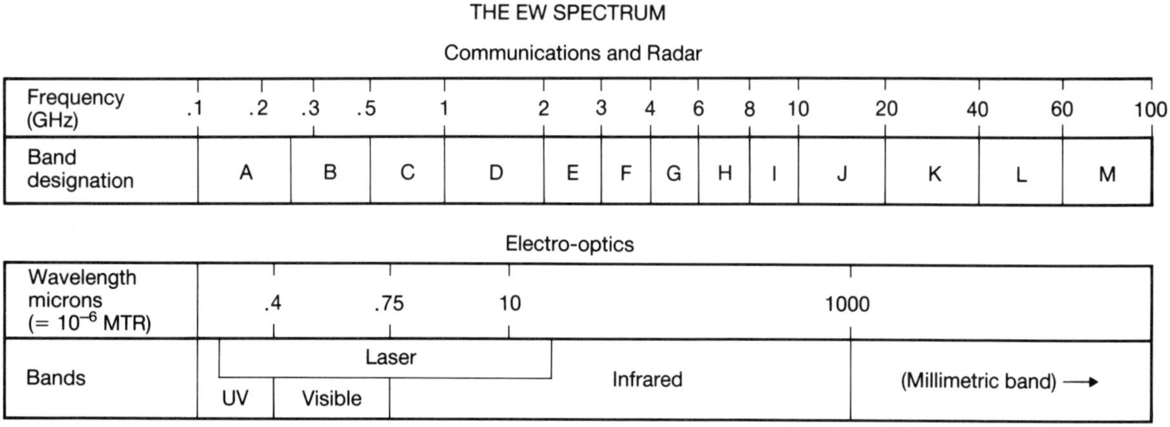

THE EW SPECTRUM
Communications and Radar

Frequency (GHz)	.1	.2	.3	.5	1	2	3	4	6	8	10	20	40	60	100
Band designation	A		B	C	D	E	F	G	H	I	J	K	L	M	

Electro-optics

Wavelength microns (= 10^{-6} MTR)	.4	.75	10	1000
Bands	UV	Visible / Laser	Infrared	(Millimetric band) →

Fig. 10.1 The electronic warfare spectrum

phones were placed in aircraft engine compartments and the resultant noise used to modulate voice channels and disrupt enemy ground controllers' communication with their fighters. Metallic radar-reflecting strips called 'window' were dropped during bomber raids over Germany to confuse ground radar operators. Simple receivers illuminated a cockpit warning light when a fighter's radar was detected. Subsequently, during the Soviet invasion of Czechoslavakia, a blanket electronic interference was used to mask events from the West. Jamming of surface–air missile radar and pyrotechnic decoys to counter heat-seeking missiles featured extensively during Israeli-Arab wars. American 'Wild Weasel' specialist EW aircraft were employed to attack and neutralise radar-controlled weapons in Vietnam.

More recently, during the Falklands campaign, Lynx helicopters protected British Task Force ships by broadcasting radar noise to confuse and decoy Exocet missiles. From these simple but robust beginnings, EW has grown to become a vital and integral part of modern air warfare.

Improvements to radars have prompted more innovative and effective ways of interfering with their use. Successful disruption techniques led in turn to further modifications and enhancements to the radar – and so on. Moving target indicators, frequency agility, pulse compression and bistatic operation are just a few of the techniques employed to make life difficult for the enemy bent on disrupting radar reception. To succeed, EW must be therefore innovative, responsive and dynamic.

Aircraft self-protection equipments are a necessary compromise between operational effect, cost, size and weight. It can be argued that a reasonable proportion of aircraft cost to allocate to self-protection is in the order of 10–15 per cent. Unfortunately available space on modern aircraft is at a premium and weight budgets are usually set with aircraft performance uppermost in the designer's priority list. Weapons, communication and navigation equipments compete for real estate and are readily recognised as essential to safe and effective operation. EW effectiveness is more difficult to quantify. How many aircraft will be saved if a jammer is fitted? How much more effective will combat tactics be if used in conjunction with a radar warning equipment? EW is often destined to receive the lowest priority of budget, space and available sites. Many pilots would rather sport a gun than a jammer. The effect of a gun is easily demonstrated and understood. It is harder to put one's faith in a guaranteed result of invisible electromagnetic

Electronic Warfare – an Overview

waves. Consequently, the introduction of airborne EW equipments has been a slow and painful business. Education, trials and demonstrations are important but most progress has resulted from the harsh lessons of attrition during combat.

Before the capabilities and benefits of EW are addressed, the practicalities of specifying airborne EW equipment must be considered. All too frequently, once the advantage of fitting EW equipments is accepted, the sky becomes the limit! A comprehensive range of equipments capable of dealing with past, present and future threats is specified. Budgets are exceeded and the aircraft takes the form of a 'one-ton budgie' (Fig. 10.2), and although extensively equipped, it would be unable to get off the ground. This results in, not the procurement of heavy and overspecified systems, but in many combat aircraft throughout the world, flying around without vital EW protection.

Fig. 10.2 The electronic warfare equipped one-ton budgie! *Shaun Singleton. Plessey Avionics*

THE EW SPECTRUM

So that radar, communications and EO systems can be identified, located, monitored, suppressed or neutralised, it is necessary to catalogue the enemy's electronic order of battle (EOB). Much of this assessment involves eavesdropping and data-gathering – signals and electronic intelligence (SIGINT and ELINT). Apart from the obvious resources needed to collect information, the agencies responsible must be backed by large analytical and technical teams to carry out the substantial task of extrapolating and evaluating emitters of the EOB.

Obviously, a great deal of valuable information is transmitted over the air waves in peace time which can be intercepted and used by the enemy. Ground radar operators need to practise intercepts; communications channels and weapon systems must be tried and tested. From these transmissions, a substantial amount of data is available to the electronic eavesdropper including frequency, pulse repetition interval, (PRI), scan type and rate, antenna beamwidth and power levels. Clearly, efforts can be made to minimise the amount and value of information released.

The EW spectrum

Availability of 'war only' frequencies and other parameters, limiting power to reduce the intercept range and coding transmissions provides some measure of protection against SIGINT and ELINT during peacetime. Nevertheless, the practical limitations on parametric variations are restricted by physical factors such as the performance of waveguides, antennas, power amplifiers, magnetrons and filters. These factors are in addition to the limitations imposed by physics which govern choice of propagation frequencies, pulse width, PRI and power required to operate effectively.

In general, equipments designed for long range use are easier to intercept and exploit than close-in systems. As well as high levels of radiated power, long-range early warning equipments have low scan rates and relatively long dwell times. This contrasts markedly with a single shot laser range finder which provides a once and forever chance of detection, and only then when within visual range. Interception of some transmissions must therefore be supplemented by less honourable tactics: espionage, photography, loan or theft of components, equipment, handbooks and documents, spoils of war or purchases through a trusted third party. Additionally, a vast amount of data is legally available through open press – newspapers, technical papers and defence journals.

During evaluation of the EOB, it is also important to consider the effects of known or potential technological advances. Some estimates can be made by extrapolating data from friendly systems. However, it is all too easy to overemphasise the enemy's capability by using the most advanced technology as a yardstick.

Intelligent use of emitter data in conjunction with the combat scenario therefore, is the first step in defining the EW requirement and developing suitable tactics. An example of representative threats is shown in Fig. 10.3. Fundamental issues governing the need for EW include: definition of the threat, aircraft role, performance, flight profile, planned altitude and mission. Having established the background and need, a detailed study of the range of EW equipment types and the technology available to provide the capability required can be made.

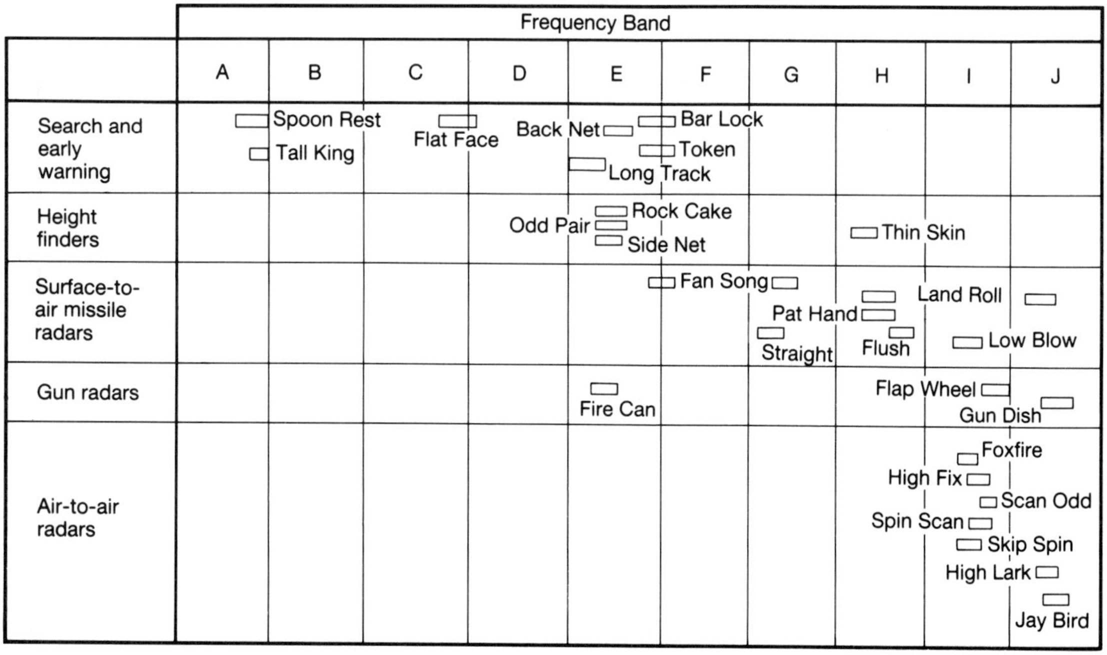

Fig. 10.3 Frequency spread of Soviet search, acquisition and fire control radars

	Frequency Band									
	A	B	C	D	E	F	G	H	I	J
Search and early warning		Spoon Rest, Tall King	Flat Face	Back Net	Bar Lock, Token, Long Track					
Height finders				Odd Pair	Rock Cake, Side Net			Thin Skin		
Surface-to-air missile radars					Fan Song, Straight	Pat Hand	Flush	Land Roll	Low Blow	
Gun radars					Fire Can			Flap Wheel	Gun Dish	
Air-to-air radars							High Fix, Spin Scan	Foxfire, Skip Spin, High Lark, Jay Bird	Scan Odd	

231

ELECTRONIC SUPPORT MEASURES

Electronic support measures

Electronic support measures (ESM) is defined as the interception, location and identification of radiated electromagnetic energy to provide threat recognition and intelligence. In this section, the characteristics of ESM equipments ranging from self protection threat warners to the specialised collection of detailed parametric data for intelligence purposes will be considered.

Probably the most widely used airborne ESM device is the radar warning receiver (RWR). The purpose of the most basic RWR is to provide timely warning of the presence of a threat, together with an indication of direction and lethality. RWRs should also provide threat identity – or at the very least threat priority (radar in search, track or lock), and category (controlling guns, missiles, ground-to-air, air-to-air) and estimated range.

Broadly speaking, the performance of an RWR is influenced by the type of receiver, sensitivity, the operational scenario, signal density, processing capability and storage – all factors which impact on cost. Even then, the performance of an installed RWR is only as good as the quality of the supporting data base.

The most common in-service RWRs are based on the crystal video receiver (Fig. 10.4). These receivers typically cover the main radar band between 2 and 18 GHz. Incoming radar pulses are received in each of the four spiral antennas at signal strengths related to the direction of arrival (DOA). The crystal detectors then convert the signals into a video representation of the incoming pulses. Although this detection removes phase and carrier frequency information, the essential inter-pulse characteristics of the pulse train are retained (Fig. 10.5).

Fig. 10.4 Basic crystal video radar warning receiver

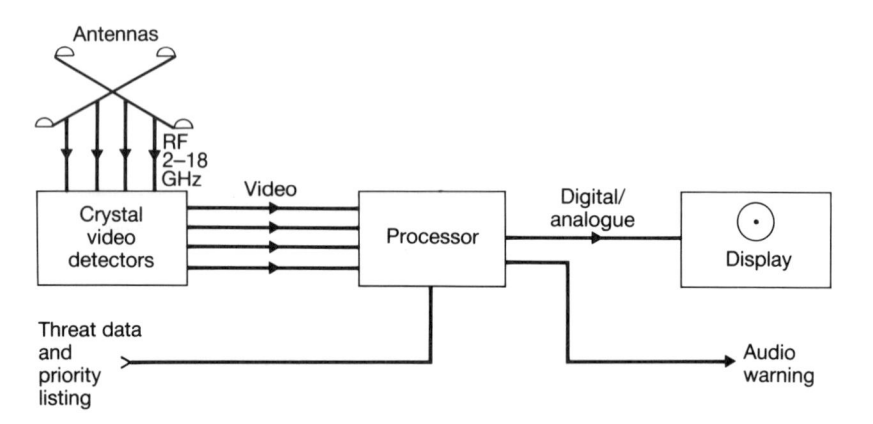

Fig. 10.5 Video pulse train, crystal video receiver

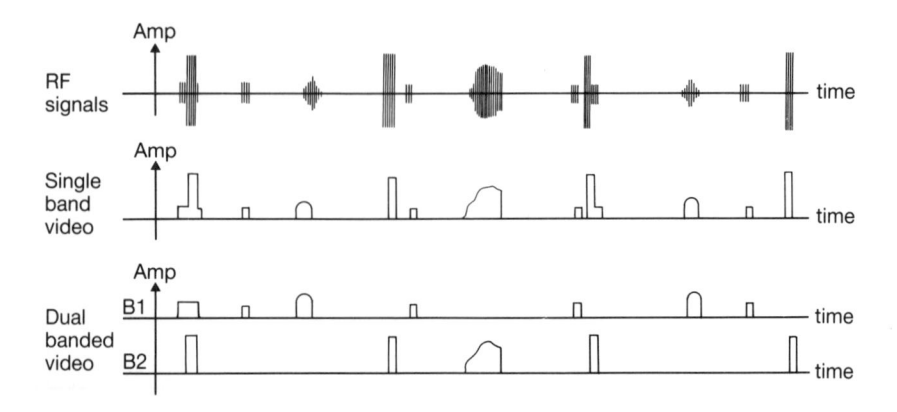

The resulting video waveform enables time of arrival (TOA), pulse width (pw) and amplitude to be derived. From this data, PRI can be deduced using the repeat pattern (for stable PRI) or the pattern sequence (for staggered/jittered PRI). The amplitude variation of succesive video pulses enables scan rate and type to be assessed. In theory, pulse width can also be derived – however, the inevitable rounding and misshaping of pulses during propagation makes PW a notoriously unreliable parameter.

From the crystal video receiver, we can therefore extract DOA, scan pattern and PRI. Separation of particular threat radars depends on matching the received sequences and patterns against known PRI and scan data. This matching is relatively simple when operating in low signal densities.

Problems occur in higher densities because signals coincide, overlap and obscure each other. Some temporary relief can be obtained by splitting the 2–18 GHz coverage into three or four sub-bands; this reduces the signal density in each band. However, as pulse densities increase still further, it becomes necessary to raise the receiver sensitivity threshold. Although this technique effectively lowers pulse density, wanted as well as unwanted signals are suppressed.

These techniques work well for pulse radars, but what about continuous wave (CW) signals? The video translation of CW produces a d.c. voltage which, instead of registering a threat, merely raises the noise level. To overcome this deficiency, the input signal is chopped at a predetermined rate; when this repetition rate is detected, the presence of CW threat is assumed.

A further limitation of the crystal video RWR is that, even when pulses are successively de-interleaved and matched, many radars share similar PRI and scan pattern, which in turn leads to ambiguities in identifying type – or even deciding if the radar is friendly or hostile. In an attempt to overcome this deficiency, some RWRs incorporate a frequency selective receiver (FSR). When ambiguities of PRI occur, this separate FSR is switched into circuit to measure frequency and so provide another parameter to assist identification. These refinements of the basic RWR are shown in Fig. 10.6.

Whatever the limitations, the crystal video RWR nevertheless provides simultaneous coverage of the radar band from 2 to 18 GHz, is low cost, robust, and operationally attractive for lower value platforms. Even with its limitations, it vastly enhances survivability by providing rapid warning of most immediate threats which at least enables some evasive manoeuvre to be carried out.

If RWRs are to be improved further, RF frequency information as a sort parameter for all signals must be retained. The benefits of reducing receiver bandwidth to cope with the high pulse density environment have already been seen. These features are included in many modern RWRs and airborne ESM equipments (Fig. 10.7). For example, the channelised superheterodyne receiver utilises a number of narrow-band channels to cover the desired radar spectrum. To achieve even greater frequency resolution, these channels can be further sub-divided. Although less economic in size, weight and cost than crystal video RWR, it has the same ability as the wide open receiver to monitor all pulses in the spectrum simultaneously.

A lower cost alternative which also retains frequency information is the swept or stepped receiver. A narrow band filter is swept across each of the receiver sub-bands. In this way, frequency is measured to the accuracy of the narrow band filter. At any instant in time, all pulses outside the coverage of the filter are ignored. This technique ensures that the receiver functions well in high pulse sensitivity environments but there is a danger that wanted signals using short

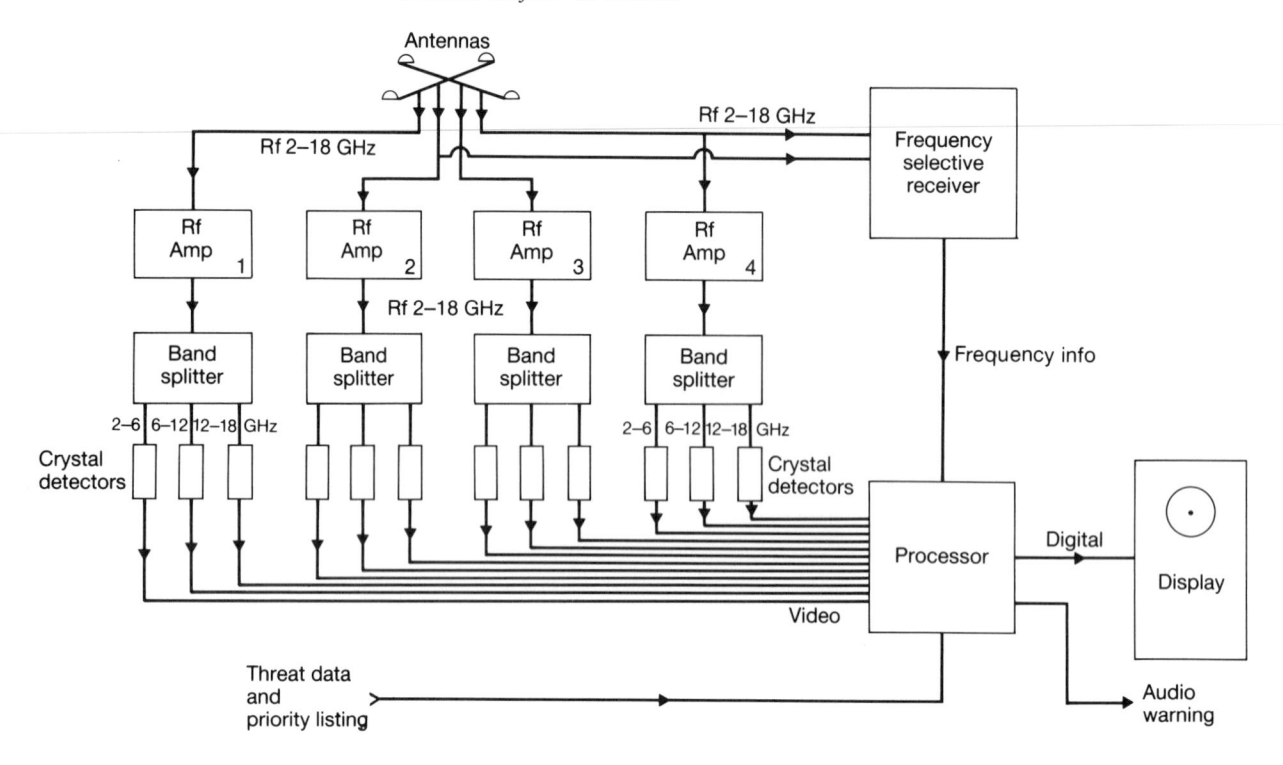

Fig. 10.6 Advanced crystal video, radar warning receiver

Fig. 10.7 Advanced radar warning receiver
Plessey Avionics

duration pulses will be missed if the sweep rate is too slow. Conversely, if the sweep is too fast, insufficient time will be available to extract the required data. Overall, the swept/stepped receiver can provide a highly efficient system if the right balance is struck between bandwidth and sweep rate in conjunction with intelligence data, scenario, role, cost and physical constraints.

There are many variations on the theme. Wide open and compressive superheterodyne receivers, instantaneous digital frequency measurement equipments and state-of-the-art receivers using acousto-optic devices called Bragg-cells exist today. The more complicated and costly systems are generally more suited to intelligence collection than threat warning. Key parameters for the threat warner are time, lethality and general direction. Information gathering systems require accurate DOA and detailed parameter data at the expense of immediacy. Nevertheless, there is an ever-greying area between threat warners and airborne ESM receivers as technology advances, the quality of available data improves and operators demand better and more capable RWRs.

ELECTRONIC COUNTERMEASURES

Electronic countermeasures (ECM) is defined as the actions taken to prevent, disrupt or reduce the enemy's effective use of the electronic spectrum. In this section the various properties, types and characteristics of radar and communications jamming will be outlined. The properties and application of passive countermeasures (such as chaff) and other off-board radar decoys will also be investigated.

The aim of radar ECM is to induce sufficient interference into a receiver to deny acquisitions, tracking or engagement. If a missile is fired, ECM may be used in the 'end game' to defeat missile guidance. The best known and frequently encountered airborne ECM device is the self protection jammer (SPJ).

The purpose of an SPJ, is to achieve a large measure of protection over the complete range of enemy radar assisted or guided missiles and guns. Parameters such as the power required, frequency and antenna coverage, type of jamming, complexity and configuration are related to aircraft type, role and the threat scenario. In particular, availability of aircraft space, flight profile and area of operation are crucial factors in deciding the SPJ specification – or even more fundamentally, whether it is worth fitting a jammer at all!

A classic candidate for the SPJ fit is the ground/attack/reconnaissance (GR) aircraft. These aircraft carry out missions deep behind enemy lines and are therefore extremely vulnerable to attack from the full range of hostile weapon systems. Even before specifying the actual SPJ fit, some measures can be taken to minimise the risk of operating in this environment. Flying sorties at very low level limits the radar horizon and hence the time of exposure to threats. Also, when engaging low flying targets radar performance is degraded due to the effect of ground clutter. Additionally, decreasing the radar cross-sectional area by structural design or use of special paints and materials makes the aircraft harder to detect, track and destroy. Strict control of communications and radar emissions reduces the opportunity for the enemy to detect and exploit the aircraft's transmissions. Nevertheless, a GR aircraft has to run the gauntlet of the extensive, diverse and formidable threat systems so even with these measures, survivability could well depend on the capability of ECM.

From the typical threat table in Fig. 10.3 it can be seen that jammers should cover the radar band from 2 to 18 GHz. However, the usual cost/weight/size constraints prevail. Frequently a single 'extended' band from 6 to 18 GHz is used to cover most of the high priority overland threats. There is a growing acceptance that future ECM equipment will need to cover threats in the millimetre radar band above 18 GHz.

Two categories of SPJs are in operational service – noise and deception jammers. In simple terms, noise jamming is a robust technique aimed at swamping the victim radar with a band of unwanted interference. Depending on the effect desired, this type of jamming can be broadband, spot or swept spot. Broadband jamming is aimed at covering a large number of threats or a frequency agile radar. However, using broadband jamming means that, as available power is spread over a large part of the spectrum, the power against any particular threat is reduced proportionally. Only jamming power within the receiver bandwidth will be effective. At the other end of the scale, spot jamming over a few MHz will maximise power against a single victim radar. A working compromise is seen in the swept spot jammer where the narrow band spot is swept over a wider frequency to cover multiple or agile threats. As usual, the laws of physics prevail – unless the spot to sweep ratio is selected carefully, during periods when the signal is unjammed, operators will be able to acquire and track the aircraft. These types of jamming are illustrated in Fig. 10.8.

Noise jamming relies on directing sufficient power into the victim radar's main beam to mask reception of the target. Figs 10.9(a) and (b) show the typical result on a PPI display. Target range is denied completely and direction is made

Fig. 10.8 Types of noise jamming

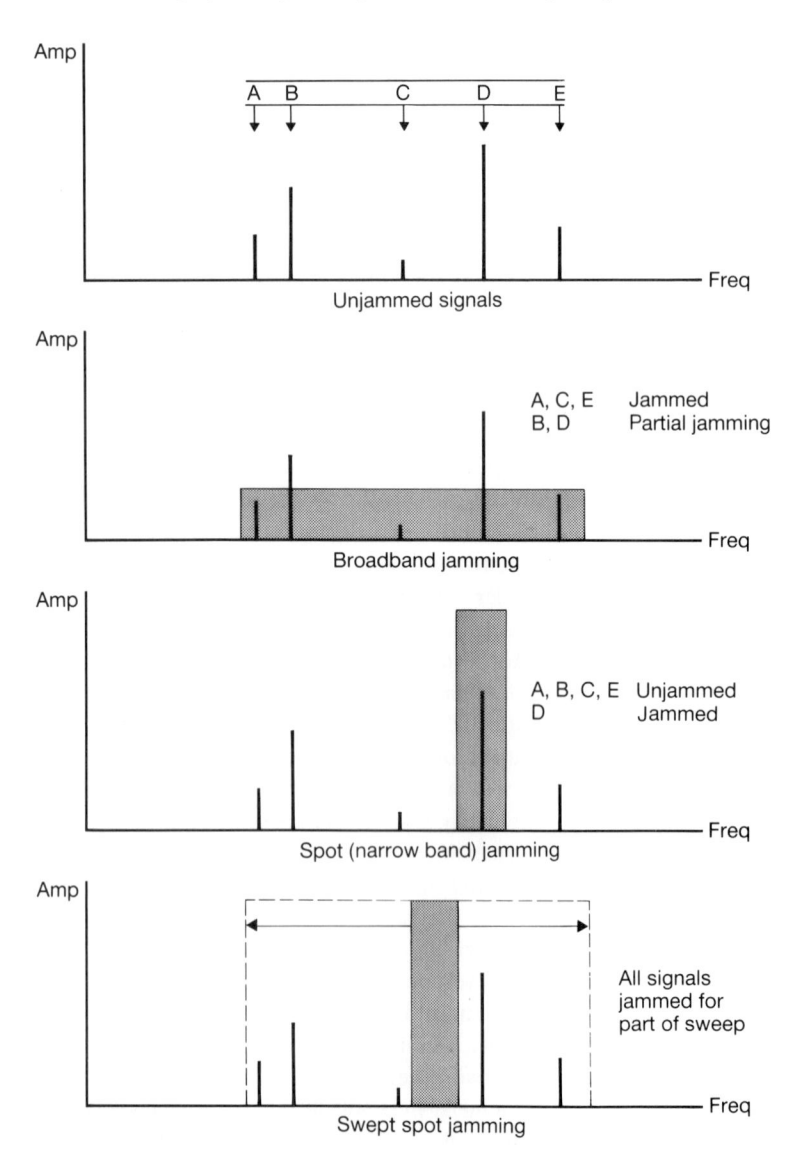

Electronic countermeasures

Fig. 10.9(a) Un-jammed PPR radar display

Fig. 10.9(b) Main beam noise jamming

Fig. 10.9(c) Main beam and side lobe jamming

237

ambiguous to the extent of the radar's antenna beamwidth. If higher levels of jamming power are used, it is possible to penetrate the victim's sidelobes and further complicate the radar picture (Fig. 10.9(c)).

If noise is transmitted continuously on the threat frequency, there is little need for in-depth intelligence on the radar's parameters. However, continuous noise is wasteful of power, particularly against pulse radars with low duty cycles. An attempt to rationalise jamming with the radar pulses is achieved in smart noise jammers. These equipments attempt to predict and jam each radar pulse but revert to the quiescent state during the pulse off period.

Ways of managing the available power even more effectively are found in the family of deception jammers. As the name implies, these jammers attempt to trick the victim into tracking the jamming pulse rather than the target. An example of such a technique is range gate pull off (RGPO). RGPO is achieved by firstly synchronising the jamming pulse with that of the radar so that a strong signal is returned. This pulse will depress the radar's automatic gain control so as to hide the real target. The jamming pulse is then 'walked off' in time by inducing a gradually increasing delay in the transmitted pulses. This series of pulses will be interpreted by the radar as a strong target, but will be at a different range to the real AGC depressed target. In practice RGPO is applied for 10–20 seconds to walk the radar several hundreds of metres away from the aircraft. The sequence is then repeated again to prevent the radar re-acquiring.

Against Doppler radars, a similar technique is used to cause velocity errors, called velocity gate pull off (VGPO). In this case, a frequency shift is applied instead of the time delay to the jamming signal to induce Doppler errors and walk off the velocity gate.

Other deception programmes rely on modulating the jamming signal in such a way as to attack the victim's azimuth and elevation tracking circuits which rely on comparing signal amplitude to assess the target position. For example, a scanning radar relies on receiving maximum amplitude return when the target is on boresight. If the jammer provides an interference modulation geared to the radar's scanning pattern, errors will be induced in the tracking gates. The optimum effect can be achieved if the scanning rate is matched by the jammer and the false signal amplitude is minimum when the radar is on boresight. This technique is known as inverse scan modulation.

Whilst deception jammers use the available power economically, they are dependent on a high degree of threat intelligence information to ensure the desired effect will be achieved. In addition to transmitter scan rate, frequency, PW, PRI information, harder to obtain details of tracking gates, AGC limits, ECCM features and receiver thresholds must be correctly assessed to ensure an effective jammer programme.

One advantage enjoyed by the jammer is that its transmissions have only to travel one way from aircraft to radar. The threat radar's pulses have to travel to the target and back again. As received signal strength is inversely proportional to distance^2, this means that the jammer pulse is attenuated by distance^2 and the radar's signal by distance^4. The point in space when the jammer's advantage is negated and the radar's higher power allows it to see the target above the jamming is known as the burn-through range.

We have seen that in its simplest form, a broadband jammer could be operated continuously to obliterate reception over a range of enemy transmission frequencies. However, we have also seen that this technique is extremely inefficient in its use of power, weight and space. Furthermore, if jamming is used indiscriminately, it merely serves to alert the enemy; aircraft jamming will be detected by ESM at far greater ranges than a radar which relies on reflective energy from the aircraft skin. Home-on jam missiles can be launched at the convenient beam of continuous noise. Intelligent-jamming therefore, relies on detecting and periodically monitoring the threat radar to respond effectively.

Recent United States jammer development has addressed the needs of multi-service and multi-platform. The system, known as airborne self-protection jammer (ASPJ) (Figs 10.10, consists of five units – two receivers, two multi-octave transmitters and a processor. The units are common to a number of aircraft, including A-6, AV-8B, F-14, F-16 and F-18.) Cooling and electrical connections are made through an interface which is unique to each aircraft.

Fig. 10.10 Airborne self-protection jammer (ASPJ) *Westinghouse/ITT, USA*

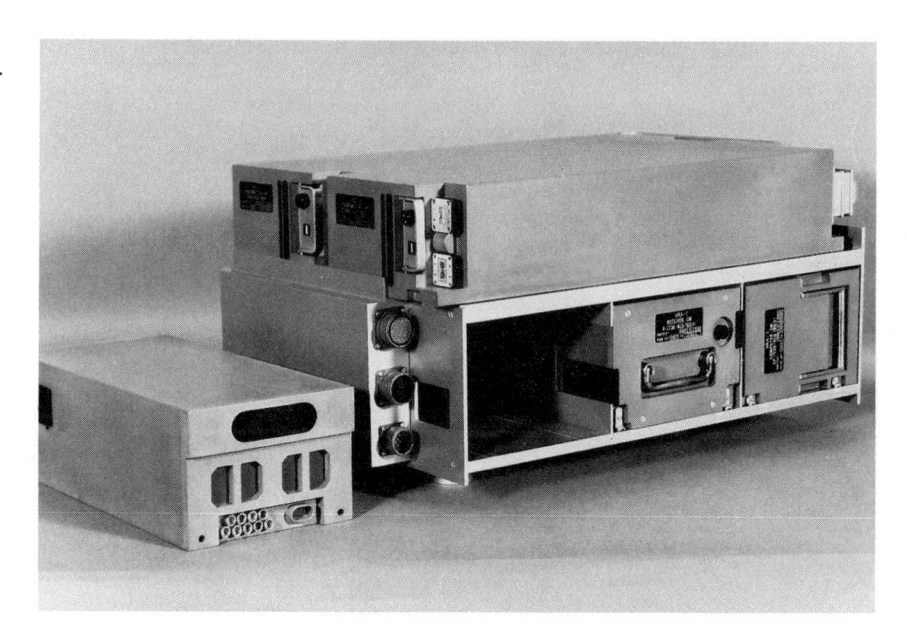

Jammers can utilise a dedicated receiver or share the information from the aircraft ESM. Whichever option is used, a necessary compromise must be made when operating a high power noise generator on the same platform as a highly sensitive receiver. Antennas must be sited to provide maximum isolation between jammer and receiver (70 dB typical). In some installations, a combination of look-through (switching off the jammer periodically to allow the receiver to operate) and look-around (where the receiver scans around the narrow frequency band of jamming) is used. In all practical installations, antenna siting is therefore a compromise between the requirements of the jammer, the RWR, aircraft obscuration and achieving satisfactory isolation.

To supplement the SPJ, or more often to give some measure of protection to poorly equipped aircraft, two other categories of jammers exist: escort and stand-off jammers. The escort jammer is used to provide an extensive capability to degrade, deceive and destroy search and weapon radars. Probably the best known example of the escort jammer is the US Wild Weasel aircraft which accompanied aircraft raids on Vietnam. In this case jamming was backed by air-to-ground anti-radar missiles which frequently induced the prudent operator to shut down the ground radar as the aircraft approached. Stand-off jamming operates on a similar principle but, in this case, the jammer provides blanket protection from a safe

Fig. 10.11 ASPJ – current applications. *Westinghouse/ITT, USA*

distance from the threats. Much of the characteristics of jamming remain the same except that the geometry of mission aircraft, jammer and radar is complex. Screening other aircraft requires much more power than that needed to cover one's own skin echo. Additionally, if a jammer is operating at long range, the power/distance^2 ratio soon begins to bite and large jammers, antennas and aircraft are needed. In general, few air forces can afford to operate sufficient stand off and escort jammers. However, if operated effectively, these specialist aircraft can have a marked impact on the success of combat missions.

Communications jamming (CJ) is also generally reserved for specialist aircraft and operations. However, CJ requires a great deal of power (it does not enjoy the SPJ's advantage over the 2-way radar path). The geometry of transmitter, receiver and jammer is also complex. Voice communications are particularly difficult to jam because of man's innate ability to recognise language in high noise environments and to make sense of broken transmissions and messages. In addition, the use of data link, frequency hopping and spread spectrum transceivers makes effective CJ more and more difficult to achieve (Fig. 10.12).

Electronic countermeasures

Fig. 10.12 'Sky Shadow' pod mounted jammer *Marconi–Plessey–Racal*

Although one of the earliest forms of airborne ECM, chaff is still fitted extensively to many combat aircraft. Known as 'window' during the Second World War, chaff consists of packages of metal strips cut as miniature dipoles so as to produce strong reflections over the radar band. Each dipole is coated to ensure clean separation from the others and weighted to produce a tumbling action during its fall under gravity. Various dispensers are available to cater for the variety of aircraft and roles. The dispensers either guillotine open flat packets, stream out the dipoles by forced air through a venturi or fire the chaff out of cartridges (Fig. 10.13).

Many EW gurus felt that chaff would become obsolete with the advent of moving target indicators (MTI) and Doppler radars. However, experience has shown that chaff is an enduring and robust, low cost countermeasure. When used in a dynamic air battle, with speed changes and manoeuvre, chaff is still considered

Fig. 10.13 Expendable countermeasures – chaff. *Chemring Ltd*

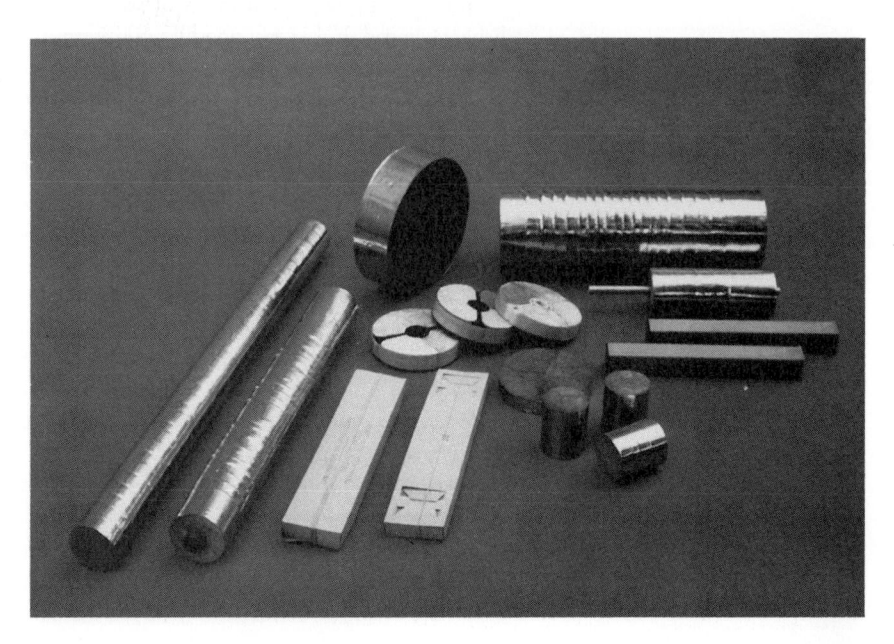

to be a valid counter to most threats. Although originally chaff was dispensed in many thousands of bundles to form safe corridors for subsequent missions; its main use today is in the self protection role. Bundles are dispensed at short intervals when under threat (as detected on the aircraft RWR). As a rule of thumb, each package of chaff should be equivalent to at least twice the echoing area of the aircraft. Radars using speed gates to distinguish chaff from typical aircraft must allow some latitude for highly manoeuvrable targets. So even if MTI is partially successful, it is likely that it will be upset by the sequence of chaff packages leaving the aircraft at discrete intervals. Although not an end in itself, chaff forms an integral and important part of modern aircraft self protection systems.

Other forms of off-board radar countermeasures are aimed at providing a similar capability to chaff or supplementing the effect of the aircraft's jammer. Mini decoys providing a few watts of power can be released from chaff dispensers against semi-active radars. These active decoys have two advantages over chaff: their echoing area will remain relatively constant whereas chaff tends to disperse, and a Doppler shift frequency can be superimposed to provide realistic aircraft velocities. More ambitious decoys can be fired ahead or towed to perform the function of mini jammers. Operating these jammers away from the aircraft negates the effect of home-on jam, prevents radars from using leading edge tracking and provides a more effective countermeasure to monopulse radars. Although off-board radar decoys are fitted to few of today's aircraft, there is little doubt that they will form a vital counter to future more capable radars.

ELECTRO-OPTICS AND INFRA-RED

Electro-optics (EO) and infra-red (IR) are playing an ever increasing part in air warfare. Ninety per cent of aircraft shot down in combat since 1972 have been casualties to IR homing missiles. Optical gun sights and night vision devices add enhanced capability to the already formidable radar threat. Lasers are used extensively to provide high accuracy weapon tracking and ranging. EW must therefore take account of this growing and often covert capability.

In this section, the applications of IR and lasers and the problems of detection will be reviewed and some possible countermeasures outlined.

One of the most insidious anti-air threats is the IR heat-seeking missile. It may be launched from land, sea or air platforms. Shoulder launched IR homing missiles are also widely available to most armies of the world. The IR threat is all the more awesome because it is usually undetectable before impact.

The most common IR countermeasure is the expendable flare decoy (Fig. 10.14). These rapidly igniting decoys are ejected away from the aircraft to produce a more attractive source for the missile to follow. However, IR missiles are frequently launched covertly (without assistance from other sensors and therefore not detectable on aircraft RWR/ESM). Unless the approach of the missile can be detected, therefore, it is not possible to assess when best to eject the flares. Missile approach warners (MAW) use Doppler radars to sense the approach of the threat and missile launch warner (MLW) search for the missile plumes. Unfortunately, few of these warning equipments are in service and aircrew cue the release of flares by less scientific means – crossing enemy battlelines, overflying troops or shipping and starting target runs.

An alternative countermeasure to the heat-seeking threat is the IR jammer (Fig. 10.15). These equipments also rely on producing an attractive IR source by using a solid (hot brick) or filament to produce the correct signature. However, whereas

Electro-optics and infra-red

Fig. 10.14 Expendable countermeasures infra-red flares. *Pains Wessex Ltd*

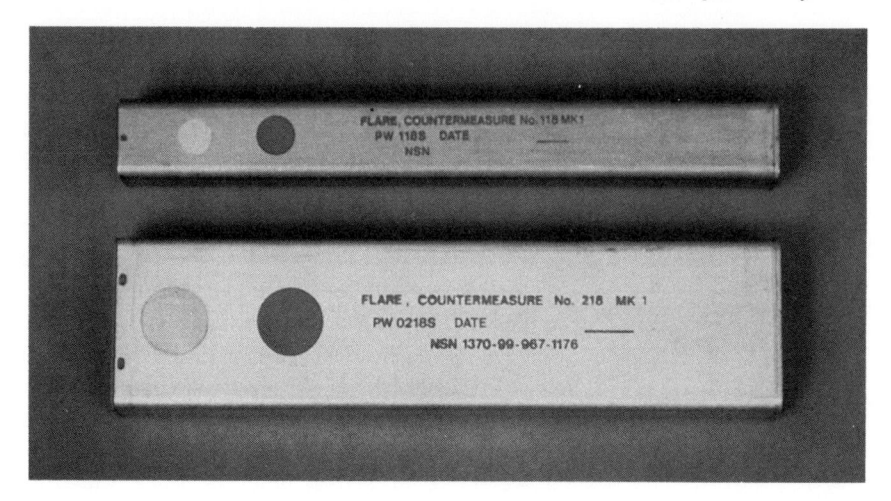

Fig. 10.15 Infra-red jammer *British Aerospace, Naval and Electronic Systems Div., Filton*

flares decoy the missile by physical separation from the aircraft, on-board systems must generate the required miss distance by modulating the IR source to cause errors in the missile guidance. As IR operates over line of sight, intelligence on these missiles is not easy to obtain. Furthermore, unless the actual kind of missile is known, a less than optimum or general 'catch all' modulation programme will have to be used. It is not surprising therefore, that with these limitations on both flare and jammer, IR missiles are such a formidable threat.

Light Amplification by Stimulated Emission of Radiation – known as the laser was developed a mere 27 years ago. Yet its application to the battlefield is already extensive and increasing year by year.

Lasers are used for highly accurate range finding, tracking and weapon guidance. They can be used to designate (illuminate) targets to provide terminal homing for laser guided bombs and missiles. Lasers can also be used to dazzle, blind and damage optical sensors – including eyes. When a high level of power is used, they have the potential to become hard kill weapons in their own right.

Unlike radar, laser beams do not diverge appreciably with range. Laser radiation possesses two important and unique characteristics – it is pure in that emissions from a single source occur at the same frequency; it is also coherent in that all waves are in step in time and space with each other. The result is a highly concentrated beam of energy.

With these properties, it is not surprising that lasers have been so readily incorporated into weapon technology. To cope with this new dimension of the EW threat, laser detection and countermeasures are finding their way into tomorrow's requirements.

Helicopters are particularly vunerable to the ground–air laser threat during their relatively slow, 'nap-of-the-earth' operations. Laser detection and threat warning is therefore becoming essential to helicopters crossing enemy lines. However, since a laser beam could be less than a metre in diameter when it strikes the aircraft skin, detection poses a particular problem. To ensure main beam detection, laser detector heads would have to be spaced around the airframe at less than metre intervals. Although such a system could be built into new aircraft, it is a complex and expensive solution with little application to retrofit programmes (Figs 10.16 (a) and (b)).

A more practical warning system is therefore likely to use one or two centrally mounted clusters of detectors to intercept the direct laser beams. Individual detectors have a limited field of view to enable direction of arrival (DOA) to be assessed. These main beam clusters are then supplemented by secondary beam or scatter detectors. Scatter detectors sense laser energy reflected from the airframe or by atmospheric particles. They must therefore have a wide field of view and DOA cannot be assessed. Nevertheless, all other parameters can be correlated with the main beam detectors to achieve a high probability of intercept in a practical and affordable equipment configuration.

Warning equipments must also differentiate between lasers and optical interference – especially sunlight. It is fortunate that the peculiar characteristics of the laser – purity and coherence – can be used as sort mechanism. Adjacent frequencies can be checked to ensure they are clear of the signal (purity) and the received waveform is analysed for coherence.

Finally, having detected and analysed the laser beam, there is still the question of countermeasures. If laser warning is provided, at least turning away from the threat and flying lower will reduce vunerability. On slow moving platforms, deploying smoke or optical chaff will inhibit the laser's operation. Aircraft optical sensors can be protected by closing shutters as soon as the beam is detected. Ultimately, attacking the lasing source with airborne high power laser promises to be the fastest and most effective means of countering this growing threat.

The future

Fig. 10.16(a) Laser warning equipment – direct detector

Fig. 10.16(b) Laser warning equipment – indirect detector. *Plessey Avionics*

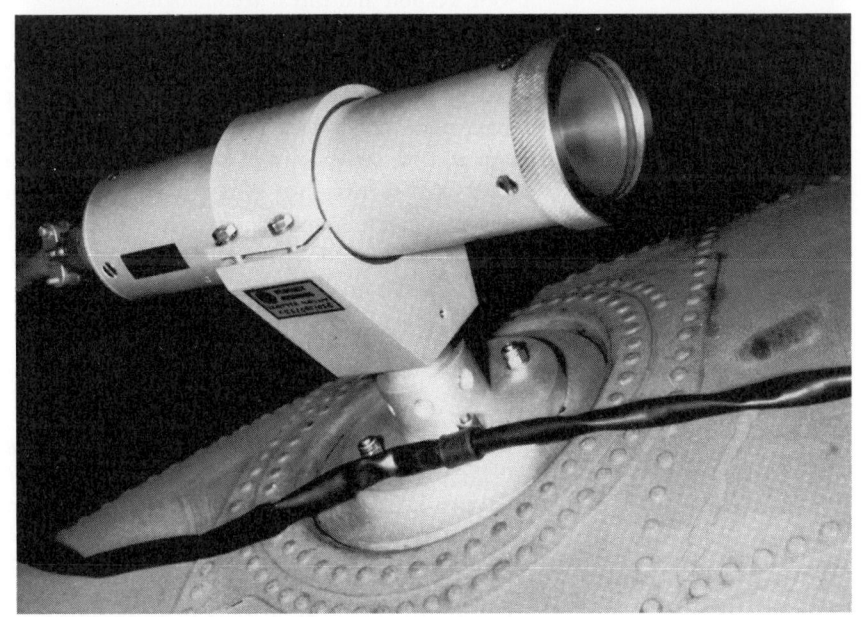

THE FUTURE

EW is, by nature responsive, innovative and difficult to predict. The speed of change in the past decade, particularly in processing capacity and capability, has been astonishing. New dimensions have been added to threat radars, weapons, countermeasures and methods of detection. Weapons are capable of pin-point accuracy over ever increasing ranges. 'Fire and forget' missiles have become the norm. EW systems have responded with more efficient detection, better jamming and increasing automation. In simple terms, EW will grow to meet the new threat, the latest ECCM measure and the newly discovered technology.

ECM and ESM must become more integrated and interoperable. RWR and jammer must work together effectively to detect, classify, counter and reassess radars with continuously changing parameters. Power must be more efficiently managed and directed at the threat. New techniques must be evolved to deal with monopulse, millimetric and more capable radars. Although even more information will be collected and processed; it must be used to ease the workload of aircrew by automatic operation and continuous updating of countermeasures. Digital radio frequency memories (DRFMs) will accurately store parametric information for instant recall and application.

The integrated electronic warfare system (INEWS) is under development to meet the needs of future United States aircraft such as the Advanced Tactical Fighter (ATF). The system is intended to achieve protection across the whole of the EM spectrum and will use state-of-the-art technology, avoid unnecessary duplication of modules and utilise artificial intelligence to ensure a high degree of integration and interoperability.

Aircraft ECM must be supplemented by innovative off-board devices. Radar absorbent materials (stealth) will play an increasing role in reducing aircraft detection ranges and enhancing the effectiveness of ECM. Better ways of detecting and countering IR and laser systems must be developed. In sum, EW must keep pace with weapon and threat technologies.

The need for EW is, by and large, undisputed. The lessons of several decades of air warfare are clear. EW is the key to survival in an increasingly hostile air combat environment. And yet, many front line aircraft throughout the world have inadequate or obsolete protection equipments. Unless the value of EW is fully appreciated, capability will continue to be assessed in numbers of aircraft rather than force survivability. Where budgets, weight and performance are crucial, EW must be carefully quantified, specified and then qualified. Above all, the capabilities and limitations must be understood if EW is not to be the Cinderella of avionic systems.

BIBLIOGRAPHY

Hirst, M. *Flight International*, Nov 1977.
Johnston, Stephen L. *Soviet Electronic Warfare. International Defense Review*, 1985.
Price, A. *Instruments of Darkness*, Wm Kimber, 1967.
Savage, J. *Laser Technology – a Background*, Miltronics, 1986.
Schleher, D. C. *Introduction to Electronic Warfare*, Artech House Inc., 1986.
Schlesinger, Robert J. *Principles of Electronic Warfare*, Prentice-Hall International, 1961.
Van Brunt, Leroy B. Applied ECM Vol. I, Oct., E W Engineering Company, USA, 1982. *The International Countermeasures Handbook*, 9th Edition, E W Communications Inc., 1984. *International Defense Review Special Series 8 – Electronic Warfare*, Interavia SA, 1978.

CHAPTER 11 FUTURE TRENDS AND DEVELOPMENTS

GRAHAM WARWICK BSc. News Editor, *Flight International.*

Graduated in 1976 from Southampton University with an Honours Degree in Aeronautical Engineering.

After two years with Hawker Siddeley Group Kingston, now British Aerospace, Mr Warwick became a reporter on *Flight International*. He has held various editorial posts on the journal since 1978.

AVIONIC INTEGRATION

Preceding chapters have highlighted the increasing integration at all levels in avionic systems from the surface of the microchip to the screen of the cockpit display. It is a trend that shows no signs of abatement.

If there is a single factor driving this trend towards increasing integration, then it is the rapid advance of microelectronic technology. This is reducing the overheads associated with data bus communication – the 'enabling technology' for avionic integration.

The electronics that must be installed in every avionic box capable of using the data bus have been progressively reduced from several cards to a single card and now to a set of chips that occupy only the corner of a single card. The benefits that cheap and easy access to data bus communication bring are enormous.

A 'first-generation' avionic architecture has many individual boxes independently supplied with power and, if necessary, cooling, and individually connected to a dedicated control panel in the cockpit. Boxes that wish to communicate must also be individually connected to each other.

A 'second-generation' architecture has each box sharing its information with several others via a one-way digital data link. Such is the principle behind the ARINC 429 data bus, where each box has a single output bus, but multiple input buses. Such a bus allows several boxes to be managed from a single multifunction control panel, however.

A 'third-generation' architecture has each box plugged into a bi-directional digital data bus along which it both transmits and receives information. This is the principle behind the military 1553B data bus in which a single (dual-redundant) bus connects up to 32 boxes, and the new civil ARINC 629 bus, which allows up to 120 boxes to communicate.

The next generation could see several changes, including an increasing trend towards distributed computing and away from individual avionic boxes, or line replaceable units (LRUs).

A degree of physical integration has already been achieved in some aircraft by combining the functions of two or more 'first-generation' boxes in a single LRU.

Future trends and developments

The combining of air data and inertial reference in a single unit in the Airbus A320 airliner is one example. The combining of flight management and autopilot functions, also in the A320, is another.

In military aircraft physical integration can take many forms. Perhaps the most common is the combination of head-up display symbol generation and weapon-aiming computation in a single box. With increasing computing power becoming available, there has been a trend to add other functions, such as air data computer, mission computer, and data bus controller, to the same box.

The result has been a reduction in the number of LRUs, but the next step could eliminate the familiar 'black box' altogether. The latest packaging trends call for individual modules to be plugged into a rack which supplies the necessary power, cooling, and communications services. The result is easier maintenance despite the higher packaging density.

The 'next-generation' avionics architecture introduces the concept of common modules and distributed processing.

Most avionic functions can be broken down into generics – input/output, arithmetic computation, data and signal processing, etc. – some or all of which are currently required in every LRU. This increases overheads.

In simple terms the common-module approach breaks down the various LRU tasks into such basic functions, for each of which a common module is developed. These modules are then assembled – in racks – into groups able to perform particular tasks. The modules are connected to a high-speed data bus which is capable of mass data transportation.

This bus enables tasks to be shared between similar modules distributed throughout the aircraft. Signal processing, whether for the sensor system or electronic-warfare suite, can be performed in any one of several common modules. Ultimately the overall system will not care where a function is performed, only that it is done efficiently.

The common-module approach provides redundancy without many of the overheads currently associated with duplicating or triplicating critical LRUs. It also allows 'graceful degradation'. In the event of a module failing another will assume its task – with no apparent reduction in the system's overall capability.

The ability of different avionic systems to communicate rapidly and fully via a high-speed data bus allows the sharing of scarce aircraft resources, such as good antenna locations. The principle of shared resources lies behind the concept of integrated communication/navigation/identification and integrated electronic-warfare systems.

In the drive to reduce drag and radar signature, there is a need to minimise the number of external antennas allocated to radio systems. Many of these operate in the same frequency band. Military IFF and civil ATC transponders, the JTIDS tactical data link, and GPS satellite navigation all share D band. This brings the possibility of sharing radio-frequency elements of the communication system, such as antennas and amplifiers.

Other resources that could be shared include cryptographic units for encoding and decoding secure IFF, JTIDS, GPS, and radio signals, and digital voice encoders for JTIDS and V/UHF radio.

The need to share antennas becomes more important if complex 'adaptive' aerials are required to counter jamming. These antennas use a technique called null steering which shapes the reception pattern to generate notches, or nulls, which can be steered to blank out any jammers.

While discussing antennas it is worth mentioning the trend towards integration of the various apertures with the airframe in the form of conformal or embedded aerials.

Already being tested are conformal radars which take the form of many small transmitter/receiver modules embedded in the leading and trailing edges of the wing and tail, or along the fuselage sides to provide all-round coverage. These modules are operated as a phased array to electronically scan the resulting beams. Figure 11.1 illustrates the principle applied to an orthodox antenna.

The array's ability to generate multiple beams tailored to different functions can be usefully employed to enable a radar to perform several functions simultaneously, from terrain-following to air target tracking. Similarly conformal phased-array antennas can be used by the electronic-warfare system to counter several different jamming sources by generating beams at specific frequencies and power levels.

Another concept for physical antenna integration is the 'smart skin'. Recent development of useable electrically conducting plastics brings the possibility that antenna patterns can in future be incorporated directly into the composite surfaces of aircraft – avoiding entirely the drag and radar signature of external aerials.

COCKPIT INTEGRATION

Integration is all about information exchange, and the advent of the digital data bus has brought with it an information explosion inside the aircraft – and the problem of how to present that information to the pilot.

The major problem is the limited amount of instrument-panel 'real estate' available in an aircraft, one which is particularly acute in combat aircraft, where panel space is severely limited.

In our 'first-generation' architecture, each avionic box had associated with it an area of instrument panel containing lights, dials; or switches – or all three. The pilot had to remember how to operate each system and, if alerted to a problem, remember where to look on the panel.

Before the advent of the cathode ray tube (CRT) cockpit display all the available panel space was occupied by controls and displays each having only a single function. Use of the CRT as a flexible display medium has allowed at least part of the panel space to be used for more than one function.

The monochrome CRTs used initially had the ability to display clearly only a limited amount of information at any one time. The advent of colour displays allows the information density to be increased, enabling the information previously presented on several instruments to be integrated on a single screen.

The shadow-mask technology of domestic televison sets has been successfully adapted to provide full-colour cockpit CRTs, but major limitations remain. One of these is brightness. While perfectly adequate for the shaded cockpit of a transport aircraft, shadow-mask CRTs are not bright enough to be viewed in full daylight. Their use in a fighter cockpit still poses problems, therefore.

Alternative display technologies are available, if less well developed. One is the beam-index CRT, which dispenses with the energy-wasting shadow mask. This produces a bright full-colour picture, but is electronically very complex. Beam-index cockpit displays are being developed, but their useful life could be limited because of the rapid advances now being made in flat-panel display technology.

There are many candidate flat-panel technologies, each with its own advantages and disadvantages, but emerging as the leader is the liquid crystal display (LCD). This is a solid-state device that is thin and light, requiring little power and in turn

Fig. 11.1 Texas Instruments' active aperture radar antenna comprises an array of small, low-power solid state gallium arsenide transmit/receive modules. *Texas Instruments*

generating little heat. It works by modulating ambient light falling on the display and therefore the brighter the sunlight, the easier it is to read. Conversely, the LCD needs to be backlit at lower ambient light levels.

LCDs have yet to match the colour capabilities of CRTs, or their resolution and display area, but it is only a matter of time. Rather than simply displacing the bulky, power-hungry, heat-generating CRT from the cockpit, however, the LCD could change the instrument panel beyond all recognition.

The installation advantages of LCDs will allow them to occupy more panel space than is now allocated to CRTs. They may even displace the conventional electro-mechanical standby instruments now used. While a single, panel-sized liquid-crystal display may prove difficult, if not impossible, to produce, the ability to butt several smaller displays together to provide a totally flexible panel is not in question.

If the CRT's display flexibility can be extended across the entire instrument panel using LCDs, then the problem of presenting information to the pilot becomes easier. A more natural form of display, including animated graphics – cartoons – can be used to convey the status of aircraft systems, or present a complete picture of the battle.

In combat aircraft, the head-up display has become a vital source of information, but its position as the primary flight instrument is now being challenged by the helmet-mounted display.

The HUD's major limitation is its field of view. Ways of increasing this both horizontally and vertically have been developed, but the HUD still only provides information over a relatively small region straight ahead. If the pilot looks outside its field of view he loses the information it provides on his aircraft, its weapons, and the target.

The helmet-mounted sight has been around for some time as a relatively simple device able to project an aiming mark which the pilot can place over a target by moving his head, the position of which is sensed by one of several means. Head position can then be used to direct aircraft sensors or weapon seekers, minimising the need to manoeuvre the aircraft.

While a simple light-emitting diode array can be used to generate the aiming mark, a CRT is needed to project the type of information normally presented on the HUD. This poses the problem of helmet weight, particularly in high-G combat. Solutions include a very small helmet-mounted CRT or fibre-optic cable which relays the picture from an airframe-mounted CRT.

The design problem becomes particularly acute when the pilot wants to fly at night. Currently he would find his way using a forward-looking infra-red (FLIR) sensor image projected on his HUD, augmented by helmet-mounted night vision goggles for all-round vision. If a helmet-mounted display is to be used, then it must somehow coexist with the image-intensifying goggles.

Alternatively the FLIR image could be projected on the helmet-mounted display, using a gimballed sensor slaved to head position so that it always points along the pilot's line of sight. All this must be achieved in a helmet that weighs no more, and preferably less, than today's unencumbered headgear.

Whether a pilot uses a head-up display, helmet-mounted display, or both, he must communicate with his aircraft efficiently. In a modern data bus-equipped aircraft the most likely means of communication will be a keyboard, or keys bordering a CRT display.

The move to data bus communication has meant a move away from dedicated, fixed-function control panels – a welcome change, but one not without its problems. Keys are often software-controlled, their functions changing with each keystroke as the operator moves down through several layers of control.

Menu-driven displays which take the pilot deeper into the system with each keystroke are relatively simple to operate. Captions displayed alongside the keys prompt the next action. But they can be time-consuming. Pilots have also been shown to be prone to errors when using alphanumeric keypads, such as those used on flight management systems.

Critical systems, which cannot tolerate operator errors, are usually given dedicated controls. In combat aircraft this has resulted in a proliferation of switches on the stick and throttle, where they are instantly accessible throughout combat.

Two developments could ease the pilot's workload by providing more natural communication with the aircraft.

The ability of CRTs, and later LCDs, to display full-colour graphics is the key to successful use of touch-screen technology. The pilot is better able to assimilate pictures than digits, and selecting a particular weapon by touching the relevant location on a cartoon representation of the aircraft is quicker and less error-prone than selecting from rows of digits by pressing the required key.

The use of touchscreens in the heat of combat, when stress, vibration, and high G degrade the pilot's pointing ability, is to be questioned, but the technology certainly has benefits in more quiescent phases of flight. The value of voice control in combat may also be in doubt but its role in managing cockpit functions in all other flight phases is not.

Speech recognition technology has made remarkable advances in recent years. In benign noise environments devices can now recognise numerous words regardless of the speaker. In aircraft, however, the noise environment is both harsh and changing and the vocabulary, if specialised, is large.

Speech recognition systems come in several forms – speaker dependent and speaker independent; isolated word, connected word, and continuous speech.

The simplest is a device capable of recognising isolated words uttered by a single speaker – a speaker-dependent isolated-word speech recognition system. Speaker-dependent systems must be 'trained' to recognise the speaker's voice. This is achieved by repeating the vocabulary several times to produce a template for each word to be recognised. On hearing a word the system then compares its library of speech templates with the utterance in an effort to obtain the best match.

A speaker-dependent connected-word recogniser waits for a pause following several words and performs the pattern matching process in the gap between bursts of speech. This allows a more natural form of speech, but presents the system with the problem of determining where each word in an utterance begins and ends.

A continuous speech recogniser does not wait for a pause, but performs the recognition process in near real time. In both connected-word and continuous-speech systems techniques are employed to limit the number of templates 'active' at any one time. This increases recognition reliability.

Syntax is used to limit the active templates to a subset of the complete vocabulary. The pilot's first word, for example 'Radio', will determine which subset is called up for the word, or words, which immediately follow, say 'Frequency' followed by a series of digits.

Totally natural communication with the aircraft, in which the relevant words are plucked from a stream of speech, will probably never be possible. More than likely the pilot will have to activate the recogniser and be disciplined in speech so as to limit confusion.

Failure to recognise a word, or worse the substitution of one word for another that is similar, are major problems, particularly in the dynamic noise environment of the cockpit. Recognition accuracies close to 100 per cent are essential if voice control is to be a help, not a hindrance, to the pilot. For that reason aircraft direct voice input systems are likely to remain speaker-dependent.

There are ways to improve recognition performance, including adapting the stored templates to changes in the pilot's voice as the flight progresses, or taking inputs from other aircraft systems to put the speech in context.

Another requirement is that voice control co-exists with other means of man/machine interaction including switches, keys, and touchscreens. The pilot must be able to use whichever mode of communication best suits his needs, and be able to move smoothly from one to the other, mixing vocal and tactile commands without confusing the system.

Voice control – and the associated technology of voice synthesis – plays an important role in the 'supercockpit' concepts now in vogue. In essence these present the pilot with a totally synthetic image of the outside world, derived from onboard databases and external sensors, which is independent of outside visibility. The pilot sees a three-dimensional 'virtual world' which is generated by computer and projected on to a helmet-mounted display so as to fill his vision.

Aircraft controls and displays are represented graphically, superimposed on the computer-generated outside-world image. Control is by voice or by reaching out and 'touching' the relevant picture.

Thought control, still in the very early stages of exploratory research, shows some indication of becoming a reality, but not in its 'science fiction' form. Instead, biofeedback could be used to monitor the pilot's health, and particularly his state of consciousness. This information could be fed to the flight control system to prevent the pilot losing consciousness – and control – at high G.

SENSOR INTEGRATION

Even today the combat-aircraft pilot has a surprising diversity of sources of target information available – preflight intelligence briefings, radar, IFF, radio, radar warning, and, of course, the 'Mk 1 eyeball'. Currently, however, the correlation of all these sources, and others, to produce a unified tactical picture must take place within the pilot's brain – at the same time as he attempts to fly his aircraft.

Sensor integration – or fusion – not only reduces pilot workload, it also expands coverage, increases confidence in target existence and identity, and is mutually supportive – different sensors having different strengths and weaknesses.

Sensor 'fusion' is the art of combining the various target data sources automatically and presenting the pilot with a unified and complete tactical picture with all ambiguities resolved. Sensor fusion produces a robust system that is difficult to jam because all the elements are mutually supportive, the advantages of one offsetting the disadvantages of another.

Radar is still the primary combat-aircraft sensor, although it is often combined with IFF to provide target identity and increasingly arduous 'stealth' requirements call for its judicious use, for example, by power management. Additional clues to target existence and identity can be provided by the radar warning receiver. Increasingly electro-optical systems are used to augment radar, both to overcome jamming and to provide covert target detection and tracking. Radar contacts, IFF responses, and other sensor information are available from co-operating aircraft via a secure, jam-resistant data link.

As sensors improve, much more information becomes available, and the correlation task goes beyond the pilot's capabilities. For example, information on target identity may be available from the radar, by close examination of the radar signature, from IFF, from passive electronic surveillance, from electro-optical sensors, and from the data link.

Particularly important to sensor integration – or blending – are the advances being made in thermal imagers. Currently these devices have limited range, but excellent imaging capability. Long-range sensors are being developed, however, requiring complex signal processing akin to that developed for radar in order to pick out genuine targets from the background thermal 'clutter'.

The imaging capability of radar is also being exploited using synthetic-aperture processing techniques to increase resolution many times, to just a few metres in some cases. Radar has the advantage that it can see though any weather, but with it comes the danger of detection.

Other candidate sensor technologies include high-frequency millimetre-wave radar, which has imaging resolutions approaching those of infra-red, but with greater ability to penetrate poor weather. Laser radar also shows promise as a short-range, covert sensor with high resolution because of its narrow beamwidth.

Image recognition has an important role to play in blending the output of different sensors and providing the pilot with a clear picture of the target.

CONTROL INTEGRATION

Integration of previously discrete aircraft control functions offers significant benefits. The integration of airliner autoflight and flight management functions has already been mentioned. This reduces pilot workload and enables fuel-saving flight profiles to be flown accurately and consistently.

In this case engine control (via autothrottle) has been integrated with flight control (via autopilot) – albeit with limited authority. Integration of flight and engine control has significant potential in combat aircraft, where engines and airframe often work close to their limits. Linking the two control systems allows the engine management system to anticipate aircraft manoeuvres and so reduce the artificial margins currently imposed to ensure that engine operating limits are not overstepped in combat.

In future combat aircraft the engine will be an integral part of the aircraft control system, with movable intakes and vectoring nozzles contributing significantly to the sum of forces and moments in flight. The pitch control power of two-dimensional vectoring nozzles, for example, is such that flight and propulsion control must be integrated.

In a vertical or short take-off aircraft where propulsive lift dominates, the integration of flight and engine control is beneficial, enabling aerodynamic and propulsive lift to be combined in the most efficient manner. Benefits of integration can include accurate touch-downs in small spaces in all weathers.

The integration of flight and fire controls also promises significant operational benefits. Currently the pilot closes the loop between the fire control system tracking a target and the aircraft's weapons – manoeuvring the aircraft until the target is within the lethal envelope of his weapon. At all times he must remember the limits of his aircraft and his weapon.

With the advent of full-authority electronic flight control it becomes possible to tie the flight and fire control systems directly together. The flight control computer does not really care whether a command signal comes from the pilot or another computer, and will not accept any demand that oversteps aircraft limits.

With target tracking provided initially by radar, then with greater precision by an electro-optical sensor, it is possible to generate line-of-sight error commands for the flight control system that will bring the weapon to bear automatically. Initial pointing could be accomplished automatically, therefore, the pilot only taking over in the final stages, or when the degree of error exceeds the control authority allocated to the flight control computer.

Integrated flight and fire control also allows weapon release in manoeuvring or evasive flight, increasing the chances both of destroying the target and of surviving against hostile defences.

In a future fighter integrated fire, flight, and engine control promises to reduce pilot workload significantly. Under direction of a weapon system which decides in which order several targets should be attacked, the aircraft could be flown automatically to the best attack positions.

This is similar to the concept of 'four-dimensional' flight management, in which the aircraft is flown automatically in three dimensions so as to reach preprogrammed waypoints at predetermined times, so freeing the pilot to concentrate on the tactics of his mission. In civil aircraft similar techniques would allow airliners to achieve preset arrival times, so avoiding delays and easing congestion.

DATABASE INTEGRATION

Advances in digital storage techniques mean that aircraft can now carry vast amounts of useful data. Modern flight management systems can store information on all the world's major runways and navigation aids, for example. Charts covering the whole of Europe can be stored in a military map generator.

Moving map displays have proved immensely useful to military pilots. Existing systems are based on filmed maps which are either projected directly upon a cockpit display or scanned remotely, with the result presented on a multifunction display.

But film maps cannot easily be changed. The digitally generated map is a far more flexible medium. Colours can be changed, scales can be adjusted, and information can be added or deleted.

Maps based on conventional paper charts, digitised and stored on magnetic tape, laser-scanned optical disc, or in semiconductor memory, are two-dimensional, however. The addition of height data brings new possibilities.

The elevation data can be derived from the contours on paper charts or from digital terrain databases available from mapping agencies. Elevation data literally brings a new dimension to digital maps. Relief maps can be generated, and colour-coded to show terrain above aircraft height – and therefore to be avoided – in red, or to highlight where intervening terrain blocks the view of hostile air defences.

Terrain databases can be used during mission planning to automatically chart a 'safe' route through enemy airspace that minimises exposure to air defences. The same database can be used to generate a three-dimensional image of the route for pre-flight familiarisation; the same database being plugged into the aircraft before take-off.

The database that is used to generate maps can be used to navigate the aircraft. As the aircraft flies along, its radar altimeter traces out a profile of the terrain directly under the flightpath. This profile is unique, and can be compared with profiles generated by the on-board terrain database. A match gives the aircraft's position, accurate to within 50m, and this can be used to keep the inertial navigation system accurate.

Furnished with extremely accurate position information the aircraft can be made to follow the terrain by looking within the database to generate a flightpath that will keep the aircraft close to, but safely distant from, the ground. In contrast to terrain-following based on radar, which cannot look behind hills to see what lies ahead, terrain-following by looking ahead within the database produces a more ground-hugging flightpath, reducing exposure to hostile defences.

Terrain databases have particular relevance to aircraft which wish to remain undetected and must therefore avoid all unnecessary emissions – including terrain-following radar.

KNOWLEDGE INTEGRATION

Despite the advances outlined in previous chapters, there are still serious doubts that the lone fighter pilot can cope with the demands placed on him in a future air battle. There are moves, therefore, to provide the pilot with an electronic associate – the so-called silicon copilot – able to handle routine 'housekeeping' tasks and to help make, and execute, tactical decisions.

Artificial intelligence is a concept which has been seriously oversold, but the scepticism the term engenders disguises the fact that knowledge-based, or expert, systems are making real progress. Computer programmes which encapsulate the experience of a human expert are already a reality in ground-based systems, such as those used for maintenance, and airborne applications are in advanced stages of development.

Several areas of avionics could benefit significantly from the application of expert systems, among them complex 'human' tasks such as health and usage monitoring, fault and failure diagnosis, route planning, and target recognition.

Expert systems are at the heart of 'self-healing' avionic concepts. The ability to diagnose failures, determine the consequences, and devise alternatives is a human one, but one which computers look capable of performing.

One example now being tested is the self-healing flight control system. Modern combat aircraft have many flight control surfaces, some of which perform similar functions. The loss of a surface through equipment failure or battle damage need not, therefore, mean the loss of the mission, or of the aircraft.

In the event of a flight control surface being damaged, an expert system would diagnose and isolate the fault and devise alternative control-surface configurations. The system would then advise the pilot if his mission could still be accomplished and, if not, suggest alternatives. If the damage was particularly severe the system would advise the pilot to abandon his mission and return to base. He would be advised of any further deterioration, allowing him to eject once over friendly territory.

The Pilot's Associate, or electronic co-pilot, is a much more ambitious application of expert systems, but one which the US hopes will be mature enough for application to its mid-1990s Advanced Tactical Fighter (ATF) (Fig. 11.2).

In ATF, the electronic co-pilot would be the interface between the human pilot and his aircraft. The pilot and his electronic associate would communicate via speech recognition and speech synthesis and touch-sensitive three-dimensional colour graphic displays.

The electronic co-pilot comprises a distributed network of cooperating expert systems, each with its own knowledge database. At the top of the hierarchy is the electronic co-pilot itself, acting as system manager under the pilot's direction.

Beneath the electronic co-pilot are a series of subsystem managers. The situ-

Knowledge integration

Fig.11.2 The Advanced Tactical Fighter cockpit will bring together concepts such as flat-panel displays, touch-sensitive screens, voice-activated controls and helmet-mounted displays plus an 'electronic co-pilot' *Lockheed California*

ation assessment manager establishes the magnitude and urgency of threats and keeps track files up to date. It cooperates with a beyond-visual-range identification manager which identifies targets by comparing their signatures with a library, and a sensor manager which controls all aircraft sensors.

The four-dimensional flight manager automatically manoeuvres the aircraft into the most advantageous beyond-visual-range attack position and exercises control over the flight control and propulsion managers. The mission planning manager automatically replans the mission after any unplanned event and suggests back-up mission objectives if the primary mission cannot be achieved.

The tactical planning manager recommends ways to deal with new threats, combining information from the situation-assessment and mission-planning managers to provide a ranked list of defensive and offensive options.

INDEX

Page references in bold indicate photographs, those in italic indicate diagrams.
These are separate indexes for names, [etc.].

accelerometer   10, 11, 166, 167
active control technology   102, **103**, *104*, 131
aeronautical public correspondence   200
aircraft motion sensing unit   166
air data acquisition   88, 168
air data computer and inertial reference unit   169
air data display   82
air data inertial reference system   25
air data module   169
Airline Electrical Engineering Committee   23
Air Navigation Order (UK)   2
airspeed indicator   2
air traffic control   67, 96, 150, 179, 200, 222, 226
air transport racking   23, 24, 137
altimeter   2
antenna
  active array   **250**
  Cassegrain   212, *213*
  centre feed dish   210, *212*
  communications   174, 181, 182, 183, 184
  conformal array   211, 249
  cosecant squared   224
  dipole   211
  offset feed dish   *212*
  planar slotted array   *213*
  radar   203, 206, 209, 210, 217, 232, 238, 239, 248
  radio *see under* radio
antenna coupler unit   163
Aeronautical Radio Inc   17, 22, 23, 29, 30
ARINC characteristics
  404   24
  429   37, *38*, 42, 43, 62, 67, 121, 127, 141, 169, 247
  529   *38*
  568   155
  629 (DATAC)   37, 42, 62, 247
  702   121
  709   155
  711   154
  712   157
artificial horizon   2, 55, 56
attitude direction indicator   54, *56*, 57, 62, 65, 71, 72, 73, 158, *159*
attitude director   69, *70*
attitude heading and reference unit   169, 170
automatic antenna tuning unit   183, *184*

automatic attitude director   69, 70
automatic configuration control processor   102
automatic dependence surveillance   196
automatic direction finding   154, 157, 177, 184
automatic landing 'autoland'   21, 71, 72, 220
automatic test equipment   29
autopilot
  Sperry   26
  Honeywell   6
avionic integration   247

Battle of Britain   6
beat frequency oscillator   157
Blind Landing Experimental Unit (RAE)   159
Boltzmann Constant   207
British Civil Airworthiness Requirements   29
British Standards Institution   20
built-in test equipment   27, 28, 46, 116, 120
bus controller   39
bus interface unit   41

canard configuration   93, 111
carrier-sense multiple access clash avoidance   42
cathode ray tube   9, 15, 52, 55, 57, 58, 59, 60, 61, 62, 63, 65, 68, 70, 72, 73, 74, 76, 79, 80, 123, 124, 249, 252
  thin   60, *61*
  shadow mask   *58*, 60, 249
  travelling wave   13, 216
central processing unit   133, 169
circuit integration, large-scale   116
Certificate of Airworthiness (UK)   20
Civil Airworthiness Requirements, British   19
Civil Aviation Authority (UK)   19, 20, 159
cockpit display unit   63, 66, 68
cockpit voice recorder   142
Colossus   31
  colour display, CRT   79, 80, **81, 82**
communication navigation and identification   197
compass
  direct reading   53
  gyro   53
  magnetic   2
  remote reading   53

258

Index

computers
　actuator drive and monitor　102
　air data　25, 168, 248
　analogue　9
　digital　9
　elevator/aileron　96
　flight augmentation　96
　flight management unit　121
　navigation　78
　spoiler/elevator　96
　weapon aiming　78
computer language　*45*
　ADA　46, 113
　　development support environment　46
　COBOL　46
　CORAL　46
　FORTRAN　46
　high-level operating　34, 44, 46
　JOVIAL　46, 113
　PASCAL　46
computer memory devices　33, 35
　bubble (data bank)　36, 37, 124, 125
　core　35
　electronically erasable programmable only　36
　electronically programmable read only　36, 124
　random access　36, 124
　read only　184
　scratch pad　35, 124
　semiconductor　35
　program UV PROM　124
　working CMOS RAM　124
computer memory, non-volatile　28, 35, 125, 169
control and display unit　**123**
controlled requirement expression　135, 139
course deviation indicator　152, 155
crystal video receiver　*232, 233, 234*

data, autonomous transmission and communication. (ARINC 629)　42, 43
data base integration　255
data bus/data modulation　14, 25, 33, 37, 51, 63, 127, 128, 131, 138, 169
data transfer unit　141
direct voice input　57, 65, 76
directional gyro　2
display
　gas plasma　61
　head-down　17, 80
　head-up　9, 17, 62, 71, 72, 76, **77**, **78**, 79, 80, 81, 143, 248, 251
　integrated　74
　liquid crystal　56, 60, 62, 251
　multi-function controls and display　68, 73, **74**, 75, 79, 80
distance measuring equipment　153, 155, 156, 202
Doppler sensors　18

electromagnetic interference　175
electronic associate, 'silicon co-pilot' concept　82, 256
electronic attitude director　69

electronic centralised aircraft maintenance ECAM　67, 73, 74, *75*
electronic counter-measures　6, 13, 216, 235
electronic data processing　9
electronic intelligence　230, 231
electronic support measures　232
engines
　civil　47, 116, *119*, 120
　re-heat　117, 118, 119
　digital control system/unit　118, 119
　full authority digital control　119, 120
　indicating and crew alert system　67, 73
　main control unit　117, 118
　military　117, *118*, 119
ergonomics　49, 50, 51
European Organisation for Civil-Aviation Electronics　19, 20, 22, 30
Exocet missile　229

fan marker　4
Federal Aviation Administration (USA)　19, 20, 21, 120, 156
fibre-optic techniques, 'fly by light'　*43*, 113, 192
flight management computer unit　121
flux valve/flux gate　165, 166, 169
flare decoys　242, **243**
'fly-by-wire' techniques　17, 18, 72, 73, 96, 98, 100
forward-looking infra-red　60, 81, 251
frequency synthesis　186, *187*, *188*, 189

gimbal　11, 167
gimbal lock　170
G-induced loss of consciousness　76
glass cockpit　15
gyro　2, 9, 11, 165, 166
gyro gunsight　9

hands-on-throttle and stick control　*145*
helmet-mounted sight　78, **80**, 251
horizontal situation indicator　53, *54*, 62, 65, 150, *153*, 155

identification, friend or foe　81, 199, 226, 248, 253, 254
inertial platform　117
infra-red picture　77
input/output unit　169
Institution of Electrical Engineers (UK)　194
Institution of Electrical and Electronic Engineers (UK)　194
instruction set architecture　45
instruments
　'basic six' panel　54
　electromechanical　51, 53, 55, 56
　gyro-mechanical　55
　stand-by　82, **83**
integrated circuits　32, 188
　very high performance　33
　very large scale　33
International Civil Aviation Organisation　20, 68, 220
International Telecommunication Union　176
inverse scan modulation　238

259

Index

jammer
  airborne self-protection 235, **239, 240**
  Sky Shadow **241**
  stand-off 239
jamming
  broad band 235, 236
  communications 240
  deception 229, 235, 238
  escort 239
  home-on 235
  infra-red 242, **243**
  noise 235, *236*, **237**, 238
  spot 235
  swept spot 235, 236
Joint Airworthiness Requirements 20, 21, 30, 52
joint tactical information distribution system 177, 196, **197**, *198, 199*, 248

klystron 209, 215, 219
keypads 63
keys, soft 63

laser 13, 37, 171, 242, 244, 254
  ring gyro 13, 26, 171, *172*
laser warning equipment 244, **245**
light emitting diode 51, 52, 57, 59, 60, 61, 65
line replaceable unit 27, 28, 57, 82, 131, 137, **138**, **139**, 169, 247
London Air Traffic Control Centre **151**
Lorenz landing aid 55
low cycle fatigue counter 140

magnetron 9, 209, 218
maintenance data panel 131
Manchester bi-phase format 39
marker beacon 12, 70
Massachusetts Institute of Technology 166
master armament selector switch 145
mean time between failures 120
memory devices 33, 35
  bubble (data bank) 36, 37, 124, 125
  core 35
  electronically erasable programmable only 36
  electronically programmable read only 36, 124
  program UV PROM 124
  random access 36, 124
  scratch pad 35, 124
  semi-conductor 35
  working CMOS RAM 124
memory, non-volatile 28, 35, 125, 169
microphone
  carbon granule 189
  electromagnetic 189
  noise cancelling 189
microelectronics 14
Mil Specifications
  MIL STD 17
  1553 39, 11
  1553B 37, *39, 40, 41, 42*, 127, 131, 133, 141, 247
  1750 A 35, 45, 46
  MIL STD-1750 146
Ministry of Defence (UK) 34, 46, 128

mission adaptive wing project 106
moving target indicator 241
multi-purpose colour display 79, 81
multi-function display 25, 65, 153
multiplex data transmission 128

National Aeronautics and Space Administration (USA) 25, 65, 163
National Physical Laboratory (UK) 7
NATO Air Force 146
Navigation aids and systems
  area nav 12, 68, 156, 180
  attitude heading reference system 169, *170*, 171
  automatic direction finding 154, 157, 177, 184
  Cossor Compass 9000 display, Gatwick airport **160**
  Decca 191
  distance measuring equipment 68
  Doppler 11, 12, 164, 165
  Gee 8, 12
  global positioning (NavStar) 12, 161, 163, 164, 248
  H_2S 8, 224
  hyperbolic 12, 16, *161*
  inertial 10, 11, 12, 65, 125, 166, 167
  instrument landing 68, 71, 96, 122, 126, 153, *158*, 159, 160, 162, 174, 177, 179
  Loran 12, 161, 162, 163
  nav vertical 153
  non-directional beacon 150, 153, 154, 156, 177
  Oboe 12
  Omega 12, 161, 163, 177
  Omega VLF 12
  plan position indicator 151
  radio direction/distance magnetic indicator *152*, 155, 156
  radio magnetic indicator 53, 150, 151, 152, 154, 157
  relative bearing indicator 150, 151, 157
  tactical air navigation 68, 156, 177, 197
  VHF omni-directional radio range 152, 153, 154, 155, 161, 177, 179, 180, 221
  VOR/TACAN/DME 68, 156
  VOR, terminal 68
Norden bomb sight 6

operating languages, computer, high level 34, 44, 46

PAVE PILLAR project 45, 137
Penetrate (Ferranti) 79, 81, **82**
pi-filter network 183
pilot associate concept 82, 256
pitot static sensors 88, 89, 168
powered flying control unit 102
program loading unit 35
pulse code modulation 193

radar
  airborne early warning 202, 222, 223
  airborne interception 7, 202, 222
  air-to-surface vessel 78
  air-sea warfare (anti-submarine) 222
  Doppler/pulse Doppler 13, 17, 207, 209, 215, 218,

219, 221, 222, 241, 242
  ground collision avoidance   225
  H_2S ground mapping   8, 224
  identification, friend or foe IFF   81, 199, 226, 248, 253, 254
    monopulse   213, 214, 223, 225
    monostatic   216
    multistatic   216
    navigation/attack   223, 224
    primary   216, 220, 221
    propagation   203
    pulse   *see* radar, Doppler
    reconnaissance   222, 226, 227
    secondary   216
    secondary surveillance   202, 220, 221, 226
    sideways looking airborne   224
    terrain clearance following   225, 226
    weather   160, 219, 220
radar scanner   206
radar warning receiver   *232*, *234*, **234**
radio
  amplitude modulation   178
  antenna   181, 182, 183, 184, 249
  antenna matching system   182, 183
  antenna automatic tuning unit   184
  plug-in crystals   186
  radio detection and ranging (radar)   7, 9
Royal Aerospace Establishment (formerly Royal Aircraft Establishment)   65, 142, 171
Royal Air Force   6, 12, 53, 54, 117, 140

Sagnac effect   171
satellite tele-communication   200
satellite, geo-stationary   200
self-healing avionics   256
semi-automatic functional requirements analysis   47, 135, 140
sensor
  inertial   90, 166
  pitot static   88, *89*, 90
  sensor integration   253
side stick control   73, **74**, 96, 97
signals intelligence   230, 231
smart probes   90
smart skin   249
Society of Automotive Engineers (USA)   45
Sperry Gyrosyn compass   166
Sperry Zero Reader   158
stability augmentation   10
standard instrument departure   67, 68, 150
standard terminal approach   67, 68, 150
static source error correction   169
strapdown inertial platform   167
symbol generation   23, 25, 61
symbology   9, 60, 62, 66, 76
systems management processor   *131, 132, 133, 134*
Systems other than navigation
  Automatic communication and recording   142, 149, 196

automatic flight control   29, 73, 152, *173*
  digital   170
automatic flight guidance   65
automatic landing 'autoland'   21, 71, 72, 220
automatic management radio and intercom   *191*
control and stability augmentation   100, *101*, 102
digital engine control   118, 119
electronic instrument   22, 26, 29, 51, 55, 57, 60, 62, 69, 70, *71*, 73, 74
electronic flight instrument   26, 29, *50*, 55, 124, 152, 153
engine indicating and crew alert   67, 73
engine management control   122
flight control   131, 254, 325
flight management control   36, 63, 65, 66, 67, 68, 121, *122*, 123, *124*, 126, 127, 150, 153, 163
health and usage monitoring   72, 116, 140, *141*, 142
integrated electronic warfare   246
'intelligent' avionics   *98*
micro-wave landing   154, 159
multi-function information distribution   197
stores management   143, *144*, 146
structural usage monitoring   142
utilities management   12, 13, 115, 128, 131, 132, 134, 135, 138
weapon/stores management   116

Telecommunication Research Establishment, Bawdsey (UK)   6
Tempest requirements   191, 192
thermal imagery   81
thermionic valve   14, 31
thin tube display   60, **61**
tolerancing   175
touch screen   63
touch technology   63, 64
transponder   155, 220, 221, 248
Triad laser gyro   171
trimmable horizontal surface actuator   96
turn and bank indicator   55
turn and slip indicator   2
Type Approval testing   176

UHF/VHF transmitter/receiver   **179**, 180
United States Air Force   34, 44, 127, 137, 190
United States Defense Department   12, 166
United States Navy   9, 12, 113

versatile digital analyser   14
voice synthesis and systems vocoder   174, 189, 193, 199, 252, 253

weapon aiming mode selection   143
Wild Weasel aircraft   229
window (chaff)   8, 13, 17, 229, **241**, 242

yaw damper   10

INDEX OF NAMES

Beamont, W/Cdr Roland 32
Bowen, Dr E. G. 7

McCurdy, J. D. A. 2

Doolittle, General James 56

Post, Wiley 5

Powers, Gary 13

Sperry, Lawrence 23

Trenchard, Lord 6

Watson-Watt, Dr Sir Robert vi, 7
Wright brothers 1, 52

INDEX OF AIRCRAFT

Aerospatiale/BAC Concorde   116, **117**
Airbus Industrie
  A300   37, 66, 69, 97
  A310   37, 66, 69
  A320   17, 28, 37, 65, 66, 69, 72, 73, **74**, *75*, 93, 95, 96, *97*, 123, 169, 248
Airship Industries Skyship 600   113, 114
Armstrong Whitworth
  Whitley   8
  ATR 42/72   25, 29
A. V. Roe (Avro) Vulcan   85

Bristol Aeroplane Company
  Beaufighter   6, 12
  Blenheim   6
  Boxkite   2
  Britannia   116
British Aircraft Corporation TSR 2   14, 31, **32**, 116
British Aerospace plc
BAe 146   93, **94**
  EAP   47, 82, 85, 100, **105, 106,** *107*, **108**, 112, 118, *130, 131, 132*, 133, 134, **135,** *136, 137,* **138,** 139
  BAe Harrier   117
  BAe Hawk   140, 145, *146*
  BAe Sea Harrier   36
Boeing Airplane Company
  B 17   6
  B 247   6
  B 707   11
  B 747   **11**, 127, 165, 200
  B 757   **15, 16,** 37, 66, 120, 159, 173
  B 767   **15, 16,** 17, 37, 66, 173
  B 7J7   43

Consolidated Vultee Aircraft
  B24 Liberator   6
  Catalina   8
Curtiss
  seaplane   2
  flying boat   2, **3**
Curtiss Wright Corpn Condor   **4**, 5

de Havilland Aircraft Co.
  DH106 Comet   171

DH98 Mosquito   6, *7*
Douglas Aircraft
  DC3   6, 10
  DC8   11
  DC10   11

Fokker Aircraft Corpn F50   29

General Dynamics Corpn
  YF16   16
  F16 Falcon   13, 39, 77, 80, 109, 239
Grumman Corpn
  A6   239
  Hellcat   9
  F14A Tomcat   113, 239

Handley Page
  Halifax   6, 8
  Heyford   vi, **7**, 217
Hawker Siddeley Aviation
  Nimrod   31
  Trident   140, 159
Hawker Aircraft Co Hurricane   6

Junkers Ju 88   6

Lockheed Corpn
  Vega 'Winnie Mae'   **5**
  U 2   13
  YF22A   112
  L-1011 TriStar   11, 121
McDonnell Douglas Corpn
  F15 Eagle   13, 109, 112
  AV-8B Harrier   39, **78, 79,** 239
  MD11   25, 28
  MD80   22, 66
  Phantom 'Wild Weasel' EWA/C   229, 239

M/D Northrop F18   39, 109, 113, 239
Messerschmitt Me 110   6

North American Aviation
  F86D Sabre Dog   143
  F100 Super Sabre   10, 113
Northrop Corpn YF 23A   112

263

Index of aircraft

Panavia Tornado   13, 36, 85, *86*, **87**, *88*, 91, **99**, *101*, 102, **103**, 117, 118, 128

Rockwell International
   A-5 Vigilante   9, 14
   B1   39, 88

Sepecat Jaguar   36
Short Brothers
   Belfast   116

Stirling   6
Sunderland   8

Vickers Vanguard   116
Vickers Supermarine Spitfire   6

Westland Lynx   229

Zeppelin airship   4

INDEX OF RESEARCH AIRCRAFT

advanced tactical fighter (USAF)   45, 119, 246, 256
advanced tactical aircraft (USN)   45

BAe Experimental Aircraft Project   47, 82, 85, **105**, 108, 128

DARPA/Grumman X29   88, *109*, 111
DARPA/Rockwell/MBB X31   109, *111*, 113

General Dynamics AFTI mission adaptive wing F111   106, 109, *110*, 111

LHX Program US Army   45

McDonnell Douglas F15 STOL manoeuvre technology demonstrator   109, *110*, 112, 119
McDonnell Douglas/Northrop F/A18   113
Messerschmitt Bölkow Blohm CCV F104   105

Sepecat Jaguar FBW active-control demonstrator (BAe)   100, 102, *103*, 106, 111

INDEX OF GAS TURBINES AND PROPULSION UNITS

Bristol
   Olympus 116
   Proteus 116

International Aero Engines V2500 Superfan 120

Rolls-Royce Pegasus 117, 118, 119

Turbo-Union RB199 117, 118, 119

United Technology Pratt & Whitney PW2037 120
Unducted fan 120